THE MEANING OF THE 21ˢᵀ CENTURY

www.**booksattransworld**.co.uk

Also by James Martin:

The Wired Society

Technology's Crucible

THE MEANING OF THE 21ST CENTURY

A VITAL BLUEPRINT FOR ENSURING OUR FUTURE

JAMES MARTIN

eden project books

TRANSWORLD PUBLISHERS
61–63 Uxbridge Road, London W5 5SA
a division of The Random House Group Ltd

RANDOM HOUSE AUSTRALIA (PTY) LTD
20 Alfred Street, Milsons Point, Sydney,
New South Wales 2061, Australia

RANDOM HOUSE NEW ZEALAND LTD
18 Poland Road, Glenfield, Auckland 10, New Zealand

RANDOM HOUSE SOUTH AFRICA (PTY) LTD
Isle of Houghton, Corner of Boundary Road & Carse O'Gowrie,
Houghton 2198, South Africa

Published 2006 by Eden Project Books
a division of Transworld Publishers

A catalogue record for this book is available from the British Library.
ISBNs (hb) 9781903919842 (from Jan 07)
1903919843
(tpb) 9781903919859 (from Jan 07)
ISBN 0593057252 (cased)
1903919851

Typeset in 11.15/15.25 pt Garamond MT by
Falcon Oast Graphic Art Ltd.

Printed in Great Britain by
Mackays of Chatham plc, Chatham, Kent

1 3 5 7 9 10 8 6 4 2

Mixed Sources
Product group from well-managed
forests and other controlled sources
www.fsc.org Cert no. TT-COC-2139
© 1996 Forest Stewardship Council

FSC

To Lillian Martin,
who organized the research and fed the author

CONTENTS

PREFACE

THE 21ST CENTURY is an extraordinary time—a century of extremes. We could create much grander civilizations, or we could trigger a new Dark Age. There are numerous ways we can steer future events so as to avoid the catastrophes that lurk in our path and to create opportunities for a better world. A revolutionary transition is ahead of us, and our children have a vital role to play in it; so there is much that we need to teach them about their future.

This book is concerned with the big issues that will make the difference—large-scale trends with a momentum like a freight train, which will affect the lives of our children and their children, and types of leverage that will enable us to make significant changes. It's a book about the main ocean currents, not the waves on the surface. It is not about conventional politics, which are to a large extent unpredictable. There will always be passionate arguments about politics, taxation and wealth redistribution, but those are not my subjects here. We can identify critical high-momentum trends independently of politics, and most of the essential changes can be achieved regardless of which political parties are dominant at the time.

Some of the solutions to problems will meet resistance for parochial polit-ical reasons, often because there are large vested interests in maintaining the status quo. Putting the right incentives into place is critical to dealing with our biggest problems.

Exploring the future calls for using logic and for an understanding of history, technology and the behaviour of complex organizations, in order to weave together many long-term trends. Much can be predicted about future technology because of lengthy time lags between research and development and between the creation of products and their application. We already know what people are working on.

When I wrote *The Wired Society* in the mid-1970s, there were no per-sonal computers, no mobile phones and nothing called the Internet. Some twenty-five years later, the book was hailed as an astonishingly accurate forecast of a world using those technologies. Encouraged by this, I devel-oped files about the world's future problems and what experts had to say about them. It became clear to me that the world was drifting into very deep trouble.

After university, I spent some time as a rocket scientist (no kidding). Then I joined IBM and was trained to design computer systems that helped provide solutions to complex problems. I migrated to the IBM Systems Research Institute—a think tank and internal university with an eclectic hornet's nest of a faculty, opposite the United Nations building in New York. In the age of Nixon, we went to cocktail parties at the UN, and it was like going to a different planet. The UN delegates had no clue about what was happening in technology, and we computer gurus had no vision beyond our own world.

In 1970, a secret set of meetings was arranged between twelve top American and twelve top Soviet computer scientists to find out whether and how they could cooperate in research. I was among the American group. The US State Department trained us how to behave—for example, how to handle the vast amounts of vodka we were given during the endless toasts at Moscow banquets (empty our glasses into the flower vases) and what to do if we got back to our hotel room and found a beautiful naked woman in our bed (I was disappointed). To my surprise, I found the Rus-sian scientists to be warm and friendly but holding a total misconception of what America was like. Most of us concluded that there was great potential for cooperation, but the game evolved into the KGB watching the CIA watching the KGB, and nothing useful happened.

After IBM, I formed a business for dealing with complex problems and spent twenty-five years travelling around the world, teaching five-day "World Seminars" that steadily grew to attract high-power audiences. This led to many opportunities to work with statesmen and business leaders. I tried to pick invitations where I might learn the most. For example, J. P. Morgan had an extraordinary advisory board, with George Shultz, Condoleezza Rice, Lee Kuan Yew (who was the first prime minister of Singapore), the Saudi Arabian minister of finance, Britain's Lord Howe and the CEOs of many global corporations.

Over decades, while I roved the world, lecturing and consulting on practical techniques for building complex systems, I took a keen interest in the world's larger problems. They were getting worse. But my training as a problem-solver led me to believe that there are solutions to all the profound problems I describe in this book. I saw that it was necessary to alert the public to the grand-scale challenges of the 21st century and teach their potential solutions.

Recognizing how much research and education this subject needed, I set out to found the James Martin 21st Century School at Oxford University, the mission of which is *to identify and find solutions to the biggest challenges facing humanity in the 21st century, and to find the biggest opportunities.* The school now has many institutes populated with brilliant researchers and teachers. It needs a very high level of integrated scholarship, because its goal is nothing less than to train future leaders who can deal with the tough problems facing humanity. Its faculty does detailed research on solutions, including those described in this book, and synthesizes the knowledge in order to understand our options for the future.

It is critical to teach the subject matter of this book to the rising generation—those who will have to cope with the situation. In a sense, there is no more important subject matter, because civilization's survival depends on it.

PART ONE

THE NEED FOR TRANSITION

1

THE TRANSITION GENERATION

AT THE START of the 21st century, humankind finds itself on a nonsustainable course—a course that, unless it is changed, could lead to grand-scale catastrophes. At the same time, we are unlocking formidable new capabilities that could lead to more exciting lives and glorious civilizations. This could be either humanity's last century or the century that sets the world on a course towards a spectacular future.

We live on a small, beautiful and totally isolated planet, but its population is becoming too large; enormous new consumer societies are growing, of which China is the largest; and technology is becoming powerful enough to wreck the planet. We are travelling at breakneck speed into an age of extremes—extremes in wealth and poverty, extremes in technology and the experiments that scientists want to perform, extreme forces of globalism, weapons of mass destruction and terrorists acting in the name of religion. If we are to survive, we have to learn how to manage this situation.

Formidable problems confront us, but this is a book about *solutions*— many solutions. With these solutions we will bring about a change in course, a great 21st-century transition. If we get it right, we have an extraordinary future. If we get it wrong, we face an irreversible disruption that could set humanity back centuries. A drastic change is needed in the first half of the 21st century to set the stage for extraordinary events in the rest of the century.

Humankind has been able to thrive for thousands of years because nature provided it with resources like topsoil, underground water, fish in the oceans, minerals, oil and wetlands, but these resources are finite, like cookies in a jar. We are using up many of these resources, and some don't have substitutes.

Nature also provided us with an ozone layer and a delicately regulated atmosphere, with forests that remove carbon dioxide from the atmosphere but are now being depleted. *Every year,* because of our misuse of the Earth's resources, we lose 100 million acres of farmland and 24 billion tons of topsoil, and we create 15 million acres of new desert around the world. An inch of good topsoil can take a thousand years to form, but when people destroy windbreaks by cutting down trees, the topsoil can be washed or blown away in months.

Water is vital for our survival and for producing food. It takes about a thousand tons of water to produce one ton of grain that, fed to cows, produces only 18 pounds of meat. Today mankind is using about 160 billion tons more water each year than is being replenished by rain and fed back into water storages. If this water were carried in water trucks, it would require a 300,000-mile-long convoy of trucks every day—a convoy length 37 times the diameter of the Earth. This is how much water we are using and not replenishing.

During the lifetime of today's teenagers, fresh water will run out in many parts of the world, making food production difficult. Many fish species will be too depleted to replenish themselves. Global warming will bring hurricanes far more severe than Katrina and will cause natural climate-control mechanisms to go wrong. Rising temperatures will lower crop yields in many of the world's poorest countries, such as those in central Africa. The immense tensions brought about by such catastrophes will occur in a time of extremism, religious belligerence and suicidal terrorism, and this will coincide with terrible weapons becoming much less expensive and more widely available.

This interconnected set of problems has an interconnected set of solutions. If we humans implement these solutions, we can gradually achieve sustainable development and a sustainable but affluent life. Working towards sustainability requires many different types of actions in different subject areas. In light of rapidly advancing technology, however, sustainability alone is not enough. We need to be concerned with *survivability*. There must be a move away from the untenable course we are on today towards a world where we learn to control the diverse forces we are unleashing.

Today's young people will be the generation that brings about this great transition. Let's refer to it as the 21C Transition. They are ultimately responsible for the changes we describe—a transition unlike any before in history. They are the Transition Generation. It is vital that they—all of them—understand the 21C Transition, so that they can understand the critical role they will play. For many, understanding the meaning of the 21st century will give meaning to their own lives.

THE 21ST-CENTURY CANYON

Think of the 21st century as a deep river canyon with a narrow bottleneck at its centre. Think of humanity as river rafters heading downstream. As we head into the canyon, we'll have to cope with a rate of change that becomes much more intense—a white-water raft trip with the currents becoming much faster and rougher, a time when technology will accelerate at a phenomenal rate.

As the world's population grows, global tensions and pollution will climb, and the danger of massive famines will increase. The population of the world will continue to rise, probably until it reaches about 8.9 billion (a figure from the latest computer models). The ability to feed such a population will steadily decrease as water tables drop, farms in poor countries disappear and the huge new consumer classes in China, India and 18 other countries change their eating habits so that they eat more meat (which requires much more grain production and, thus, depletes much more water).

Demographers, who map out the course of world population growth, expect the population to decline slowly after it reaches its peak in midcentury. It has been falling in most First World countries; the birthrate in these countries has dropped so far that they are below the replacement rate of about 2.1 children per woman. The planet's population growth is occurring mainly in poor countries. At the narrow part of the canyon, the world's population will be at its highest and the world's resources under their greatest stress.

The decades in which we are swept towards the canyon bottleneck will be a time when we unlock extraordinary new technology—nanotechnology, biotechnology, extreme-bandwidth networks, robotic factories, regenerative medicine and intense forms of computerized intelligence. Although much harm may be done to the climate by midcentury, the fuels that cause global warming will be largely obsolete and will be replaced by various forms of clean energy. Eventually, there will be an endless supply of energy that does little environmental harm. Instead of smokestacks and carbon-based fuels, we'll have clean industry and a hydrogen economy. The insidious chemicals that interfere with our health mostly will be banned. The oceans and the rest of the global environment will be better understood and protected, and the damage done by our abuse of nature will slowly start to be corrected.

The job of the Transition Generation is to get humanity through the canyon with as little mayhem as possible into what we hope will be smoother waters beyond. Solutions exist, or can exist, to most of the serious problems of the 21st century. The bad news is that the most powerful people today have little understanding of the solutions and little incentive to apply them. Politicians are anxious to find votes—the next election dominates their thinking. Powerful business executives are eager to achieve profits—it is their job to increase shareholder value, and shareholders will judge them by this quarter's results. So for the powerful people who control events, *the desire for short-term benefits overwhelms the desire to solve long-term problems.* We are heading towards the canyon, but our leaders are not preparing to make the passage smoother for us. There are major institutional roadblocks.

We may see a post-canyon world with smoother sailing, but a different set of events will be taking us into a different type of turbulence. Twenty-first-century technologies will give us the ability to change life and transform humans, computer intelligence will race far beyond human intelligence, and new science will take us onto a slippery slope that will change very fast. We'll want, somehow, to stay in control.

MOMENTUM TRENDS

I don't sit at my desk with a crystal ball. Crystal balls don't work. So why is it sensible to write a book about the future?

There are trends that are foreseeable because they have unstoppable momentum. Many of these *momentum trends* will have profound consequences and seem either inevitable or very difficult to change. For example, the growth of the Earth's population can be estimated, and from these estimates we can examine the demand for food, water and other basics. Some large-scale trends are inevitable because of technology. We know that telecommunications bandwidth is set to grow inexorably because of the inherent bandwidth of optical fibres. The most famous long-term prediction is Moore's Law. In 1965, Gordon Moore, a founder of Intel, the great chip company, stated that the number of transistors on a silicon chip would double every year, and it has done so for decades. Moore was able to make this prediction because he had a deep understanding of chip-manufacturing technology.

There are about a hundred momentum trends that can help us understand aspects of the future. Some are obvious, like the decline of ocean fisheries. Some are clear to experts in the field. Some are long-term, like the spread of greenhouse gases in the atmosphere. Some are relatively short-term changes to a new form of behaviour, such as how the Internet is changing the way corporations buy goods. Together, these high-momentum trends form a skeleton of the future. We can put flesh on that skeleton in different ways. We can't predict the future in detail, but we can study the alternative directions it can take and how to influence it.

When we look at the momentum trends in aggregate, it is clear that we are in deep trouble. Every six weeks the planet's population has a net increase equal to the population of New York City. Extreme differences between rich and poor nations widen even more dramatically, with the wealthier lifestyles being flaunted in the faces of the poorer via television and the forces of globalism. The age of terrorism reflects new types of tensions. Much of the planet's water, essential for growing food, comes from large underground aquifers and dates back to many ice ages ago. When this ancient resource is used up, we'll have to live mainly on rainwater. There will be wars over water.

When environmental shortages become severe in poor countries, civil violence can erupt. In 1994, during a drought, Africa's two most densely populated countries, Rwanda and Burundi, suffered an explosion of atrocities and the genocide of nearly a million people.

SURPRISES

To map the world in terms of trends having unstoppable momentum suggests a substantial level of predictability, but even among predictable trends, major surprises occur suddenly. If a surprise is going to slow the inexorable momentum, it must be very big. America was shockingly jolted by the attacks of 9/11. Even though 9/11 changed the lens of world affairs, the momentum trends of its time continued and only a few showed blips on their curves. By contrast, in the early 20th century, the grandly civilized Belle Epoque era, with Paris as its flagship, was largely swept away by the unexpected nightmare of the First World War. In 1990–91, a major economic/political surprise changed the map of history: The Soviet Union collapsed, and people's lives were left in tatters in one of the world's best-educated countries. Capitalist societies have a pattern of booms and recessions, sometimes with economic collapses. The Asian crisis of 1997 brought much of the Third World to the brink of financial meltdown.

America has such an impressive economic momentum that it sails through turbulent weather like a grand ship. Could there be surprises big enough to upset the United States? Like every country, it's affected by China, but China could evolve into a bubble economy—history's largest bubble economy—in which the bubble eventually bursts and China has what market gurus call a "hard landing." This would play havoc with many global economies unless they had prepared for it. Some surprises may be very unpleasant. If a terrorist organization set off an atomic bomb (not just a "dirty bomb") in an American city—a Hiroshima in Manhattan—the aftermath would be devastating.

One of the worst surprises might be something that at first appears far less dramatic: a flu epidemic. Twice as many people died from the flu of 1918 as in the First World War. A new strain of flu, H5N1 avian flu, has a set of characteristics that would make it very dangerous if it mutated so that it could spread from person to person. There is no vaccine for it yet, and we can't make a vaccine until we know the particular strain of flu. If such a flu starts spreading, some governments are likely to quarantine people and shut down air travel. Because economies of wealthy countries depend increasingly on international business and supply chains, the disruption to multinational business could be devastating. A worse scenario could be a pandemic that surprises nature, because humans have created an artificially modified pathogen.

Peter Schwartz, a long-term professional in creating scenarios of the future, says that surprises large enough to change the momentum-trend skeleton of the future appear inevitable when one examines the underlying patterns of behaviour. A major terrorist attack on the American homeland by Muslim extremists was predicted well before 9/11—so much so that the destruction of the World Trade Center might conceivably have been avoided. Similarly, a Chinese bubble economy or lethal flu epidemic has an aura of inevitability, and it is sensible to take precautions.

Often surprises occur because politicians don't listen to scientists. Before Hurricane Katrina caused destruction in New Orleans, the situation had been modelled in detail. The models indicated that a Category 3 hurricane could result in broken levees and extremely dangerous flooding of the city. Katrina was Category 5 for a time, but the city was not evacuated. President Bush said he "didn't think anybody expected" the levees to break. A message that turns up repeatedly in this book is that we'd better listen to the scientists.

LEVERAGE FACTORS

In this book, I use the phrase *leverage factor* to refer to relatively small and politically achievable actions—such as minor changes in the rules—that can have powerful results. There are many examples of leverage factors: A tiny catalyst can cause a major chemical reaction, for instance. Antitrust laws have a major effect on the evolution of capitalism's tendency towards creating monopolies. Injecting a tiny amount of a vaccine into our blood can trigger our immune systems to produce enough antibodies to make us immune to a disease. Most harmful momentum trends have leverage factors that could help us avoid much of the harm.

Here's one dramatic example: When women in poor countries are taught to read—a relatively easy and inexpensive project—they tend to have fewer children. In fact, the most effective way to lower the birthrate is to provide birth control and teach women to read. This helps women improve their lives, and it is a solution that brings dramatic results.

Hay was not used to feed animals in the Roman Empire. Hay wasn't needed in a Mediterranean climate because grass grew well enough in winter that there was no need to cut the grass in autumn and store it. Hay came into use during the Dark Ages. The simple idea of using hay was a leverage factor because it enabled populations in northern Europe to make widespread use of horses and oxen. Hay eventually permitted cities such as London to grow and become great centres of activity.

Smallpox has been humanity's most dreadful disease. In the 20th century, 300 million people died from it—terribly. In 1966, a totally uncompromising leader, Donald A. Henderson, set out to eradicate smallpox from the planet. When there was an outbreak of smallpox, everyone in a ring around the outbreak was vaccinated. Slowly, in one country after another, in a programme called the Eradication, smallpox was wiped out.

The last natural case of smallpox infected a cook in Somalia on 27 October 1977. No human has had smallpox from natural causes since then.

Evil leverage factors exist as well as good ones. After 1977, the only smallpox virus existed in a maximum-containment laboratory at the CDC (Centers for Disease Control) in Atlanta and in a similar repository at the Moscow Institute. Different strains of the disease were kept in little plastic vials in a liquid-nitrogen freezer. The CDC's entire smallpox collection would fit in a lady's handbag and weighed about one pound. The USSR realized that the Eradication gave it a unique military opportunity. In total secrecy, it created a facility that could make up to 100 tons of weapons-grade smallpox *per year*. Intercontinental ballistic missiles were modified so that they could carry large quantities of smallpox virus to target destinations and release them.

To address the many difficult problems in this book, we need to identify effective leverage factors. Some leverage factors are unexpected, such as the introduction of World Wide Web software, which made the Internet user-friendly. It cost relatively little, but it ushered in the mass acceptability of Internet use and electronic commerce. This simple software was an extraordinary leverage factor.

We need to separate in our minds the momentum trends and leverage factors from the overwhelming noise of smaller issues. By identifying them we can think about how to make the future better. There is an enormous amount that can be done to transform the journey ahead.

ECO-AFFLUENCE

Sooner or later, we must realize that we have to live within the planet's means. A globally sustainable civilization is not one that's poor or without joy. On the contrary, we can have spectacularly affluent civilizations in which we don't use more resources than the environment can provide. I call this "eco-affluence." There can be new lifestyles of the grandest quality that heal rather than harm our global ecosystem.

Adopting a quality of life that doesn't damage the environment doesn't mean going back to nature. You don't have to live like Thoreau (unless you want to). It could mean living in a superbly sophisticated city, near family, with the excitement of creative work, cultural diversity, elegant parks and superlative entertainment. Cities can be both beautiful and ecologically correct. Having a good lifestyle may mean developing a connection to religion, beauty and community. Future civilizations will be anything but simple, and they will offer a wide variety of lifestyles.

There are many ways to be affluent without harming the environment. Some involve the love of nature, some involve high technology, and some involve opera, football, theatre or jazz. The Earth will have large protected areas of ancient and immense biodiversity, and some people will be passionate about understanding this biodiversity. Some will be crazy about ocean racing, paragliding, bird-watching, breeding orchids, hydroponics, cricket, camping or three-dimensional chess. Digital technology will bring complex computer games with diverse virtual-reality experiences to people around the planet, and by using earphones and goggles, we can take state-of-the-art entertainment anywhere.

A vast new buying class is growing around the world (particularly in China and India) that wants to consume the way Americans do. Before their numbers grow to billions of people—soon—they should have a hydrogen-based-energy economy, fuel-cell cars and efficient use of water. To avoid wreaking havoc around the planet, we need eco-affluence to be globally fashionable.

The future will be characterized by rapid growths in knowledge and in new techniques for putting knowledge to work. Routine work will continue to be done by machines, leaving humans to focus increasingly on jobs that demand human feeling and creativity. The 21st century will bring extraordinary levels of eco-affluent creativity.

One important momentum trend is the ongoing increase in human productivity caused by improving technology and better management. From 1995 to 2005, productivity in America rose by slightly more than 3% per year—an adaptation to increased automation and computing. Some economists expect a long-term 2.5% growth in productivity. If this rate were to continue for a hundred years, society would be 12 times wealthier in real terms. China (starting from a lower base) is likely to be 20 times wealthier in real terms. It is possible (probable, I think) that the era of nano-technology and robotics will sweep us to higher growth rates than those predicted by today's economists. It is vital that this major new wealth be spent in ways that heal rather than harm the planet.

An important statement is that the world's *increase in wealth will be very much greater than its increase in population.* This conjunction of momentum trends offers hope that the world will be made a more decent place for most of humanity. The increase in wealth will be very unevenly distributed, however. The new wealth will be based largely on intellect; so the vast majority of it will go to countries that are already richer, while many of the poorer countries will skid into deeper poverty unless a well-planned effort is made to prevent this.

The forces of the near future are so large that inevitably they will change civilization. Part of the 21C Transition is a change in civilizations—different types of changes in different cultures. What's the point of ever more extraordinary technology if it doesn't build better civilizations? Human survivability and creating new concepts of civilization are inextricably linked.

We need to ask fundamental questions about civilization. What sort of a world would you like your children to live in? What should be the principles of a civilization in which biotechnology can change human nature? In Thomas Jefferson's world, constructive debates raged about future civilization. We need something similar. What *principles* are right for the 21st century, when so much will change?

Society needs visions of a better future. We need a broader vision of the future's diverse possibilities, because civilization is certain to become more multifaceted and complex than it is now. As in the grand epic legends like *Lord of the Rings,* progression towards that vision may be blocked by catastrophes, bureaucrats, battles and distractions so seductive that we can't resist them.

The 21C Transition, if we get it right, not only will steer the planet away from a course leading to mayhem but will also set the stage for an extraordinary evolution of civilizations very different from what we know today.

DICHOTOMY

The 21st century presents an extreme dichotomy. In the stronger countries, it will be a time of great increase in wealth and a massive increase in
what humans can achieve. In the weaker countries, there will be a cycle of
steadily worsening poverty, disease, violence and social chaos. Many are
actually *destitute* nations, or *failed* nations, not *developing* nations. Developing
countries are on a ladder to improvement: Step by step, they can improve
their lot. Destitute nations are so poor that they cannot reach even the bottom rung of that ladder. Below a certain level, poverty is so crushing that
there is no way out; hunger and disease get only worse. There is no way to
educate one's children, no way to create better farming, no way to enter the
world of trade.

It is entirely possible for the wealthy nations to stop this vicious cycle
in the poorer nations. Almost half the world's people live in places where
they earn less than $2 per day. Almost a billion live on less than $1 per day.
Between now and midcentury, most of the 2 to 3 billion increase in population will be in the poorer countries. Such destitution leads to desperation.
Extremist Islam, which advocates terrorism and jihad, is spreading in some
of these countries.

We can look at the future in two ways. We can ask: What is *the right thing
to do*? Or we can ask: What is *the most likely thing to happen*?

Without visiting the poorer nations of the world or destitute shantytowns within not-so-poor nations, it is almost impossible to understand
the true horror of what is there. If we ask what is *the right thing to do*, there
are clear, fundamental answers. End poverty. Eliminate disease and squalor.
Educate children. Teach women to read. In short, clean up the mess. Take
a set of actions that lift the destitute society to the bottom rung of the ladder of development, from which they at least have a chance to progress. To
get to the bottom rung of the ladder, they need help from the outside.
Without that help, the destitution will get worse. This is not an impractical
ideal. It does not need a large amount of financial aid; it needs basic know-

how put into place along with low-cost actions. The cost to the rich nations would barely be noticed.

If we ask what is *most likely to happen,* however, we observe that the United Nations sets goals, but not much happens. Politicians in rich nations make speeches and feel good about their words, but their television-watching constituency mostly doesn't care about faraway poverty. Not enough money is spent by rich countries to help poor ones. Powerful governments (like the Chinese) bring about powerful changes in their own countries, but destitute nations have no such government. They desperately need help, which they are not getting. So, for these nations, if we ask what is *most likely to happen,* the answer is an inexorable spiral into worsening conditions. To answer "More of the same" would not be correct, usually, because below a certain level of poverty, the situation slowly deteriorates even further.

Michael Porter, the Harvard superstar of business gurus, told me forcefully, "We have all these countries that are failing, all these people in these countries that have no opportunities, no sense of self-worth. This is creating very divisive forces. We're caught in a conundrum. We want to respect the citizens of countries to make choices. We believe deeply in democracy. We want people to guide their own destinies. Yet what if that keeps not working, and we have these long-term, planetwide consequences? What do we do about it? That is a discussion that the world, right now, is not prepared to have."[1]

The richer parts of humanity will continue to spend huge amounts of money improving their lives, while the poorer parts of humanity live an almost subhuman existence. The rich kids will play video games full of virtual violence while the poor kids live in shantycities full of actual violence. People in rich societies will strive to live long, vigorous lives, while the world's poor have short, brutal lives ruined by AIDS, sporadic warfare, political anarchy and the growing threat of starvation.

In entirely different aspects of the story we have to tell, there is a stark difference between *the right thing to do* and *the most likely thing to happen.*

A SICK PLANET

The most dangerous consequence of our activities may be that we upset the way our planet regulates itself. The Earth is a small but immensely complex entity sitting in endless black emptiness. Its great forests, oceans, weather patterns and ecology enable it to regulate itself, as it has done for 3 billion years. Its complex mechanisms of regulation provide a breathable atmosphere, a pleasant climate and conditions in which diverse plants grow. The Earth is like a living thing—a green and beautiful sphere with immensely complex biology and weather. It is generally stable, but we can interfere with it. If we interfere with it gently, it will adjust and return to stability. However, if we interfere with it too much, it will change to a different state, and this would be disastrous for a population so excessive that only a portion of it can survive in decency.

The discipline of earth system science studies the behaviour of the Earth and treats the Earth's geosphere and biosphere as an integrated entity. James Lovelock, the legendary authority on this subject, calls the Earth's complex system *Gaia*. Like other complex systems—such as city traffic or world finances—Gaia has a behaviour of its own. We can change the behaviour of city traffic or world finances, but if we change the behaviour of Gaia, we are messing with enormous forces, and the consequences would be devastating for our life on Earth. A grand-scale concern today is that we are inadvertently putting too much pressure on this delicately adjusted system.

Over many thousands of years, the Earth hasn't stayed constant, and won't. It has experienced ice ages and periods of global warming, changes that occur over long periods of time. It is awesome to reflect that during the last ice age, Britain was buried under ice almost 2 miles deep. At present, this totally isolated blue planet is in a period of natural warming. The Sun is slightly hotter than usual. It is bad luck that this is the time when human civilization is causing artificial warming. We are pumping large quantities of greenhouse gases into the atmosphere at the very time Gaia is already feeling the heat. To make this situation worse, we are interfering with the mechanisms that are part of the Earth's elaborate control system. We are cutting down tropical forests and changing the surface of the Earth with industry and highly mechanized farming.

Every living thing can enjoy good health or bad health. We now have computerized centres around the world that study the climate—rather like the pathology labs of a hospital checking a patient. Some climate specialists see the Earth as being ill and running a temperature that is rising fast.

Lovelock points out that some types of interference with the Earth's condition become self-amplifying. In the Arctic regions, for example, ice normally reflects most of the sun's heat back into space. Global warming is now causing this ice to melt, so that much of it will be gone in a few decades. Then, instead of the reflective icy surface, there will be dark earth and sea that *absorb* the sun's heat, and the planet will get even hotter. That will only make the ice melt faster. Much permafrost contains methane. Man-made greenhouse gases make the permafrost melt and release its methane, which itself is an exceptionally powerful greenhouse gas. When methane is released, it adds to the warmth and so increases the melting.

Again, the oceans help us by absorbing much of the carbon dioxide that we generate, but our pollutants make the oceans more acidic. They then absorb less carbon dioxide.

The following scenario is particularly alarming: Tropical forests play a major role in absorbing carbon dioxide from the atmosphere and releasing oxygen. Forestry experts say, however, that if the temperature of the atmosphere around them were to rise by 4 degrees Celsius, the forests would start to die. When they die and decay, they emit carbon dioxide instead of absorbing it. Computer models of the climate predict that the temperature will, indeed, rise more than 4 degrees Celsius during this century. The forests will lose part of their capability to absorb carbon dioxide, and that will cause the temperatures to rise still higher, exacerbating a bad situation.

All of these are self-feeding processes that make the Earth get steadily warmer, as if running a fever. Cutting down the rain forests feeds the fever. As the rise in temperature kills the forests, it makes the fever even worse.

Much of the Earth's surface is now cities, industrial zones and farmland, which doesn't help Gaia's role. This land no longer contributes to the control mechanisms that have kept the planet healthy for hundreds of millions of years. Similarly, much of the tropical zones of the planet will become scrub or desert. Gaia has been there before, Lovelock says, and recovered, but it took many centuries. If we continue to pump greenhouse gases into the atmosphere with "business as usual," we'll have a roasted planet with feedback mechanisms that make it automatically warmer.

Lovelock, a highly knowledgeable and respected scientist, paints a grim scenario. Most of the Earth will become too hot for farmers to grow food. We don't know exactly how fast Gaia's automatic feedback mechanisms will warm up the planet. Nor do we know to what extent the warming is irreversible. If Gaia is sick, we must clearly try to cure it. That will require a massive effort to lower the content of carbon in the atmosphere and to make sure the rain forests and other means of disposing of carbon are sufficient.

We are messing with forces on a grand scale, and we need to understand those forces.

TECHNOLOGY'S AVALANCHE

While America's founding fathers were debating what their future society should be like, a handful of similarly thoughtful men in England were meeting at one another's homes (when there was sufficient moonlight to ride by). They were practical men—neither aristocrats nor scholars but manufacturers who came together because they were excited by new ideas. They built new types of machines, like the loom and the steam engine. Together they set in motion an avalanche of technology that became the Industrial Revolution. Like all avalanches, it moved slowly at first, but each wave of technology brought with it new ideas for improving things, and the waves picked up speed and followed one another increasingly quickly. Two and a half centuries later, the avalanche is thundering down the mountainside with awesome power. As a consequence of technology, the 20th century saw population and consumption multiply furiously, heading to levels that the Earth could not sustain.

The avalanche will continue to accelerate. Theoretical research in science indicates that technology probably will increase in power for centuries. To stop it would take a catastrophe of extreme form. Almost all technology can be used for good or for evil, and as technology becomes more powerful, the potentials for both good and evil become greater. The spectrum from good to evil expands and will become extremely wide as the avalanche continues. The larger this range, the greater the need to accelerate the better technologies and suppress the worse. We need the wisdom to recognize that some new technologies are godsends and others could wreck civilization. New energy technologies that will lessen damage to the climate are vital; technologies that facilitate the spread of weapons of ever more mass destruction should be stopped if possible.

When writing this book, I did a series of in-depth interviews with the most interesting authorities I could find on the subjects I wanted to explore. For example, I spoke with Lord Martin Rees, the Astronomer Royal of the United Kingdom and president of the Royal Society, an institution steeped in scientific history since 1660. He could hardly seem more civilized, living and working as he does amid the ancient magnificence of Trinity College in Cambridge, overlooking gardens sloping down to the river Cam. Despite the calm, Lord Rees has profound reasons for believing that civilization could experience an "irreversible setback." A deeply thoughtful and broad-ranging scientist, he says that we have so many dangers ahead that he rates the odds of *Homo sapiens* surviving the 21st century at "no better than fifty-fifty." He spelled out this reasoning in detail in his book *Our Final Century*.[2] He is concerned that some big-budget scientific research will become too dangerous and that one low-budget maverick could trigger something uncontrollable. Current technologies already raise questions about whether we can control technology, and far wilder technologies are not on our radar screen yet.

If you think Lord Rees's claim sounds far-fetched, imagine the accelerating avalanche of technology continuing for a thousand years. Ultimately, it will become too dangerous to live with. At some point in the future, humankind will not survive unless well-thought-out action is taken to ensure human survivability. That time probably will occur in the 21st century. This is the first century since our caveman days in which *Homo sapiens* could be terminated.

Even if *Homo sapiens* survives, civilization may not.

HUMANITY'S GRANDEST CHALLENGE

A vital task for the 21st century is to learn how to cope with the avalanche we have started and its consequences. You should think of the 21st century as setting the stage for future centuries—taking us through a driving test and then establishing a highway code so that we can be reasonably safe with the forces of technology and globalism that we are unleashing. This is the century when we learn to control what we are doing.

Today's young people will live during a time of extraordinary opportunities and immense problems. How do we help the poorer nations of the world transform themselves? How will the world cope with fully transparent globalism, mass-destruction weapons and terrorism? How do we take advantage of the accelerating avalanche of technology while at the same time preventing it from wrecking our world? If we survive this formidable century, we will have acquired the wisdom to survive long term.

The main theme of this book is an idea that should be taught and talked about everywhere: that the 21st century is unique in human history in that it will produce a great change—the 21C Transition—which will enable humanity to survive. This transition is spelled out in detail in Chapter 13, "The Awesome Meaning of This Century." Some aspects of the transition probably will occur with revolutionary suddenness. They may be triggered by a catastrophe or by a sudden change by a government that realizes that desperate action is needed. What started with the Industrial Revolution now needs another revolution—the 21st Century Revolution. The Industrial Revolution was destined to change the whole future of humankind. The 21st Century Revolution will also change the whole future of humankind. If we get it right, we will make the planet sustainable and manageable. If we get it wrong, we will see our civilization being steadily, or suddenly, destroyed. If we establish an appropriate highway code for the future, the 21st century and centuries beyond it can be more magnificent than anything we can imagine, because technology will enhance human creativity and culture in ways enormously beyond anything that is generally realized today.

The generation now at school is the one slated to bring about this momentous transition—both the parts of it that are revolutionary and the parts that are more gentle. We refer to them as the Transition Generation. Collectively, their task is awesome.

Part 1 of this book explores the trouble we are running into and indicates that there are solutions—many important solutions. But it emphasizes that if we continue to delay taking action, the consequences will be long-term catastrophes on a grand scale. Part 2, "Technologies of Sorcery," describes technologies that will give us extraordinary new capabilities but that (increasingly in the future) can get us into new types of trouble. Part 3, "Through the Canyon," opens with a chapter spelling out the meaning of this very critical century. Humankind with appropriate training is impressively resourceful; so, once the canyon is visible, humankind will find ways to deal with it. The Earth probably will suffer some serious damage, and part of our task will be to make the most of a damaged planet. Part 4, "The Gateway to the Future," describes a new world we are heading towards. Can we create new lifestyles that will usher humanity towards higher levels of civilization? Can we cope with technologies that are enormously disruptive? Can we escape from the vested interests and obsolete ideas of the 20th century? Can we stop the evil side of our nature from burning down the house or blocking the way to what could be unimaginable human advancement? Will the 21st Century Revolution be relatively gentle, as the Industrial Revolution was, or will the inevitable changes occur with revolutionary earthquakes?

If we understand this century and learn how to play its very complex game, our future will be magnificent. If we get it wrong, we may be at the start of a new type of Dark Age.

So, let us begin.

2

WHAT GOT US INTO THIS MESS?

GREEK TRAGEDIES

The damage done to the Earth has not been done because individuals or organizations had evil intent. It has happened because they were caught up in a Greek tragedy. In tragedies of classical Greek theatre, the hero does not know that his actions will lead to disastrous consequences. It is man's miscalculation of reality that brings about the tragedy.

The purpose of Greek theatre was to ask questions about the nature of man, his position in the scheme of things and his relation to the powers that govern his life. The audience is aware of forces in the world powerful enough to topple even the most admirable of men. It sees the spectacle of human greatness—of man daring to reach out beyond reasonable limits in quest of some glorious ideal. The sin of the Greek hero is hubris— excessive self-pride and self-confidence that lead him to ignore warnings from the gods and, thus, invite catastrophe. The 21st century presents such themes on the grandest scale.

Some chapters of this book discuss tragedies that could have been avoided, that have harmed the Earth, caused cancer, brought disastrous population growth and damaged our ability to survive on a small planet. Other chapters discuss tragedies that haven't happened yet. We are in the middle of the play and should ask how we can change its ending.

UNIMAGINED COMPLEXITY

Humankind, until recently, had a simplistic view of the world. It didn't have the science to understand the complexity of nature. In the last few decades, several areas of science have demonstrated that nature is incomparably more subtle and intricate than we had realized. Our mind and body, our immune systems, the ecosystems of nature, the evolution of viruses and the interactions of the planet's ecology are of diabolical complexity, as are subatomic structures and cosmic-scale physics.

The technology that humans are so proud of is crude and primitive compared with nature, but it can be brutally powerful. When the bulldozer meets the rain forest, nature is destroyed. When DDT is sprayed on the countryside, many species die. Nuclear radiation can rip apart the fragile mechanisms of life. These are blunt confrontations, but there are other more delicate ones, such as invisible synthetic chemicals that send false messages to our endocrine systems—the body's manifold system of internal communication by means of chemical messages. Human-made chemicals can accumulate in the body, eventually causing cancer, birth defects and other problems.

Because evolution has gone through billions of years of trial and error, nature has learned to protect itself from nature, but it hasn't learned to protect itself from the artificial works of man. Nature on its own is astonishingly robust. When confronted with human technology, however, some aspects of nature are remarkably vulnerable.

UTTERLY ALONE

Every society has its popular delusions. A delusion of our time is the space-travel delusion. In films and television series, the people from Earth keep encountering civilizations elsewhere in the universe—no end of them—full of interesting characters. Unfortunately, there's no life elsewhere in our solar system, except possibly for very primitive creatures, or within 1,000,000,000,000,000 miles of our solar system. The harsh reality is that we won't trek to other civilizations, and they won't visit us, at least not in the 21st century. An awesome aspect of our world is that we are totally isolated.

Our mythology desperately hates the idea that we are alone. Children are fed an endless stream of stories about people from outer space. In reality, this beautiful planet with its abundant life is, for all practical purposes, utterly, absolutely alone. It's desirable to feel this aloneness so that we can feel the consequences of our planetary destruction.

Our planet is becoming a pressure cooker, at the start of a period of intense change, grossly overpopulated, sharing what will be extreme-bandwidth networks and media, but increasingly tense because of the growing extremes between rich and poor. Humanity's capability for destruction has reached dangerous proportions. We are shrinking many of the resources that humanity needs for survival. The results may cause grand-scale famines, anarchic violence and wars of unparalleled brutality. Catastrophic climate change has no national frontier, nor does nuclear winter or new strains of flu or plagues. Human-made deserts are spreading; they are visible from the moon. Numerous species are disappearing; they will never come back. We may wipe out half the species on Earth, terminating their role in evolution over the next million years.

In 1998, Patriarch Bartholomew, an Eastern Orthodox prelate, made church history by proclaiming a new class of sins: those against the environment. He wrote, "For humans to cause species to become extinct and to destroy the biological diversity of God's creation, to degrade the integrity of Earth by causing changes in its climate, by stripping the Earth of its natural forests or destroying its wetlands, for humans to contaminate the Earth's waters, its land, its air and its life, with poisonous substances—these are sins." They need to be sins in other religions also.

The challenge of the 21st century is for this beautiful, utterly isolated planet to become well managed.

MISCONCEPTIONS

Because humankind underestimated the fragility and complexity of nature, it also had various grand-scale misconceptions about its own supremacy:

We believed nature's resources were unlimited.

It seemed, until the 20th century, that we could plunder the environment at will. When colonists landed in a new place, they killed its creatures for food until some became extinct. If they exhausted the resources of one area, they moved on to another. By the end of the 20th century, it was clear not only that the Earth was limited but also that it had been seriously damaged and depleted by human mistreatment. We and our technology had become powerful enough to wreck the planet.

We thought nature could absorb unlimited pollution.

It once seemed that there was no limit to the pollution that could be pumped into the world's rivers, oceans and air. Now, in some cases (like the once-beautiful Aral Sea in Russia), the damage that has been done is shocking. We have holes in the ozone layer and melting polar ice caps. Man-made pesticides, weed killers, waste materials, fertilizers and exotic chemicals polluted the atmosphere, waterways, soil and oceans.

We didn't expect to destroy nature's species.

In the last half-century, not only has humankind inadvertently exterminated vast numbers of species, it has turned richly diverse ecosystems, stable after millions of years of evolution, into less diverse ones dominated by aggressive nonnative species that can withstand our herbicides and pesticides.

We thought our bodies were immune to the products we made.

Rates of birth defects are rising rapidly, as are cancer rates and other problems. Human sperm count has been declining seriously over the last

25 years, and much sperm is damaged. These and other problems are associated with man-made chemicals that are now in the bodies of most creatures. These chemicals interfere with our endocrine system, especially during the early weeks of foetus development.

We thought technology could replace what nature does.

We failed to understand the extraordinary complexity of nature's topsoil and depleted much of its capability with powerful herbicides, fungicides, fertilizers and chemical waste. Now, with gene modification, we seek not the survival of the fittest but the survival of the most profitable. New technology is essential to our future, but we must use it with appropriate respect for the deep complexity of nature. We should not try to replace the wisdom of billions of years of evolution with our own cleverness but, instead, should build thoughtful partnerships with nature.

We thought we could manage society in simplistic ways.

Over the course of human civilization, totalitarian governments of different types imposed crude rules for running their societies. The results were catastrophic. Much of humankind's history is the story of dictators, bureaucrats and kings who had no idea how to manage their domain. Today we understand better what forms of governance are likely to work well.

What led humankind into deep trouble in the 19th and 20th centuries was the assumption that nature could be plundered freely and endlessly. What will get humankind into trouble in the 21st century is a similar attitude about technology—that there are endless good ideas to be discovered in the laboratory and that they are for us to use. Thus, corporations race to make a profit from each technological discovery or invention.

The problem with this attitude is that we are becoming increasingly capable of ripping nature apart. We can spread genetically modified life forms without knowing the consequences. We can spread artificial chemicals that subtly interfere with the mechanisms of our bodies. As we learn not to destroy nature directly, we continue to damage it in invisible ways.

It doesn't make sense to have a blanket antitechnology view. Better technology is essential, and future generations will shudder at the crudeness of much of what we work with today. The applications of technology will become increasingly advanced, and we will need to be much more careful about avoiding the attendant risks.

The 19th-century colonist would have been furious at the suggestion that regulations should slow down his innovations, especially the profitable ones. The scientist or corporate leader today is similar in that the race for profits causes an absence of caution. In many situations, the desire for short-term benefits overwhelms the desire to solve long-term problems.

TRAGEDIES OF THE COMMONS

In England, the village green used to be shared by everyone in the village. People would graze their sheep on it. Land that was shared in this way was called "the commons." It supplied just enough grazing to support all the sheep. An individual, perhaps with a spirit of initiative that elsewhere could seem admirable, might decide that he can benefit by putting more sheep on it. If many people did that, the commons would become over-grazed and everyone would be worse off. In a village, it is obvious what is happening and the situation is easy to control. A town meeting would decide on the maximum number of sheep that anyone could graze.

The tragedy of the commons is a term used by economists to indicate a shared resource that is overexploited. Because the resource is free, people increase their use of it until its use is destroyed for everyone. Many resources in our global environment are a shared "commons"—the oceans, the rivers, the fish, the atmosphere and unseen parts of our ecology such as the water table and the ozone layer. The modern world has created such new "commons" as motorways, the radio spectrum, the Internet and the geosynchronous satellite orbit (in which satellites appear to be stationary).

Use of the old village commons was regulated, and use of these resources should be regulated, and regulation can take on different forms: You buy a fishing licence to catch trout. A hunter is allowed to shoot only one bear per season. A telephone company can buy the right to use a defined and limited portion of the radio spectrum in a specific geographical area.

Until recently, we took many of the Earth's "commons" for granted because they seemed so vast and inexhaustible—the atmosphere we breathe, the jungles, rain forests and oceans—but now technology gives us the capability to lay waste to these resources. As population, affluence and the drive for profits grow, great damage can be done. The "tragedy of the commons" used to be about the village common land; now it's about the whole planet.

Many "commons" we depend on have become global issues. One of the biggest "commons" is the Earth's oceans.

Seen from space, the Earth looks like an ocean planet, dominated by blue seas and most of it shrouded in clouds of water vapour. The oceans cover 71% of the Earth's surface. Their greatest depth is more than the height of Everest. The amount of land above sea level is a fraction of the volume of water below sea level. When one sails across the oceans, they are so vast that it seems inconceivable that we have destroyed 90% of the edible fish in them—but that is the case, and the fish that are left are much smaller than their ancestors. This harm done to the ocean is not directly visible to the public, so less attention is paid to it than if it were something we could see clearly.

DEATH ON THE GRAND BANKS

At the beginning of the 17th century, the Grand Banks off Newfoundland held unbelievable quantities of fish. The skippers of English fishing boats reported cod shoals "so thick that we were hardly able to row a boat through them." Cod were the most numerous of many fish living in profusion in the Grand Banks, but there were also halibut, haddock, pollock, flounder and plaice. Sturgeon 8 to 12 feet long choked New England rivers, and children used hand rakes to collect buckets of 10-to-20-pound lobsters to use as pig feed.

Cod was then a very hardy fish that grew to be between 5 and 6 feet long, and it was incredibly fecund. Today, a typical cod is about 18 inches long. A female used to produce 9 million eggs in a single spawning and spawned 10 to 15 times during her life. The species has existed for millions of years, through ice ages and spells of global warming that changed the ocean levels by 300 feet. It has adapted to all *nature's* changes but not to the challenge of man's modern fishing technology.

In 1951, a curious ship flying a British flag arrived on the Grand Banks. It had tall funnels and was the size of an ocean liner, but at its stern was a massive ramp, like that on a whaling boat. This was the world's first factory-freezer fishing ship. It had been built by a whaling firm that realized the stocks of whales could become commercially extinct and so reapplied its technology towards the world's fish. This one ship could catch more cod than all the other boats combined that were fishing in the North Atlantic.

Soon there were hundreds of such ships. They used radar, sonar, fish-finders and echograms to pinpoint and capture whole schools of fish and hauled up their giant nets even in winter gales. They could be recrewed and serviced by oceangoing tenders; therefore, they could fish almost indefinitely, 24 hours a day, 7 days a week, without visiting a port. Such ships started strip-mining the sea, rather like clear-cutting a forest.

By the 1970s, the stocks of fish on the Grand Banks were nearly depleted. Factory ships from Russia, Europe and Japan had wiped out an area as prolific as the Grand Banks in a decade. It was clear that the New-foundland fishing industry would be destroyed unless something was done.

In 1977, Canada extended its territorial limit to 200 miles, as Iceland had done, in accordance with the United Nations Law of the Sea. Most of the fish of the Grand Banks were inside this limit, and Canada acted aggressively to keep foreign fishing ships outside the limit. Newfoundland fishermen cheered, and the government of Canada assured them that now they had a great future. The Canadian government was in a position to save the life of the Grand Banks. This was no "tragedy of the commons" because the Banks (except for a small area) were under the control of one government department that had excellent marine scientists.

But the declining Newfoundland economy drained tens of millions of dollars a year from Canada's treasury. The government's solution was to encourage Newfoundlanders to catch more fish. Canada built its own deep-water trawler fleet and provided generous fishing subsidies that encouraged thousands more people to join the Newfoundland fisheries at precisely the time when fish stocks were collapsing. The number of inshore fishermen grew from 13,736 in 1975 to 33,640 in 1980. Atlantic Canada's fish-processing sector grew by two and a half times in the late 1970s. This growth meant that more fish had to be caught to keep the fish-processing industry busy. Because of this, joint ventures were set up between fishermen and the foreign factory ships that had caused the devastation in the first place, allowing them to fish inside the 200-mile limit if they handed over part of their catch to Newfoundland fish-processing plants. Both long-experienced fishermen and scientists knew that it was a serious mistake.

By the mid-1980s, the Grand Banks catch was declining rapidly and the fish that were being caught were noticeably smaller. By 1988, the marine scientists' models showed that the stocks of fish in the Grand Banks were on the brink of collapse, with cod being in the worst shape of all. They insisted that the fishing quotas be cut by half, but nervous politicians compromised and cut the quota by only 10%.

The mass of spawning cod had been about 1.6 million tons when the first factory ships had come in 1951. By 1991, it was only 130,000 tons. Refusing to face reality, the Canadian government set a quota of 120,000 tons *per year* to be fished. Soon the mass of spawning cod fell to only 22,000 tons. Large quantities of juvenile cod, too young to spawn, were being caught.

In July 1992, the Canadian government did what it should have done years before: It closed the Banks to cod fishing so that the stock of cod and other fish could recover, but it was too late. A codfish doesn't reach sexual maturity until it is six to seven years old. Various scientific surveys of the Banks showed that not a single generation of juvenile cod had survived to age three, let alone to breeding age. The cod were not coming back.

The social and economic costs of closing the cod fisheries were enormous. Hundreds of small communities were decimated. The Canadian government had to spend billions of dollars to support them, and 32,000 fishermen were thrown out of work.

THE PARABLE OF THE BLACK SEA

The story of the Black Sea is a parable of our time with serious lessons.

People of the Republic of Georgia tell a story that when God was dividing up the Earth, the Georgians showed up late because they had been partying all night. God was out of land, but he gave them the piece of land he had been reserving especially for himself on the coast of the beautiful Black Sea with its warm beaches, green surroundings and snowcapped mountains.

The Black Sea is deep, more than 7,000 feet in places. For thousands of years, it was a prime fishing area that over the centuries sustained ancient Greece, Byzantium, the Ottoman Empire and Imperial Russia.

The Danube flows from the Black Forest to the Black Sea, but it is hardly the *blue* Danube of Johann Strauss. It crosses several Eastern European countries, all of which dump into it unprocessed sewage, oil, pesticides and toxic industrial waste.

The Danube has a delta of 2 million acres that filters its water before it reaches the Black Sea. This delta used to sift out the river's toxins and algae. Unfortunately, Nicolae Ceaușescu, Romania's murderous dictator, ordered the delta drained and developed because he thought it was a waste of Romania's real estate. At a time when many countries were increasing the pollutants being pumped into the Danube, Ceaușescu destroyed the filter that protected life in the Black Sea.

The Danube and lesser rivers carried excess fertilizer from large numbers of farms into the Black Sea. This made algae bloom spectacularly and caused an explosive growth of the zooplankton that feed on the algae. These creatures began to consume most of the oxygen in the sea, which led to a condition called "eutrophication."

The large underwater meadows of seagrass and enormous forests of kelp in the Black Sea were a major source of oxygen, and they fed and sheltered 170 animal species—sponges, anemones, crabs and other creatures that were essential parts of the food chain. The pollution killed the seagrass and kelp forests, along with the creatures that lived there. Eutrophication and industrial pollutants wiped out most of the fish, and the sea became as green and bad-smelling as a stagnant pond. The stench of dead fish descended on the streets of the once-fashionable resorts of Odessa and Yalta.

When an ecosystem is weakened, an aggressive predator can take over. In 1982, Soviet scientists noticed a creature in the Black Sea that they hadn't seen before, and it took time to identify it. It was a bell-shaped comb jelly, *Mnemiopsis leidyi,* which was native to brackish-water estuaries on the East Coast of the United States. It must have travelled across the ocean in the ballast water of a ship. This jellyfish has voracious eating habits, and it found a profusion of food that it liked in the Black Sea. The creature opened its wide jelly jaws and vacuumed up the dense concentration of microorganisms. It sucked up fish larvae, baby shrimps, crabs and molluscs, grazing until the sea was almost devoid of fish life.

As they had seemingly endless food and no predators, the comb jellies multiplied with almost unbelievable fecundity. By 1990, there were more than a billion tons of them—more than the weight of all the creatures that all the fishermen in the world land in a year. The Black Sea beach resorts then had no usable beaches. The fishing towns had no fish. In a few years, the Black Sea had gone from good health to a state of collapse. The élite dachas of top-level officials of communist Russia were abandoned because of foul American jellyfish.

An important part of this and other such stories of our time is that marine scientists had computer models that showed them that the Black Sea was being destroyed. They knew how to stop the destruction (although they would have argued about the finer details), but governments took almost no notice of the scientists' findings. What happened with the Black Sea was an entirely preventable catastrophe, but it played itself out to a devastating conclusion.

Once the economies of the Black Sea towns were in ruins, the politicians were under pressure to find ways to correct the problem. Only then did they start to cooperate. On 31 October 1996, six Black Sea countries—Bulgaria, Georgia, Romania, Russia, Turkey and Ukraine—signed the Black

Sea Strategic Action Plan. It is a blueprint for restoring and protecting the Black Sea. It contains the most comprehensive set of measures ever undertaken to restore a sea. Specific measures are defined for reducing and monitoring pollution, managing the resources still living in the sea and controlling human development. It also calls on the Danube River basin countries to reduce pollutants and nutrient loads from agriculture, industry and domestic sources.

The Black Sea is a sobering object lesson in halting the damage in other "commons" before it leads to similar, or far worse, tragedies. There could be similar collapses of larger marine ecosystems.

Scientists know how to limit further damage, but governments know that the necessary changes could cause job losses and declines in tax revenues. Although a democracy needs sound public knowledge to help enlighten political actions, the public is spectacularly ignorant about many large-scale scientific issues.

The head of science at the British Environment Agency told me that the biggest threat to our well-being is our limited ability to use all that we know to make policies that yield practical sustainable benefits.

LETHAL OCEAN TECHNOLOGY

The latest technology deployed by fishing companies is devastating. Supertrawlers pull nylon nets thousands of feet long through the water, capturing everything in their path—400 tons of fish at a single netting. Recently, a net was introduced with a mouth the size of 50 football fields. Bottom-scraping nets scour all life from the bottom of the ocean like giant machines clear-cutting a forest. Roughly a third of what they catch is not sufficiently profitable to use, and it is chopped up and pumped back into the ocean. It is referred to as "bycatch." These supertrawlers stay at sea for months at a time, processing and freezing their catch as they go. This efficiency gives them a massive advantage over traditional fishing boats.

The supertrawlers can fish as deep as a mile, catching species that wouldn't have been considered edible a decade before. They catch squid, skate, black oreos, rattails, black scabbard, red crabs, chimeras, slackjaw eels and spiny dogfish. Creatures not previously harvested are cooked into fish sticks or processed into fake crab meat for seafood salads and sushi.

New trawlers that fish in deep waters can catch those fish that spawn late in life. In the 1990s, fashionable people in the United States and Europe were told in smart magazines that the thing to do was to eat high-priced orange roughy. This fish is caught in cold seas a mile deep off New Zealand. The orange roughy lives to an old age and only begins reproducing at thirty. The factory trawlers caught orange roughy that were well below that age, and the poor things never lived to spawn. The principal stocks of them collapsed.

Because factory trawlers are so expensive to operate, their owners need to keep them busy. The financial stakes are high. The ships keep fishing until the major fish are gone, and then they catch juvenile fish that have not yet spawned. As these are depleted, the technology is used to catch fish lower in the food chain. As the lower levels of the food chain decline, the chances of revival at the top of the food chain are destroyed.

THE CATASTROPHE-FIRST PATTERN

Where there are warnings of environmental danger, it is often the case that humans don't pay enough attention. Sometimes it is only after a catastrophe happens that appropriate precautions are taken. This catastrophe-first pattern is observed in many different areas. I'll refer to it repeatedly in this book. It describes the pattern in which we tend to deal with severe problems only after an unspeakable catastrophe forces us to take them seriously.

In 1962, the thalidomide tragedy came to public notice. Eight thousand children were born with appalling deformities because their mothers had taken this prescription drug when they were pregnant (to treat morning sickness). Babies were born without arms or legs. Some babies had hands sprouting directly from their shoulders. The photographs in the press shocked the world and caused the US Food and Drug Administration to take tough action. All drugs were subsequently tested thoroughly for their effects on pregnancy.

The catastrophe-first pattern is not a good way to run the planet, but it will become increasingly prevalent as the 21st century progresses because the possible catastrophes will become larger. We can't afford a catastrophe-first pattern. An essential part of the 21C Transition is the modelling and science that enable us to anticipate a catastrophe and prevent it from happening. We must make sure that the authorities don't ignore the science as they did with Hurricane Katrina in New Orleans.

To avoid a catastrophe-first pattern, the politicians and the public must listen to the scientists. Some future catastrophes are highly probable, but the public is indifferent, and this public indifference changes to fear when the catastrophe starts or becomes inevitable. By then, however, it is too late to implement preventive measures or control procedures. This may be true with severe climate change or with the spread of a massively deadly pandemic.

The evidence of overfishing is indisputable. We are catching far more fish than the oceans can replenish. It is like a once-wealthy person continuing to spend more money than he has—and winding up filing for bankruptcy. If similar destruction occurred on land, there would be an enormous outcry. Because we can't see the ocean destruction, we say nothing.

OCEAN RECOVERY

The oceans need not be destroyed. Very slowly they can be recovered and fished in a well-managed, sustainable and profitable fashion. A key to ocean recovery is to set up no-fishing zones called marine protection areas. These have been set up in many countries, and scientists around the world have measured the results. Often, fish populations increase and reach equilibrium in approximately three years. If fish have been caught before they reach sexual maturity, it takes much longer to return the population to equilibrium. Some fish species migrate, and the migration routes between marine protection areas also need to be protected.

Scientists estimate that, in order to have sustainable fish populations, at least 20% of the oceans need to be marine protection areas. At present, the figure is less than 0.01%. As with many of the problems described in this book, we know what actions to take, but they are being applied on a scale that is hopelessly inadequate.

Under the United Nations Law of the Sea, all coastal nations have near-total jurisdiction over the seas within 200 miles from their shore. Ninety per cent of fish stocks and most of the world's breeding grounds are within these 200-mile limits. If all countries enforced their 200-mile limits and controlled fishing within them, much of the ocean fish could be slowly brought back.

In 2005, Britain produced a highly detailed and scientific plan for how to run its fisheries. To do so requires a carefully planned network of marine protection areas. Britain's fishing industry has begun to accept marine protection areas as a way to save the industry. It's also necessary to ban bottom trawling and some other types of fishing. To make Britain's fishing industry sustainable, the capacity of its fishing fleet has to be limited. Such measures would enable ocean life to slowly recover and would allow British fishing to be a viable and profitable (although reduced) industry.

Britain will encourage the rest of Europe to follow the same rules. In principle, such rules could be made enforceable everywhere. The marine countries of the world should agree on a global fishing treaty. Every fishing boat, worldwide, over 8 metres in length must be registered and required to transmit its position—as determined by GPS (the Global Positioning System)—to an international fishery-management computer. Violators would be fined.

Because it is vitally important to protect marine food chains, many protected areas need to extend on shore and include wetlands, mangrove forests and river deltas. Where practical, it is important to restore wetlands that have been damaged. Rivers taking pollution to the ocean need to be cleaned up. Woodlands should be planted by rivers to keep farm runoff from flowing freely into the rivers. Where the sources of pollution flowing into a river have shut down, the river has usually cleaned itself with fresh water in a year.

Local and national governments ought to ban fishing technology that is excessively harmful. Trawling equipment has been designed that glides several feet from the bottom of the sea, which avoids scraping the seabed, and fishing equipment has been developed that allows small fish to escape, which avoids killing so many dolphins and turtles.

If the seas are ill-managed and overfished, as they are today, fishing is not profitable. To compensate for this, governments have given their fishing industries outrageous subsidies. In 1995, according to the Worldwatch Institute, $124 billion was being spent annually to catch $70 billion worth of fish. These numbers sound insane—no normal business could be run that way. The $54 billion difference is government subsidies, most of which foster overfishing.[1] The world has twice as many fishing boats as are needed for the entire world's sustainable fish catch, but governments are spending many billions of taxpayers' money to build yet more super-trawlers. Factory boats capable of scooping up all life on the ocean bed and pumping back the dead bycatch are subsidized by top politicians. It is outrageous that politicians, in return for contributions or other favours, hand out massive public funds to help destroy the environment.

In the 1990s, the world formed an agreement, with difficulty, on how to stop the slaughter of whales. Southern right whales were almost wiped out; only 1.5% of their population remained. Once the slaughter stopped, their population gained strength, and large numbers of these gentle giants are now seen happily frolicking off the shores of South Africa in the spring. Their numbers are increasing by 20 or 30% per year. Two or three decades from now, there will be as many as there were before man started killing them.

If reasonable controls are put in place, the oceans by the end of the 21st century will be healthy, vigorous and manageable. If we continue as we are now, the oceans will be totally destroyed. We have other stories to tell with a similar choice of endings.

There is a core of deep intelligence in humankind amid the broiling commercialism and greed. We now have excellent science of ocean fisheries, and it is getting better as the oceans and their fish become better researched and instrumented. Some scientists who rise to powerful management positions are wonderfully articulate at explaining their science. One such scientist is Robert Gagosian, head of the impressive Woods Hole Oceanographic Institution in Massachusetts. He commented in my interview with him that the only way the ocean "commons" problem can be solved is for the countries of the world to get together. He thinks that will happen when many of them have almost no fishing business left. They'll come to their senses and say, "OK, we've got to really sit down and resolve this issue." Gagosian adds, "It takes people five times of getting hit with a two-by-four across the back of the head before they really realize what's going on."

Establishing the mechanisms for protecting our "commons" facilities is part of the meaning of the 21st century, and as we will describe, extraordinary new commons facilities will be created.

3

RICH KIDS AND THEIR TRUST FUNDS

THE DESCENDANTS of a wealthy person sometimes have a trust fund—a large endowment that is being managed to provide them with an income. If it is managed well, the capital remains in place so that the income can be used year after year. A wanton individual might, of course, spend so much that the capital declines. He might reduce the capital so much that there will be little or none left for the next generation.

We humans have a spectacular trust fund provided to us by nature. We have fish in the oceans, grasslands on which we can graze cattle, nutrient-rich soil, an abundant water supply for growing food, grand forests that help keep the atmosphere clean by absorbing carbon dioxide and producing oxygen, and a planet whose intricate ecology provides us with a beautiful place to live.

In financial terms, the value of this trust fund is enormous, but like the wanton rich kid, we are using it up. We are overfishing the oceans, depleting the topsoil, lowering the water tables, polluting the environment and overgrazing the grasslands so that they are sometimes reduced to desert. Every year mankind burns an amount of fossil fuel that took 10,000 years to create. One-third of the world's forest areas has disappeared since 1950, and the destruction is accelerating.

We can't go on doing this for much longer because using up this trust fund would spell disaster on a grand scale. Running out of water means starvation in many countries. So does overgrazing the grasslands, destroying the soil quality and destroying life in the seas.

Living beyond the means of our trust fund would be inadvisable even if the population of the Earth remained constant, but the number of people added in the next 20 years will be more than the entire population of the Earth at the start of the 20th century. In the lifetime of most readers of this book, the population will increase by 3 billion people, almost all of that increase being in the countries least able to conserve food-growing resources. The Earth's population will have grown from roughly 2 billion to 9 billion in the hundred years from 1940 to 2040. If the world's people are to be reasonably nourished, food production will have to double in the next 30 years.

From 1980 to 2000, the US Dow Jones Index climbed from 839 to 11,000, but during the same time, every measure of the Earth's trust fund was seriously down. The once grand Soviet Union had an economy that hid the truth, and this deception led to its massive collapse. Capitalist economies are hiding the truth in a different way, and unless we change, that will also lead to a massive collapse.

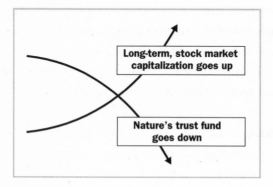

It is difficult to put a precise value on the trust fund. In the 1990s, a group of economists estimated that 17 of the services we receive from nature have an average value of $36 trillion per year with a high estimate of $58 trillion (1998 dollars).[1] The global economy was then estimated to be $35 trillion. Although there could be many squabbles about how to do such a calculation, it's commonly accepted that the trust fund is larger than the global economy.

RAIDING THE COOKIE JAR

It once seemed that the Earth's resources were boundless. Topsoil would renew itself, rain would keep the aquifers full, and fish would breed in profusion. Now we know that they are not boundless. Today many of our forms of development are not sustainable. The Earth has finite resources, like a cookie jar—and we are raiding it.

No one disagrees that it is morally right to bequeath to our children the planetary resources we ourselves have had. If we deplete the water supply, damage the topsoil and cause global warming, we are, in effect, taking vital resources from future generations. Preserving nature's resources for future generations is a desirable part of *sustainable* development. We are nowhere near sustainability, however. We are leaving future generations an increasingly depleted planet. We are stealing their cookies.

We are also leaving them technology with spectacular new capabilities. Our legacy to the next generation is genetic modification, nanotechnology, an ultrafast Internet, fuel cells, new nuclear energy technology and better medicine. Each generation lives in a world with fewer natural resources but more advanced technology. Each generation is falling deeper into a technology trap—it can't survive without technology, and its technology is becoming ever more advanced. We are changing the kind of capital we need.

We are doing something worse to our children if we *make the planet unmanageable*. Degrading natural resources and causing global warming leave the next generation with serious problems. Allowing weapons of mass destruction to fall into terrorist hands is asking for trouble. Perhaps the worst thing we can do, however, is to allow the Earth's population to grow unnecessarily.

NATURAL CAPITAL

The term *capital* refers to accumulated wealth. Human-made capital is in the form of investments, factories, cars, houses, equipment, software and so on. Natural capital refers to nature's resources: water, air, oil, minerals, natural gas, coal and living systems such as forests, grasslands, wetlands, estuaries and the oceans. Some of these are very important to us. Human-made capital is produced by human activity; natural capital is not. Much natural capital is nonrenewable, and we are depleting it.

Natural capital is often so natural that we don't think about it, just as a fish doesn't think about the water it swims in. We don't think about what makes our air breathable, why we need insects and microbes, what the wetlands do for us or how the detergent-filled runoff from our dishwasher might harm the wetlands.

In the 21st century, the obstacle to prosperity is not a lack of *human-made capital* but a lack of *natural capital*.[2] Our economy is totally dependent on a depleting supply of natural capital. It is estimated that, in the last half-century, the Earth has lost a fourth of its topsoil and a third of its forest cover. We are losing fresh water at the rate of 6% per year. A third of the world's natural resources were consumed in the last three decades.[3] Most were consumed by the billion people in the rich countries. This is alarming when you reflect that the heavy-consumer countries will soon include China and India and others. Three billion more people will join the heavy-consumer club.

The fish, topsoil, trees, water, fertile land and nature's resources in general have immense value to humankind, and every month about 2,000 species disappear from the planet, but on the books of the corporations, they have no value. As we deplete topsoil and lower the water tables, farmland is being converted to towns or factories. The acceleration is alarming. When Britain had its Industrial Revolution, it took a century for it to double its income. After America began industrialization, it took 50 years. China took less than 10 years.

There are two types of natural capital: commodities and services.

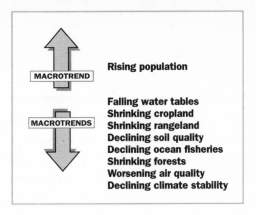

Commodities are the goods provided by nature—air, water, wood, oil, coal, minerals, fish and so on. Services include such things as the maintenance of a breathable atmosphere and the maintenance of an environment in which lush vegetation can grow—the rain, sun and winds; numerous microscopic creatures that help make the soil fertile; insects for pollination; and seeds that travel, blown in the winds and carried by insects.[4]

We can slowly find substitutes for many of the commodities of natural capital. As oil runs out, we will find other forms of energy. If iron ore ran out, we could make car bodies of carbon fibre. But there's no substitute for sunlight or water or the air we breathe.

While we can find substitutes for many of nature's *commodities*, it is often too difficult or expensive to find substitutes for its *services*. The pollination services that wild bees provide to blueberry farmers, for example, are 60 to 100 times more valuable than the honey the bees produce. Blueberry farmers describe wild bees as "flying $50 bills." In some areas, the bee population is seriously declining. There are many other ecosystem services. A river's delta filters out toxic chemicals before the water reaches the sea. The ozone layer protects us from harmful ultraviolet-B radiation. Nature decomposes our organic waste and regulates our climate. We need air pure enough to breathe, water pure enough to drink and the services of the floodplains and wetlands.

FALSE ACCOUNTING

The most common number used to rank a country's economy is the GDP (gross domestic product)—the annual revenue generated by the domestic economy. This number is divided by the population of the country to give the much-quoted figure GDP per capita. Using the average revenue generated per person annually, countries can be compared in terms of how much people earn. For example, for 2004, the GDP per head of the United States was $41,530; that of Egypt was $1,030.

Although these are the accepted numbers for describing the size of economies, we have an accounting anomaly of enormous proportions. The calculation of GDP ignores natural capital. Many countries proudly announce that their GDP per capita is growing but, in reality, if the depletion of natural resources were considered, the GDP for almost all nations would be declining.

Corporations, in most cases, don't pay for the natural capital they use, nor do they figure it into their accounting. Fishing companies pay the cost of *catching* fish but don't pay for the depletion of fish stocks. Farmers don't pay for the water they take from an underground water supply. Oil companies pay for drilling and refining but pay nothing for the black liquid they take from the Earth. When humanity lays waste to a forest, we count the value of the timber as an increase in wealth but don't account for the damage done to nature's ability to absorb carbon dioxide. Corporate balance sheets put zero value on the Earth's resources.

Today's market system is remarkably effective in allowing people to decide what goods and services they want and having corporations meet those needs, but the prices don't reflect nature's trust fund. As a result, organizations waste precious resources, clear-cut forests and dredge the sea with life-destroying nets. Capitalist corporations are intensely driven to improve earnings. If resources are free, management takes advantage of them. "Pump up as much water as you want out of the aquifer; it's free!"

Not including natural capital in corporate or government accounting gives us a false view of our current balance sheet. Today's capitalism neglects to assign any value to the largest stocks of capital it employs—natural capital. It liquidates natural capital and calls it income. Chief executives have been given prison sentences for doing this with other capital but not with natural capital. Somehow we must account for the costs of natural capital. It's difficult to assess exactly what these values are, but approximate assessments have been made by many organizations. Any realistic value would be better than treating natural capital as having zero value.

If we proposed the use of accounting that reflects natural capital, there would be the most vociferous protests from business executives everywhere, but there may be more acceptable ways of achieving a similar result—fishing licences, for example, or a tax on carbon emissions or charges for removing water from an aquifer. In many places, the laws relating to aquifer use are chaotic or nonexistent. Eventually, if there are appropriate forms of payment for using natural capital, entrepreneurs would then invent new ways to profit within the guidelines of the new principles and benefit from the 21C Transition. In the long term, major new profits will be generated by products and services geared to planetary well-being.

Placing zero value on nature's capital encourages us to trash the Earth rather than trying to look after it. Like a primitive society that eats its seed corn, we destroy natural resources that are essential to the future. If we don't correct this and recognize the magnitude and consequences of natural capital depletion, we'll drift into deep trouble. Ultimately, it's essential that we employ policies that protect natural capital.

In addition to not paying for the natural capital they use, corporations often don't pay for the damage they do to the environment. Corporations didn't pay for the damage they caused by creating a hole in the ozone layer. Manufacturers don't pay if they pollute rivers. Bottom-scraping fishing fleets are destroying much of the oceans' web of life, but nobody pays. The damage caused by global warming will become immense. Air-conditioning systems are a major contributor to global warming, but people don't pay for the cost of that. Because they don't pay for it, they never normally think about it.

The technology used by mining companies has become so increasingly powerful that machines can grind up mountains of ore. Poorer-quality ores can be used, and this appears to make metals more abundant. It fails to account for the costs of stripped forests, the mountains of slag, destruction of villages nearby and toxic chemicals spilling into rivers. None of these are factored into the costs of production. A cleanup at *abandoned* mining sites in the United States is costing U.S. taxpayers an estimated $33 billion to $72 billion.[5]

EVIL SUBSIDIES

To make the false accounting even worse, huge subsidies are handed out by governments, and they often have nothing to do with how taxpayers want their money spent. How many U.S. taxpayers today, for example, would vote to give tobacco farmers $800 million a year?

Certain subsidies are important in order to make a complex society work well. There needs to be money for educating people who can't afford to pay for education themselves. Sometimes subsidies are a good investment, like the United States' DARPA (Defense Advanced Research Projects Agency), which developed packet-switching networks that eventually evolved into the Internet. Although such investments sometimes produce impressive results, most of them don't, but there is sometimes a strong case for government funding of research that private enterprise wouldn't fund because it has too long a payoff time.

The total subsidies for agriculture in the world's eight richest countries greatly exceed those in the 160 poorest countries. They add up to $350 billion per year. In Europe, subsidies and quotas are used to suppress surplus food production and stop "lakes" of milk or butter from being sold, at a time when parts of the world are close to starvation. One sometimes sees fertile fields growing only weeds. Subsidies to farmers in rich countries make workers in poor countries poorer. They leave both the environment and the economy worse off.

Subsidies are supposed to help people, industries or regions weather financial or other disadvantages, but many subsidies leave the environment or the economy worse off than if the subsidy had never been granted. Environmental scientist Norman Myers refers to these as *perverse* subsidies. He set out to do an approximate accounting of the world's perverse subsidies.[6] He found this difficult because governments often refused to reveal information about payments; they are often concealed. Subsidies are not officially tallied in the United States or in most other countries.

Myers's list of perverse subsidies is enormous. They total $2 trillion per year—larger than all but three of the economies in the world. As Myers was being televised in drenching rain, he said, "It's insane. If governments got rid of even half of their perverse subsidies, they could get rid of their budget deficits at a stroke, they could increase their health and education spending by at least 50% and, with the amount of money left over, they could throw a week-long party for the whole country."

The average American family pays $2,000 per year in subsidies but doesn't know it. Since many of those subsidies damage America, the taxpayer ought to know the facts and be able to express outrage.

Some subsidies are thought of as being necessary to help the poor, but many help the rich at the expense of the poor. The rich know how to manipulate the political system; the poor don't. The total overseas aid given to developing and least-developed nations is only 2 to 3% of the money wasted on perverse subsidies.

It's difficult to imagine a worse job than being a coal miner—to hew coal with a pick deep underground in cramped tunnels in an atmosphere that causes lung disease. Germany used to pay $6.7 billion ($73,000 a year per worker) to subsidize the hopelessly uneconomical coal mines of the Ruhr Valley.[7] It would have been much more economical to close all the coal mines and send the workers home at full pay. Imagine what $73,000 a year could do to train a man to do a worthwhile, enjoyable job. Some perverse subsidies are being reduced, and this is one example.

Gasoline prices at the pump in the United States are a third of what they are in most of the world. American subsidies to fossil-fuel industries that help cause global warming exceed $20 billion per year. Subsidies for fuels that will help avoid global warming are less than $1 billion per year.[8]

There would be huge leverage in stopping harmful subsidies everywhere. All of today's subsidies should have planetary-correctness ratings. The public could have access to a list of all subsidies sorted by this rating so that the most harmful ones are at the top of the list.

The world's multinational energy companies feel the need to tell the public how much they care about the environment. One of them invited me to visit its corporate headquarters in the Southern Hemisphere, which was heralded as a *green* building. The building used only a few solar panels (its architect explained that it was cheaper to buy electricity from the electric company), and it caught almost no rainwater. Both electricity and water in that region were artificially cheap, thanks to government subsidies. The profitable business decision was not to be ecologically correct but to use public relations to persuade the public that one *is* ecologically correct. This is referred to as "green wash."

DESTROYING OUR HOME

In Roman times, you could walk through trees for the whole length of the North African coast. The rich landscape has since been turned to desert by overgrazing, deforestation and salination caused by irrigation.[9] In our fathers' time, there seemed to be unlimited water, soil and grassland, and it was natural to assume that such bounty was free. Nature was grand enough to absorb our pollution, and the forests seemed to have little use other than to provide us with wood for building and burning. Now we realize that forests absorb the carbon dioxide we breathe out and replace it with oxygen. If there were not enough plants to do this, the Earth's atmosphere would become toxic. We take this service for granted because it's invisible and free, but we are close to exceeding the capacity of nature to recycle all the carbon dioxide we produce.

Seven percent of the people on Earth use 80% of the available energy. Many others want to catch up, but if everybody used as much energy as the 7%, the strain on the planet would be intolerable. Media-driven consumption patterns are taking root in Asia and India, outstripping the resources the Earth can provide.

Twenty years ago, I took photographs of a happy-looking area in Indonesia that seemed a model of village bliss. Hillside villages had rice paddies that had been there for many centuries, with fruit trees, flocks of ducks and lush jungle with colourful flowers. Recently, the residents were persuaded to sell the colourful batik they made, and they cut firewood to heat the vats in which the batik was dried. Dreaming of profits, they stripped too much wood. Then the hillsides were washed away by rain. It took only three years for their little piece of paradise to be destroyed.

ECOLOGICAL FOOTPRINT

The term *ecological footprint* is used to give people an idea of how much they consume of nature's resources.[10] A person with a 10-acre footprint uses the equivalent of 10 acres of the Earth's resources.

There are 5.3 acres of land, on average, for every person in the world. In 2000, the average person in the world used 6.9 acres' worth of resources and ecoservices. The average American uses 24 acres' worth. A typical American consumes energy, water and other natural resources equivalent to the consumption of 140 people in Afghanistan or Ethiopia. The average person in Britain uses 11 acres; the average person in China uses 4 acres, but this number is poised to rise rapidly.

Four decades from now, the average number of acres per person will have dropped to 3.5. So, our world ecological deficit is rising rapidly. Three main factors contribute to this: declining resources, growing population and lifestyles of increasing resource consumption. Of these, the last is, by far, the largest contributor. It is vital to develop a broad understanding of how better-quality lifestyles can be achieved with less impact on nature's resources.

A country is said to have an *ecological deficit* if the number of acres needed to support its lifestyle is greater than the number of acres that exist in that country. The United States has a deficit of 11 acres per person. In Japan, the deficit is 10 acres per person, and in the main European countries, the deficit ranges from about 5 to 9 acres per person (and 10 in Holland). China has a deficit of 1.2 acres per person, but that will rapidly increase. Poor countries like Pakistan, Bangladesh and Nigeria have a deficit below 1 acre per person. Some countries with a low population density have a *surplus*: New Zealand, for example, has a surplus of 23 acres per person, Australia 17, Brazil 14 and Indonesia 1.7.[11]

The Earth's ecological deficit can't last. We are using more water than the rain can renew, catching more fish than are spawned, cutting down more timber than can regrow, pumping more carbon dioxide into the atmosphere than can be absorbed and depleting topsoil that took tens of thousands of years to accumulate. Even if people with good management skills stop doing this, billions may not. We can fight the ravages described by Malthus if we have good management, but the world's massive population growth will be in the countries least capable of managing their water, agriculture, fisheries and forestry. Ultimately, managing these sustainably is a nonnegotiable condition for life.

THE PONDWEED PATTERN

Geometric growth is the mathematical phenomenon that causes many aspects of our world to become catastrophes with astonishing suddenness. Unless we understand how geometric growth works—and works in nature—we can be deluded into thinking that there is no cause for alarm.

For instance, a farm pond has a fixed size, and a farmer can be surprised how suddenly it can be choked with weed. He looks at it one day and observes that most of the surface has clear water. Suddenly, it becomes choked. Duckweed spreading on a pond can double its surface area in a day. Suppose that it takes exactly a thousand days to cover the whole pond. This is the coverage of the last ten days:

DAY	COVERAGE (%)
990	0.0976
991	0.1953
992	0.3910
993	0.7810
994	1.5630
995	3.1250
996	6.2500
997	12.5000
998	25.0000
999	50.0000
1,000	**100%**

Suppose it was your responsibility to keep the pond free of duckweed. On day 990, you might observe the duckweed and say, "I've been watching it for more than two and a half years, and it still covers less than one-thousandth of the pond's surface; so, don't worry about it. I'm going for ten days' holiday." When you came back, the pond would be wiped out, and the fish would be dead because of lack of oxygen.

An example of this type of danger is that a sea has a fixed size. Geometrically growing pollution in the sea may not be obvious to most people until, suddenly, it's unstoppable. Its geometric growth is deceptive because it reaches the fixed limits unexpectedly. The Black Sea weakened slowly at first and then collapsed with shocking suddenness. Overnight, one of the most beautiful places on Earth became so foul that its luxury resorts had to close. A foetid pond can be dredged but not a foetid sea.

If we continue on a deficit course, nature will take care of the problem, as nature always has. The deficit will end in the lifetime of most readers of this book. It will end either because we manage the situation well or because nature ends it with starvation, catastrophic climate change and shocking conditions for much of humanity.

The Earth Day organization describes it as "Heating our house by burning the furniture. Then the walls. Then the roof. Then the floor."[12] Running into ecological limits is not like a sudden train wreck; the limits are not obvious to most people, and they are easy to exceed. There may be no serious shortage of food or raw materials in the affluent countries. Rich, ingenious nations will even manage to maintain an affluent lifestyle. The limits will be masked by advances in technology. Grain prices will become very high, but this will benefit America. The richest countries will try to hold back floods of immigrants from the victim countries and may increasingly adopt a fortress mentality.

THE WAGES OF SPIN

Carl Sagan commented that the most scary thing of all is to live in a society dependent on science and technology where almost no one knows anything about science and technology. But perhaps one thing is more scary:

to live in a society dependent on science and technology where PR companies have become superbly skilled at making the public believe lies about science and technology. Corporations can hire them to do this for profit reasons. One of the early examples of this was tobacco companies setting out to create false science—phony experiments that "prove" that nicotine is harmless and nonaddictive, and expert witnesses paid handsomely to testify under oath to falsehoods.

Respectable science has a peer-review process. A paper for publication in a responsible scientific journal has to be reviewed by scientists not connected with the author before it is accepted for publication. The public is unaware of this when cleverly persuasive advertising feeds them phony science.

Politicians have become highly skilled at representing events to the public with a well-designed spin. Washington created "spin doctors," who became masters of deception. Some of the solutions that urgently need to be implemented would harm corporate profits—so, corporations also have spin doctors. Expensive PR campaigns persuade the public that profit-depleting actions are unnecessary. Fishing and coal mining represent two excellent examples: It is deeply cynical to use fleets of factory supertrawlers to clear-cut the oceans, knowing they are destroying the fisheries, and then use clever advertising to persuade the public that it's the best solution to the world's food problem—or to persuade the public that coal is the key to future clean energy, knowing that it would be a primary contributor to climate change and lung disease.

A democracy based on high technology and advanced media skills needs mechanisms for preventing the clever falsification of science. Could the falsification of science actually be made illegal? We need the brilliant sunshine of science and tutorial clarity to penetrate the cobwebby caves of vested interests, perverse subsidies, false PR, ignorance, misgovernment and corruption around the world.

4

TOO MANY PEOPLE

UMAN ACTIONS ARE OFTEN FILLED with deep ironies. From 1950 to 2000, the medical profession found ways to eradicate some dreadful illnesses and to keep people from dying, especially in the Third World. This was one of the great achievements of its time. Its consequences, though, were catastrophic for the planet because, while the death rate dropped, the birthrate did not. The world's population soared from 2.5 billion in 1950 to 6.5 billion in 2005 and seems set to rise to 8.9 billion a few decades from now. Excessive population growth leads to poverty, starvation, disease, squalor, unemployment, pollution, social violence and war. Many poor countries have become so destitute and violent that the social misery seems irreversible. In an in-depth interview, former Congressman Peter Kostmayer commented, "You really can't separate overpopulation from the environment, the economy or the political system. You can't separate it from political instability or injustice. Until we begin to deal with the population problem and begin to stabilize world population, we're not going to be able to solve any of these other problems."

In a few countries good governance and sound economic policies overcome the effects of population growth, but then another problem looms. If the population were to become 8.9 billion people, and most of them want to live like the new consumer class in China, the pollution, global warming and stress on the planet would be extreme. That is just too many people for the planet if they all want to live high on the hog.

Seasoned engineers quote a Law of Unintended Consequences: "A complex project will always have consequences that are unintended." The population explosion caused by better medicine is a grand-scale illustration of this.

A key task of the 21st century is to correct the situation. To do that, two things have to be achieved: first, lower the average birthrate; second, achieve eco-affluence in which wonderful lifestyles can be achieved with a sustainable use of resources.

AVOIDABLE DEATHS OF CHILDREN

In a poor country where a high proportion of children die, people usually choose to have many children, hoping that some will survive. If there is good health care, they generally choose to have fewer children. Good health care is one of several social factors that correlate with a drop in fertility rate.

Every year about 3 million children in poor countries die of diseases that could have been prevented by basic health care and vaccinations. About 30% of the world's children who ought to be vaccinated are not—a total of about 50 million vaccinations per year. The cost of getting a package of basic vaccines to a child is about $30. A death that is a heartrending tragedy for the parents often can be avoided for the price of a night out at the movies in the West.

Bill Gates discovered these numbers in the World Bank's World Development Report of 1993. At first he didn't believe them. He worked out that these deaths are equivalent to about 80 civilian jet crashes *every day*, each crash killing a hundred children—surely, he thought, that can't be true. He had the numbers verified and found that they were, indeed, correct. He knew that an expenditure that was a small fraction of his net worth could prevent most of those deaths. The realization that such leverage was possible was a factor in the creation of the Bill & Melinda Gates Foundation, which set out to establish many measures to improve health in poor countries. The foundation is demonstrating actions that could be taken by governments. By not taking such action, Gates maintains, governments are treating a human life in the world at large as being worth only a small number of dollars.

A SUSTAINABLE POPULATION

In the mid-1970s, the world's population was rising at an almost exponential rate. Demography experts then had models that forecast that it would reach 15 to 20 billion before it levelled out.

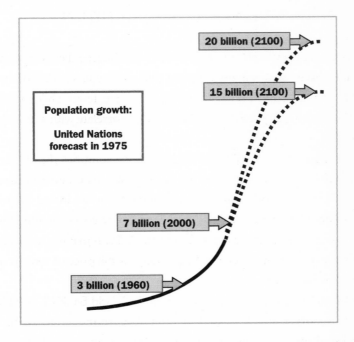

An intelligent observer from outer space would have looked at the Earth in 1975 and said, "You idiots. Find some way to put the brakes on. A world population of 15 to 20 billion would be unspeakably harmful. If you try, you can find ways to make people have fewer babies."

There has been substantial success in lowering the rate of population growth:

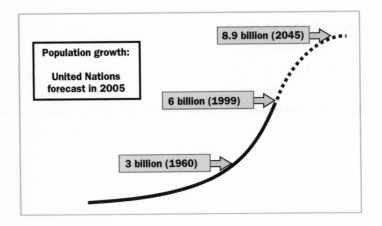

Geometric growth, like compound interest, if continued long enough produces very large numbers. If women on average continue to have more babies than the replacement rate, the population will grow relentlessly. Today's birthrate (much lower than two decades ago) produces a population increase of 1.33% per year. If that increase continued, after four centuries (as far in the future as Shakespeare was in the past), the Earth would

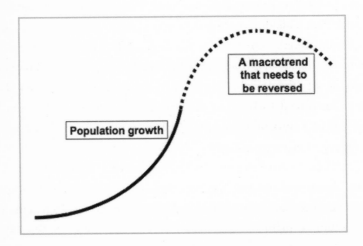

have over a trillion people. Clearly, that won't happen. Either we must bring down our birthrate or nature will decimate humankind in brutal ways, as it does when fish overbreed in a pond. We will eventually have to live within the resources of the planet.

CHINA'S ONE-CHILD POLICY

Until the 1970s, the Communist Party in China vigorously encouraged population growth. Then it vigorously encouraged the opposite.

The population growth had put a strain on the government's efforts to help its people. After the "Great Leap Forward," there was massive famine and widespread poverty. In 1979, when China had a quarter of the world's people, a *one-child* policy was adopted. This policy was not formally written into law, but was enforced by family-planning committees at local and county levels—all of which were under orders to achieve results. The policy advocated the following three main points:

- delayed marriage and delayed child bearing
- fewer and healthier births
- only one child per couple

In 20 years China reduced its population *growth* by about 300 million people. In 1999, China claimed credit for delaying by four years the birth of the world's 6 billionth person on 12 October 1999.

During China's "Great Leap Forward" and "Cultural Revolution," it became very brutal in the way it implemented social policies. This brutality continued with the one-child policy. Women pregnant with a second child were often forced to have an abortion and be sterilized. Reports have come out of China of pregnant women being handcuffed, thrown into hog cages and taken to operating tables in rural clinics.

The Chinese government claimed that its one-child policy was "voluntary," but it was an Orwellian definition of voluntary. Local authorities could fine a pregnant woman, lock her up, subject her to sleep deprivation and morning-to-night brainwashing sessions, fire her from her job, fire her

husband from his job and her parents from their jobs. As long as a pregnant woman walked the last few steps to the local medical clinic on her
own, the abortion was said to be "voluntary." In some cases, the enforced
abortion was so late that, when it happened, the foetus gave a little cry.
China is a very different country today, without such brutality.

Not surprisingly, women pregnant for the second time were clever at
avoiding the authorities. They left town, went to stay with friends or rela-
tives and had their second baby secretly. The resulting child was unregis-
tered; so it would be difficult for the child to be educated, but it might be
able to support the family.

Fertility rate is defined as the average number of children born per
woman. The fertility rate at which the population stays constant is 2.1. In
1970, the fertility rate in China was 5.9. It dropped to 2.9 by 1979, before
the one-child policy was introduced, because of a "late, long, few" policy,
meaning have the first child late, have a long time between children and
have few children. With the one-child policy, the fertility rate didn't drop
to 1. It eventually stabilized at about 1.7 on average. It is lower in the cities,
and higher in rural areas, but many Chinese are moving to the cities.

After the one-child policy was introduced, some Chinese people
decided to have a male rather than a female child. Ultrasonography made it
possible to determine the sex of a foetus and abort a female foetus. This
was made illegal, but somehow unwanted female babies were disposed of.
In 1997 the World Health Organization issued a report claiming that
"more than 50 million women were estimated to be 'missing' in China."
Fifty million Chinese men won't be able to find a wife, and prostitution is
illegal in most of China. In some countries the sex ratio is much higher than
in China. In Qatar there are 179.6 males, and in the United Arab Emirates
189.7 males, for every 100 females.[1]

In China, many second-generation children of the policy will have no
siblings, cousins, aunts or uncles. By 2020, one in four Chinese will be over
60 years old. In the streets of China today, one sees many beautifully
dressed and happy little girls—*only children,* pampered and loved by their
parents.

POPULATION AND POVERTY

Almost all of the world's population growth is happening in poor countries. The rich get richer and the poor have more babies. Many affluent countries are experiencing a decline in population, for example, Japan and many European countries, including Ukraine and Russia. Population growth in the United States, Canada and Israel is mostly from new immigrants.

Starving populations, strange as it seems, usually grow faster than well-fed populations. A dying plant produces more flowers. Poor people, expecting their children to die, make more children. More children mean more workers to care for parents in their old age. The world population is expected to grow by 2.5 billion by midcentury—about twice the population of China will be added to the planet. Tragically, almost all of this increase will be in areas such as shantytowns, with the least capability to feed, care for or find employment for the increase. The terrible conditions of the shantycities will get worse.

Many women in sub-Saharan Africa before 1990 had eight children. The HIV figures are dreadful, but in spite of AIDS, the population in most of Africa continues to grow. Medical care has declined woefully in much of the region. A country with widespread deaths from AIDS and threadbare medical care is hell on Earth.

I had a friend in India who was once the Minister of Population Planning. I once stayed with him when he was white-haired, 80 and long since retired. He had the wonderful ability to tell stories that some Indians have. He normally didn't drink alcohol, but he had a secret supply of Scotch that would come out after midnight when the conversation got interesting. With a tilting head and a musical Indian accent, he said seriously, "Do you know, I was the most unsuccessful executive of all time."

He told me that, from 1960 to 1995, India increased its average food per person by 17 percent and changed from being a major food importer to being self-sufficient—a gigantic achievement. However, during that period, India's population increased by half a billion. Providing the necessities for this population put an immense strain on the economy. Meanwhile the water table dropped, and it will continue to drop so that Indians will face major crises caused by water shortages. If India had managed to control its population growth, it would be incomparably better off now.

In the last hundred years, India's population grew from 300 million to over 1.1 billion; most of this has occurred since 1950. When developing countries have access to Western medical technology, fewer people die. India has a thriving middle class, but hundreds of millions of poor people earn less than one dollar per day. India's GDP per head (annual gross domestic product divided by the number of people) is $384. A small part of India is very high-tech today, with superbly skilled software developers and entrepreneurs. The government wants to expand such wealth-generating activities to as much of the country as possible. The United Nations has estimated that by midcentury the population of India will be about 1.6 billion. India, like China, already has severe pollution. Less than one Indian in a hundred has a car today. Imagine a future in which most Indians want a car and a home with air-conditioning.

The worst scenario of all could be that envisaged by James Lovelock—a planet with damaged control mechanisms and positive feedback that makes its temperature keep rising so that much of the Earth becomes uninhabitable or unfarmable. Such a planet could sustain a population that is only a fraction of today's.

LEVERAGE FACTOR

One of the world's vitally important leverage factors is the effect of illiteracy on the world's population growth. There is a high correlation between female illiteracy and fertility rate. Around the planet, the highest fertility rates are found in countries where most women can't read. When women are taught to read and have birth control methods available to them, the fertility rate drops dramatically. In the 1980s in many of the world's poorest countries, only 3% of the women could read, and the average number of children a woman had was seven, sometimes eight. In the United States around 1800, the average family size was seven children.

As female literacy spreads, fertility rates drop. For example, in Thailand, 87% can read, and the fertility rate is 2.6. When almost all women can read, the average number of children a woman has is often below two. Then the national population will start to decline.

There are comprehensive databases containing population figures and literacy rates. The numbers show a remarkably consistent correlation worldwide between increase in literacy and decrease in fertility rate. A chart of literacy rate against fertility rate for the countries of the world doesn't have smooth mathematical curves, but its message is unmistakable. Teaching women to read in poor countries is not expensive, and it slows the population growth.

The process described here is only partially complete. Female literacy percentages can be improved until they are over 90%, and as this happens, fertility rates will drop further, perhaps approaching those in developed countries. A goal ought to be that *no country* has a fertility rate higher than the population replacement rate of 2.1 births per woman. Fifty-one nations now have a fertility rate below that level today. Several of these are developing nations, among them Brazil, Bulgaria, China, Croatia, Cuba, Georgia, Lebanon, Kazakhstan, North Korea, Romania, Slovenia, Sri Lanka, Thailand, Tunisia and Turkey.

This impact of teaching women to read surprised people. As fertility rates started to drop, the United Nations adjusted its population growth forecast. In the late 1980s, it predicted a world peak population of about 12 billion people. In 1990, it forecast a peak of about 10 billion. Now the demographic models indicate a plateau of 8.9 billion between 2040 and 2050, and then population will decline, very slowly at first.

In 1989, the state of Kerala, in the far south of India, began a statewide campaign to make its entire population literate.[2] Kerala was one of the poorest parts of the world; the average person earned 89 cents per day. The campaign used the notion of the Brazilian educator Paulo Freire that the immediate problems in people's lives provide the best teaching material.[3] The Kerala readings centred on hunger, poverty, safe drinking water, housing and employment. The health lessons were coordinated with an immunization campaign to protect against common diseases. The goal was to make everybody feel involved.[4]

Thousands of literacy activists travelled through the state, stopping to perform songs and street dramas, holding public discussions and issuing calls to illiterate people to come forward and learn to read. They used the slogan *"Sakshara Keralam, Sundara Keralam"*—a literate Kerala is a beautiful Kerala. Classes were held in fields, courtyards and cowsheds. They took place on the seashore for fishermen. Volunteers collected 50,000 donated pairs of old spectacles for people with poor eyesight. They taught leprosy patients to hold pencils in stumps of hands with rubber bands. Nobody was left out.

In a decade, Kerala achieved a literacy level among both young and old people almost as high as that in the developed world. The birthrate has fallen to 18 births per thousand people per year, whereas it averages 40 births per thousand people in the world's 54 poorest countries.[5] In much of Africa it is over 60.

Kerala demonstrated that people in very poor communities could live longer, healthier lives, have fewer children and participate in democratic debates.

Kerala, so far, has remained poor. It's interesting to reflect what could happen if the Kerala-style drive for literacy were connected to a drive for people to start their own businesses. The reading matter of the drive for literacy could tell people the benefits of entrepreneurship and owning a company. The combination of Kerala-style literacy with micro-loans, entrepreneurship and appropriate technology can put poor nations on a staircase to a better world.

LIBERATED WOMEN

There are four stages in the evolution of the role of women that relate to fertility rate: First, when women learn to read and have birth control methods, the fertility rate declines. Second, when women have jobs, the fertility rate drops further. Third, when they become "liberated," it goes down further. Fourth, when women become ambitious and aspire to the most impressive jobs in a society, the fertility rate falls to well below the replacement rate.

As women become educated, they seek out interesting jobs and contribute to the economy, and women with interesting careers tend to marry later in life and have fewer children. Statistics from the developing world show that women with no education tend to have about twice as many children as do women with seven years of education. It's common sense to give women in society basic education, but many of the poorest areas in the world still don't do it.

After women become independent, as many are in the developed world, they can be ambitious and aspire to being executives, doctors, researchers and professionals. Top law schools may enroll more female than male applicants. Liberated women can enjoy sex, as men do, without having children—and because bringing up and educating children is far more expensive than it used to be, many couples are motivated to have small families.

Japan is a society with deep traditions that were thought to be unchangeable, but now increasing numbers of Japanese women are reluctant to marry; they want careers, financial independence and personal freedom. Japanese college students, both male and female, have started to have multiple sex partners. Japan's younger women are less willing to stay at home and be traditional wives. They question whether they want what old-style Japanese marriage offers them. Japan's fertility rate, now at 1.32, will probably fall lower in coming years. Japanese demographers now have serious worries that the nation will have too many old people.

In Spain in 1975, very few university students were women, and the average fertility rate was 2.9 children per woman. By 2000, 60% of university students were women, and the fertility rate had dropped to 1.12 children per woman. In Russia it is 1.14; in Singapore, it is 1.04.

Interestingly, although the Chinese were brutal in enforcing their one-child policy, many countries with no population-control policy achieved a birthrate substantially lower than China's. In Italy, the Catholic Church decreed that there should be no birth control, and Italian men thought of themselves as icons of sexiness, but the number of births per woman today in Italy is 1.2, as opposed to 1.8 in China. In Switzerland, it's 1.46, Greece 1.38, Russia 1.35 and Germany 1.35.[6]

These figures indicate that freedom of choice among liberated women lowers total fertility rate more than brutal methods of enforcement. One of the most extraordinary examples is Hong Kong (which is now part of

China). Although there was no attempt to lower the fertility rate, Hong Kong women had far fewer babies than those in mainland China. Taiwan used to have a fertility rate above 8, which now has dropped to 0.7—the world's lowest. Fertility rates are lower in the cities than in the countryside. As vast numbers of people migrate to the cities, fertility rates will drop. For cities that become as exciting as Hong Kong, they will drop well below the rate at which population declines.

In the Third World, soap operas on radio or television can be designed to help lower the fertility rate. Miguel Sabido, a former vice president of Mexico's national television network Televisa, pioneered this approach. Sabido created a series of soap opera segments on illiteracy. The day after a character in his show visited a literacy office wanting to learn to read and write, a quarter of a million people showed up at the literary offices in Mexico City alone. Eventually, 840,000 Mexicans who watched the show enrolled in literacy courses.[7] Sabido found that soap operas can change people's attitudes on a variety of social issues. He addressed the issue of contraception in a soap called *Acompáñeme,* which translates as "Come with Me." This serial ran for over two years. It was sexy and became very popular. It featured a poor young family with whom viewers could identify. The mother wanted to stop at three children but didn't know how. Her macho husband resented her efforts to try the rhythm method. As the show continued, it changed people's views on family size and the role of the woman in the family. Mexico's fertility rate fell 34% over 10 years. David Poindexter, at that time with The Population Institute and now honorary chair of Population Media Center, promoted Sabido's model in China, India, Brazil, Pakistan, Nigeria and elsewhere.[8]

In Kenya, radio serials provided an ideal vehicle for changing people's views on a variety of topics relating to population growth. Soap operas on radio and television can be inexpensive and remarkably effective in changing attitudes and behaviour.

It's a fortunate fact that we can lower fertility rate by increasing the quality of life of women. Give women interesting lives, and they have fewer babies. It is likely that world population will decline in the second half of the century, but it won't do so early enough to avoid the terrible conditions, starvation and civil violence caused by gross overpopulation.

5

THE GIANT
IN THE KITCHEN

POLITICIANS AND CIVIL SERVANTS usually tell me, "Food is no problem. We can feed the world." They quote an influential study by Amartya Sen, Nobel Prize winner in economics, that showed that almost all famines can be traced to poverty, not food shortage,[1] and modern famines are far more likely to result from armed conflict (such as those in Sudan, Ethiopia and Somalia) than from food shortages. This study, however, was published in 1981, when the world population was below 5 billion. In the future, not only will the world population be billions larger, but the rising consumer class in China wants to eat meat, which needs large amounts of grain, and China can't grow enough grain. Many parts of the world will have catastrophically declining water supplies. The Green Revolution, an extraordinary effort to increase food production from 1965 onward, is sputtering. Global warming will slowly reduce the farmers' yield in many areas; in certain countries where farming is already difficult, climate change will wreak havoc on food production. The comforting popular wisdom of the Nobel Prize winner has given governments a false sense that no action is needed.

Much food production won't be in the same areas as food consumption; so, key aspects of the question are, "How can the food be distributed to where it is needed? What happens if 2 billion or more people cannot pay for the food they need?"

Well over a billion people on Earth are spending about 70% of their income on food. Unfortunately, world prices for grain are likely to become so much higher than today that poor countries won't be able to afford them. The situation is exacerbated as half the world's population live in countries where the topsoil is being depleted, water is running out and good farmers are leaving.

Higher prices will lead to substantial increases in grain production in places like Brazil, Argentina and Ukraine. Large areas of land that have been uneconomic to farm in Brazil will be brought into production, and Argentina will produce more cattle, but that won't help countries where the people are too poor to buy food at inflated prices.

WATER WARS?

Perhaps the most critical resource of all is something so simple we don't think about it—fresh water. When I was young, good water seemed to be in infinite supply. I occasionally visited homes in the American Midwest, and my hosts would say, "We're lucky here; we've got the sweetest tap water." Now, people in the same homes assure you that they're giving you bottled water—because their tap water has been poisoned by insecticides, fungicides and herbicides that seep into groundwater and wells, often from crop-dusting planes. In my interview with Senator John McCain, he said, "I think water is going to be the prime issue of the 21st century."

Each day we drink about 4 to 5 litres of water in tea, juice, beer and other beverages, but to produce the food we consume that day requires more than 2,000 litres of water. If there are water shortages, they translate into food shortages. Ninety percent of all the water we use in the world is used to produce food. It comes from two sources: rainwater and natural underground water called aquifers, where it may have existed for millions of years. In many parts of the world, aquifers are being run dry. The annual depletion of aquifers worldwide amounts to at least 160 billion tons of water per year. This is the equivalent of the 300,000-mile-long convoy of water trucks every day that I mentioned earlier.

When an aquifer runs dry, the community that used it will have to live largely on the water it receives from rain. There will be a time when most of humanity will have to live on rainwater. We must use it much more efficiently. (We can create fresh water from seawater, but to do so is very expensive, and the process currently requires fuel that contributes to global warming.)

Because it takes about a thousand tons of water to produce 1 ton of grain, the water taken each year and not replaced could produce 160 million tons of grain—enough to feed about half a billion people at today's average rate of world grain consumption.

Several trends further explain the worsening situation. For example:

- Migration to cities is increasing. By 2030, 80% of the population of Africa will live in cities. As urban areas grow, their demand for water increases. Consequently, river water is being diverted at an increasing rate from croplands.
- The use of powerful diesel and electric pumps, which drain the aquifers faster, is spreading.[2]
- Much of the cropland that rain falls on is being destroyed by salination or soil erosion.
- About 40% of the world's food comes from irrigated cropland, and this percentage will only increase. But as the population increases and the water supply shrinks in many developing countries, they won't be able to maintain their level of irrigated agriculture.

The Punjab is widely regarded as India's greatest agricultural success story, but underground water is being pumped there at twice the rate at which it is being replenished by rainfall. Water tables throughout India are falling an average of 3 to 10 feet per year. The International Water Management

Institute estimates that India's grain harvest could be reduced by a quarter in the next few years because of aquifer depletion, but India's consumption of grain per person is increasing, as is its population.

A depleted aquifer is not visible to the public, but when a great river dries up, it is highly visible. It would be difficult for Americans to think of the Mississippi not reaching the sea. The great river of Northern China, the Yellow River, is almost as large as the Mississippi and had flowed for thousands of years into the Yellow Sea. Shandong Province is like the Kansas of China; it produces a fifth of China's corn and a seventh of its wheat. Half the province's water used to come from the Yellow River. One day in 1972, the river failed to reach the sea. Now it runs dry on most days. China is building canals to take water from the Yellow River to its growing cities. It is possible that almost no water from the river will reach Shandong, which is on the ocean. This province makes up for the loss by pumping more water from an ancient underground aquifer, but the water level in the aquifer is dropping steadily, and its contents will soon run out.

As the soil in north-west China has dried out, vast dust storms are increasing in size and frequency. Satellite photographs of the region show dunes of dust covering villages and roads and huge dust plumes reaching the cities in north-east China. Once fertile parts of north-west China are turning to desert, and the Gobi Desert is spreading uncomfortably close to Beijing. The top of tall pylons that were once part of the electricity grid now stick out of sand dunes. To hold back the desert from Beijing, the government is planting a billion dollars' worth of evergreen trees.

China has terrible memories of the famines caused by Chairman Mao's Great Leap Forward and Cultural Revolution, in which at least 30 million people died of starvation in four years. (A recent book puts the figure at 70 million.) There will be determined measures to prevent another famine.

America is the preeminent provider of grain to the world, and China has built up an enormous quantity of dollar reserves. This is its insurance against its own domestic grain shortages. There'll be a constant line of ships crossing the Pacific from America to China. Lester Brown, the global environmentalist, in my interview with him, describes this in detail as being like an umbilical cord tying the two societies together.

A food minister in India used to use the saying "The food reserves of

India are in the silos of Kansas," and that made sense in 1980. In future famines, the breadbasket of America will be far too small to provide reserves for the world. India and much of the rest of Asia will have surging populations that are eating ever more grain.

China's grain production increased from 90 million tons in 1950 to 392 million tons in 1998—one of the outstanding economic achievements of the last half-century. It peaked in 1998, but since then it has dropped by 70 million tons. This drop alone exceeds the entire grain harvest of Canada. To cover its shortfall, China has been drawing down its once vast stocks of grain, but they're now largely depleted. China is buying grain from world markets at a time when these markets don't have much grain to export. China, the giant in the pantry, will be competing with about a hundred countries that already import grain. As the world's demand for grain exceeds the supply, the prices will rise, and the demand from the new consumers in China and India will eventually make the price rise large.

WATER PRODUCTIVITY

As much of the world runs short of water, it is essential that water be used more efficiently. A huge amount of rain hits the Earth every day, but humankind wastes most of it (just as we waste most of the energy from the sun). The water supply can be enhanced if we put into place inexpensive means of capturing rainwater before it runs down the streets or goes into drains and unproductive earth. We need to collect the water and pipe it to where it is needed.

I feel particularly close to this subject because I live for part of the year on an island off the coast of Bermuda that has no wells or springs. The rock of the island is porous to the sea, and we have to live entirely off rainwater. When I bought the island, it had sparse scrub vegetation; now it has dense, exotic vegetation, with masses of flowers. The water that used to run down the rocks into the sea is now channelled into storage areas and ponds for computer-controlled irrigation. The drinking water is caught on the roofs and kept in dark tanks.

Although Bermuda is a tiny country with no wells or aquifers, it has beautiful gardens, and though its population density is 45 times that of the U.S., it lives on rainwater. When Bermudians travel, they are amazed at the waste of water they see everywhere—house roofs that don't catch rain and property that allows all the rainwater to run away. Most rain soaks uselessly into the ground or runs down gutters into the street. Houses in Bermuda, by law, have clean roofs designed to catch water and underground cisterns to store it. Hotels have large rain-catchment areas.

The use of irrigation can increase the yield of most crops up to four times. In China, 70% of food is produced on irrigated land, but in Ghana, Malawi and Mozambique the amount is less than 2%. In sub-Saharan Africa, only 4% of arable land is irrigated, and many farms are using less than 2% of the available water.

Approximately 70% of all water worldwide is used for irrigation. The answer to this crunch situation is to switch to crops that are less water-intensive and to use water much more efficiently. Rice that yields 4 tons per acre uses little more water than rice that yields 2 tons per acre. Wheat typically produces 50% more calories per unit of water than rice. More controversially, most protein crops can be genetically modified so that they produce a high yield with less water.

Rotating sprinkler systems waste a huge amount of water. The best water productivity comes from drip irrigation systems, which use thin hoses with holes in them to take water directly to the plants that need it. An electronic system attempts to take the right amount of water to the right plant at the right time. Because water goes directly and only to the plants, such systems can reduce water use by 70% while doubling crop yields. Many of the water-stressed areas of the world have not installed efficient watering systems. As elsewhere in the story I have to tell, great increases in efficiency are possible. These often come from computer-controlled systems. They need intelligent engineering and disciplined management. The bottleneck is a severe shortage of skills and discipline. Skills can come from good training, but in many places, discipline is a more difficult cultural issue. Unlike most of the world, some parts of Africa have large untapped reserves of groundwater. As Africa learns to use water resources better, its people need to learn from the bad practices of other areas. They need to learn how to avoid salination, where water evaporates on the surface of irrigated land and leaves an accumulation of salt, as well as how to prevent the creation of waterlogged land. An important part of the solution is to deliver the right amount of water to the right place at the right time.

THE GREYING GREEN REVOLUTION

The Green Revolution is one of the remarkable achievements of mankind. From the mid-1950s to the mid-1980s, it provided a major increase in the world's capability to grow food. The world population grew by 2.5 billion during that period, but the increase in food more than kept pace with the increase in people—the yield per acre almost tripling in many areas. By 1985, the rate of improvement had slowed down. In the 1990s, crop productivity fell in some countries as water became scarcer. Further improvements in crop productivity will be more difficult to achieve.

The Green Revolution created a switch from growing many species of crops to growing only the most efficient varieties. It went from biodiversity to monocultures. A problem with monocultures is that they are much more susceptible to harm of different types, including the evolution of predators. As a result, in some places, farming is reverting back to greater biodiversity.

When the Green Revolution started to sputter, the world population was 5 billion. Since the Green Revolution depends on intensive watering, often more water was taken from underground aquifers than nature could replenish. When water starts to run out, crop yield drops. A good case in point is Saudi Arabia: From 1980 to 1994, Saudi Arabia increased its wheat production 20 times. Then, suddenly, the aquifer on which it depended became almost empty. Within two years, wheat production dropped by more than half.

DESTRUCTION OF TOPSOIL

Good topsoil is astonishingly complex. One teaspoon of good grassland soil may contain 5 billion bacteria, 20 million fungi and a million protoctists.

In a square yard of such soil, there are thousands of spiders, ants and wood lice, beetles and fly larvae, 2,000 earthworms, 20,000 pot worms, 2,000 millipedes and centipedes, 8,000 slugs, 40,000 springtails, 120,000 mites and 12 million nematodes.[3] The life forms below ground weigh more than those above ground—the equivalent of a dozen horses per acre.[4] In each gram of soil, there can be 4,000 distinct genomes, and these differ greatly from one location to another. This rich organic life has been destroyed in farmlands by the use of herbicides, pesticides and crude fertilizers.[5] In the US, farming has been a great success story, but a third of the original topsoil has gone, and much of the rest is degraded. After a century of farming in Iowa, which has the world's highest concentration of prime farmland, half the soil has gone, and the rest is half dead.[6] Soil productivity in the Great Plains fell by 71% during the first 28 years of farming it.[7] In much of the rest of the world, the soil loss per ton of food produced is worse than in the United States.

Although major efforts have been launched in the United States to restore soil and create new soil, 90% of American farmland is losing topsoil on average 17 times faster than new topsoil is being formed.[8]

In dry climates, salination causes a drop in the productivity of the soil, and if it is allowed to continue for long, the land becomes useless for crop growing. In India, which desperately needs to feed its people, about a third of the irrigated land is damaged by salination, and more than 12% has been abandoned. One can fly over such land and see large expanses of glistening white salt. One-tenth of the world's irrigated croplands suffer from salination.

IMPROVING PRODUCTIVITY

The amount of land available for crop growing is declining because of the spread of cities, roads and industrial buildings. Many new cities are coming into existence, and many people without a car would like to have one—in China, for example. At the same time, much cropland is being slowly degraded. More land can be made available for crop growing, but it is of poorer quality and more expensive to farm. To feed 8.9 billion people, the world is going to have to greatly improve farm productivity.

Three types of productivity improvement are important: improving the yield of plants, improving cropland productivity and improving water productivity.

In the last half-century, there has been a dramatic increase in the amount of calories that grain crops can produce, but there is a limit. Many of today's staple crops—such as rice, corn and wheat—are close to that limit because there is a maximum to the products of photosynthesis that can increase the yield of the crop. There is much scope for choosing other plants with greatly improved vitamin content, however.

There are many ways in which the productivity of land can be improved. In well-developed countries, much cropland has nearly tripled in productivity since 1950, and some countries have quadrupled productivity for wheat or corn. Little further improvement is possible. In much of the least developed world, however, there is major scope for improvement. Many poor areas are not even using fertilizer.

Today, about two-fifths of the world's harvest is lost in the fields. Of the world's crops, 13% is damaged or destroyed by disease, 15% by insects and 12% by weeds.[9] Careful use of chemical sprays, fungicides, herbicides and insecticides can reduce this, but such chemicals can degrade the soil, accumulate in aquifers, wash into rivers and cause harmful pollution. The use of genetically modified (GM) plants, however controversial, is having much success in helping farmers in their battle against insects and weeds.

At the end of the 20th century, the world was using 10 times as much fertilizer as at midcentury. The downside is that excess fertilizer degrades the soil and washes into rivers. In some areas, it has destroyed wetlands.

In many farmland areas, it is possible to increase the food produced by "multicropping"—moving from one crop per year to two, or sometimes three. A winter crop of wheat can be produced as well as a summer crop of corn or soyabeans. In India, winter wheat and summer rice are grown on the same land. In the US, government subsidies discourage multicropping; it is much more common in China and other countries.

Often, when crops are harvested, there is a great deal of waste—for example, there are major residues when a cornfield is cut that can be saved and used as roughage for animal consumption. India used crop residues to expand its milk production from 20 million tons to 79 million tons.

Eating beef consumes far more precious resources than eating chicken, and eating chicken consumes more than eating fish. We may like steak, but a field of cows produces less than a tenth of the nutrients of a field of vegetables. In modern farming systems, 25 gallons of water are used to produce a pound of wheat, but over 5,000 gallons of water are used to produce a pound of beef. Nonetheless, world meat consumption climbed from 44 million tons in 1950 to 217 million tons in 1999 and is now climbing higher.

Americans consume 800 kilograms of grain per person per year; in Italy the figure is 400; in Japan it is below 200. Yet life expectancy in Japan is substantially higher than in the US. In Tokyo, men smoke like chimneys, have high-stress jobs and live in cramped quarters, but they survive an average eight years longer than Americans. The United States spends far more per person on health care than other countries. What accounts for its lower life expectancy? The main factor appears to be the diet. Americans eat red meat extensively. Italians have a diet rich in starch, fresh fruit and vegetables, but with fewer livestock products. The Japanese eat raw fish and nutrient-rich seaweed.

THE BLUE REVOLUTION

A rapidly growing form of food production is fish farming in freshwater ponds, referred to as "aquaculture." World protein production by aquaculture is growing at about 11% per year. Aquaculture output grew from 13 million tons in 1990 to 31 million tons in 1998, and it was predicted that it will exceed the production of beef by 2010.[10]

China produces two-thirds of the world's pond fish. During the 1990s, it almost doubled the yield per acre of its ponds. Knowing the best types of fish to grow is critical, and China has learned how to boost productivity by growing multiple types of fish in the same pond. On millions of acres of rice land, it grows fish in the rice paddies. The rest of the world could practise aquaculture, like China.

To judge from China's experience, aquaculture has such massive global potential that its future has been called the "Blue Revolution" and compared with the past Green Revolution. Two decades from now, more fish may come from aquaculture than ocean fishing. It could be an effective way for people in poor countries to obtain the nutrients they desperately need.

Daniel Pauly, a leading fisheries biologist, emphasizes that there are bad forms of fish farming. Farmed salmon and sea bass have to be fed on wild fish caught in the ocean, and it has been calculated that several kilos of wild fish are needed to feed one kilo of salmon and sea bass; so, this type of fish farming does more harm than good. Shrimp farms have had a surprisingly harmful effect on ocean ecology. Shrimp grow best in coastal estuaries, and shrimp farmers often carve out areas in mangrove swamps to create shrimp farms. The Worldwatch Institute states that shrimp farms raise 120,000 tons of shrimp per year but cause an annual loss of 800,000 tons of harvestable wild fish per year.[11] Overusing antibiotics has also been a problem. Also, specially bred fish sometimes escape the fish farms and breed with ocean fish, contaminating the gene pool of the wild fish. In some Norwegian fjords, 90% of the fish in sea farms have escaped.

On the other hand, fish farms using freshwater ponds don't harm the oceans. Tilapia is a freshwater plant-eating fish that grows and multiplies profusely; it could become the "chicken" of fish farming. It can eat waste from agricultural operations; so, fish farms can be integrated into conventional farms, and farmers can use agricultural wastes to fertilize fishponds. Researchers are breeding fish with higher growth rates, greater fertility and better resistance to disease.

Aquaculture farmers need to know how to maximize the net yield of fish. Local governments have the means to monitor and control aquaculture because it occurs mostly in their province. As is not typical for deep-ocean fishing, there's no out-of-control tragedy of the commons.

HYDROPONICS

I was travelling in a sparse, water-stressed area of Africa where the landscape was brown and goat herders tended thin animals. I was reflecting how little use was being made of the land, when I came across a set of crudely erected plastic greenhouses. Inside was an astonishing lushness of crops: tomato plants 10 feet high, each with six or seven trellises of large succulent tomatoes. This was one man's hydroponic farm.

Hydroponics is a well-developed technology for growing plants with their roots in liquid. The farmer measures the contents of this liquid and maintains an optimal mix of minerals for the plants. The liquid nutrients are recycled, with sensors helping maintain the best mix of nutrients. A hydroponic farm may have different types of plants in different types of nutrient troughs. The farmer learns to recognize from the appearance of their leaves, buds and roots what adjustments to the nutrient mix need to be made.

In an increasingly water-short world, an extraordinary aspect of hydroponics is how little water it uses. Both the nutrient solution going to the roots of plants and the water sprayed on plants as artificial rain are recycled. Such farms can use from a fifth to a tenth of the water used in conventional farming, depending on the type of crop. As water and cropland per person continue their steep decline, hydroponics will be increasingly useful.

People ask, "Wouldn't food grown in chemicals taste awful?" I have eaten wonderfully sweet strawberries from hydroponic farms, better tasting than from any other method of strawberry production. Some restaurants that rank highest in the Michelin Guide serve strawberries grown using hydroponics.

Hydroponics has the advantage that the mix of nutrients going to the plant roots can be measured and optimized under computer control. Highly nutritious vegetables can be grown with hydroponics, and the yield per acre can be much higher than with conventional farming. Some hydroponic farms growing tomatoes produce 18 times the yield per acre of a well-managed conventional farm.

Successful hydroponic farming is not just about chemicals and water. When I filmed a beautiful hydroponic farm in Africa, lush with lettuces, I expected to find that it could be a solution to growing food in water-stressed areas, but the farmer disagreed. He said that the operation required skilled know-how and nonstop attention to detail. If the water failed to run down the trough of plant roots for an hour, the crop would be ruined. He said the problem in Africa is not shortage of water; it's shortage of discipline. Famines, he said, will happen because of this lack of management.

I once grew orchids hydroponically in the sunlight in a Manhattan office. They liked the controlled office temperature and did better than most suburban orchids. Cities of the future could have diverse hydroponic farms on the sunny sides of glass skyscrapers or under sloping glass roofs—close to the restaurants they supply.

Hydroponic farms are not used to produce the massive quantities of rice, corn and wheat needed to feed vast numbers of people. The mass production of calories will continue to be the job of Green Revolution–style agriculture. The role of hydroponics may be to produce fruits and vegetables high in vitamins and essential nutrients rather than calories. In doing so, it could have a huge effect on the health in poor countries. Places like the Canary Islands export vast amounts of hydroponically grown tomatoes, cucumbers and salad greens. Puerto Rican and Mexican growers ship large quantities of hydroponic fruits. Simple hydroponic pots could be designed for poor homes and the public taught how to supplement their nutrient-poor diets.

The hydroponic farmer is almost the opposite of the traditional farmer ploughing cow manure into his fields. The traditional farmer is exploiting the immense richness of nature, using soil millions of years old and praying for good weather. The hydroponic farmer is more like a laboratory worker, making sure every variable is under control. Whereas nature's topsoil is too complex to be explained in a textbook, the nutrient baths of a hydroponic farm are simple. Growing plants with hydroponics is scientific and measurable.

ORGANIC FARMING

There is a strong move back to the subtle art of organic farming and to shops selling organically grown foods. Where possible, organic farming seeks to bring back, with efficient composting and animal fertilizers, the quality of soil that has been degraded.

Take monoculture, for instance. The Green Revolution rests on an extraordinarily narrow genetic base. It concentrates on the few most productive varieties—those with high-volume production. India had 30,000 native varieties of rice, but they are in the final stages of being replaced with one super variety. Centuries of farming knowledge and breeding are lost when such a change occurs. Losing all but a few varieties of rice is a mass destruction of DNA—and of knowledge that might have proven valuable in the future. The world has 200,000 species of wild plants, but three-quarters of our food supply comes from only seven species (wheat, rice, corn, barley, potatoes, cassava and sorghum). Man-made monocultures are replacing nature's genetic diversity.

Plants are constantly under attack by disease and insects, and nature has learned to protect itself by avoiding monocultures. In a monoculture, a disease or insect might do widespread damage. Janine Benyus, a science writer, comments that a monoculture is like every house in a neighbourhood having the same key—so a burglar can go through them all.[12] A monoculture species can be harmed by soil degradation or wiped out by climate change. Nature is robust because it has evolved through biodiversity and experimentation for hundreds of millions of years. To needlessly eliminate that diversity is folly. Organic farming seeks to restore nature's complexity. Unfortunately, we can't feed 8.9 billion people with organic farming.

As the world population grows and a greater proportion of people lives in cities, we will exist in an increasingly artificial world, with artificial food-growing. Such food may be less expensive than organic food, but it would be folly to allow the whole of agriculture to be based on monocultures and chemicals. Our hubris can persuade us that our chemicals and pills are better than nature, but our survival as a species depends on respecting nature, with its living, self-creative, self-maintaining ecosystems of extraordinary complexity.

FRANKENSTEIN FOODS

The subject of GM (genetically modified) food crops has generated an extreme, almost hysterical reaction in Europe. In 1999, the European Union placed a moratorium on the importing of genetically modified crops. In contrast, most Americans have eaten a large amount of GM corn products and other GM foods, and many don't know that their breakfast cereal is probably genetically modified. They seem to be unaware of any harm from this and seem happy to continue. Some scoff at European resistance to GM food.

There is no worse way for society to make critical scientific decisions than to have hysterical crowds in the streets shouting slogans and for corporations to try to combat this by using expensive public relations firms skilled at spin control. This is the situation in Europe with GM foods— "frankenfoods," as they are sometimes called there. Europe has strong financial reasons to protect its farmers from low-cost American farm products, but doing so would be illegal under the normal rules of the World Trade Organization. When oversubsidized European farmers see crowds demanding that American farm products be banned, they smile all the way to the bank.

To address the controversy, we need top-quality scientific research to study whether GM foods are safe. We ought to be cautious until we have enough evidence. We need to know which GM plants will spread by pollination and what the possible side effects of that might be.

It already appears that some GM crops seem to have a very important role to play. In addition to producing plants modified to give high yield in water-scarce conditions, the practice of growing GM crops lessens the amount of poisonous herbicides and insecticides that need to be used. It will almost certainly be demonstrated in the future that GM foods could lessen the scale of future famines.

The public has no reason to buy GM foods if they cost the same as or more than non-GM foods, as they do now. This will change. GM foods will become cheaper on supermarket shelves. A few types of GM crops are saving a great deal of money for farmers and could save money for shoppers. Many "frankenfoods" are of questionable value, however. With some crops, the benefits are small and could be obtained by conventional non-controversial plant breeding, though this takes place slowly. In such cases, it is folly to rush to market without the most thorough testing and evaluation. In the future, if GM farming becomes a bulwark against starvation, then these cautions may justifiably be sidelined.

The worst risk that some farmers face is that a predator might wipe out their crop. Genetic modification has produced plants that protect them from much of this threat. It has produced wheat that thrives in drought conditions and bananas that don't rot on the way to market. It can make it easier for cash-poor farmers to grow crops in marginal conditions. Other GM foods are designed to give the grower a higher yield. Existing GM seeds could double the yield of the sweet potato crop in Africa, for example.

Because GM seeds make the growing of certain crops more profitable, GM plants have been making major inroads into farming. In 2004, $44 billion worth of genetically engineered crops were planted in the United States, China, Argentina, Canada and Brazil, and their use was expected to grow rapidly in Asia and South America. Half of the world's soyabeans were GM, 30% of its cotton and 15% of its corn and oilseed rape. The use of GM crops is rising rapidly.

Nevertheless, there are vitally important reasons for caution. Pollen from GM plants can spread in the wind and might fertilize non-GM plants, including wild plants. Farmers use GM plants that are resistant to herbicide. If their pollen spreads to weeds, it might create weeds that are also resistant to herbicide. Only certain plants have pollen that spreads; we need to know which can be contained.

A long-term concern about GM crops is that the modified organisms may turn up where they were not intended. Although Mexico banned the planting of GM seeds, they turned up anyway six years later. GM corn from the American Midwest was found in remote mountainous regions of Mexico. It turned out that the Mexican government had imported cheap, highly subsidized, unlabelled GM corn from the US as food, and some farmers had used it as seed.

On the positive side, it is now becoming less expensive to sequence the

DNA in crops, and with research, it is easier to find out when DNA has been transferred or damaged.

When I was young, I used to tape-record the magnificent English dawn chorus. Now those birds don't sing. The song of the lark at dusk was magical, but it is no longer heard. It is because of simplistic use in the past of DDT. Even today, farmers spray numerous lethal pesticides that kill off beneficial insects, wipe out microbes that make soil highly fertile, and decimate bird populations. In some areas, conventional farming may be more harmful than GM farming.

GM plants have been designed to produce toxins so that their leaves kill *specific* insects. Corn crops have been produced, for example, that are resistant to the European corn borer. Organic farmers use an effective natural insecticide, *Bacillus thuringiensis,* which they claim is safe. Some plants have been genetically engineered to have the same substance in their leaves so that they kill predator insects. Great care is taken to ensure that the toxin in the leaves does not reach humans eating the grain of any edible crop. There are those who argue that certain insect-attacking GM crops kill beautiful monarch butterflies, but field research shows that butterflies are at greater risk from the mass spraying of conventional pesticides. There is a worry that if GM plants kill a specific predator, the predator might evolve until it becomes resistant to the pesticide, but that also happens when conventional pesticides are sprayed.

The image of the crop-dusting plane may become a cinematic icon of the past, like steam trains with smoke billowing above them as they race along the tracks. Poisons sprayed by crop-dusting planes almost certainly do more harm than GM plants.

We are at an early stage of the learning curve of how to use genetic modifications in crops. The accomplishments so far suggest that GM techniques could make crops less vulnerable to insects, weeds, droughts, salination and poor-quality soil. GM seeds may have a major effect on improving nourishment. These efforts need to be integrated with other fundamental changes—such as the widespread use of electronically controlled drip irrigation, the growing of high-vitamin vegetables, organic home plots and hydroponic greenhouses. Genetic improvements will be one of many weapons in the war against starvation and malnutrition.

To feed people well, we need to give them not only proteins, fats and carbohydrates, but vitamins, too. Almost a billion people are seriously undernourished despite eating enough protein. Many people are so poor that they eat a few bowls of rice a day and little else, and the rice is largely devoid of vitamins. Vitamins come from different types of plants—for example, vegetables with thick green leaves. Some plants rich in vitamin A were wiped out by the Green Revolution's mass use of herbicides. The Green Revolution lowered biodiversity in many places by converting mixed cropping systems to intensively grown monocultures of wheat and rice.

Scientists tried to deal with this problem by creating a genetically modified form of rice that became known as "golden rice." It contains beta-carotene, a nutrient that serves as a building block for vitamin A. This contains snippets of DNA borrowed from daffodils and from bacteria. It can be crossed with strains of rice that are grown locally and suited to a particular region's climate and growing conditions. The creators of golden rice want it to replace traditional rice so that it can add vitamin A to the diets of vast numbers of people who are malnourished today. Golden rice became the subject of a massive PR campaign using expensively produced television ads to promote the idea.

The simplest way to provide vitamin A to people in poor countries is to teach them how to grow the right vegetables. Plants rich in vitamin A are carrot, pumpkin, amaranth, mango, jackfruit, curry—and especially, coriander, which grows like weeds. Amaranth and coriander are an incomparably richer source of beta-carotene and vitamin A than golden rice. Other sources of vitamin A, such as vegetables, fruit or cod-liver oil, can easily be made available. We should calculate whether the simple growing of natural vitamin A is easier and better than highly artificial GM crops. The quantity of vitamin A in golden rice is much less than that needed for healthy human growth.

Villages in poor countries vary enormously in their capability to grow nutritious food. In some, you see skeletal, ill-nourished children; in others, healthy-looking kids full of vitality. Villages everywhere could have classes teaching everyone how to build up the soil, catch rainwater, collect animal droppings, create windbreaks and grow plants that have the most vitamins, like coriander.

FOOD SECURITY

The dreadful famines of the past will seem small compared with the likely famines of the future. The traditional grain reserves won't go far when China is aggressively buying grain to feed its own population. As the dust storms in China become worse, its water tables will continue to drop. A huge new consumer class will be eating more meat, and China will buy vast amounts of grain on the world markets. If the oil-rich countries of the Middle East fail to control their population growth and use up the water in their ancient aquifers, they will also buy increasing quantities of grain. These trends will force world grain prices up significantly. Today over a hundred countries import grain because their own agriculture can't provide enough for their own people. As grain prices rise, many will be barely able to afford to buy it.

The world needs to achieve food security—the building up of food reserves to tide populations over during dry spells. As the situation is now, there may not be enough food to feed the world if there are two or three years of bad harvests. If that happens, there'll be intense competition for limited supplies of grain, which will cause major increases in global prices. The solution is to build up large enough food reserves and spare farmland. At present, there seems little desire to address such models or build up the reserves necessary to prevent grand-scale catastrophe.

An essential element of food security is population control. Whatever actions are taken to improve farming and food supply, the situation will ultimately result in even worse catastrophe unless population growth is drastically cut back. If producing more food allows the world population to grow and remain unprepared for famine, then famine will be larger in scale when nature's correction inevitably happens. It is essential to make everyone understand the consequences of high birthrates. Using modern media, we can show people everywhere the unspeakable horror of starvation of tens of millions (as happened in China) and contrast it with a rising quality of life from literacy, good education, interesting jobs and low birthrates.

We need to ask, "Can we keep companies from taking actions that are highly profitable in the short term, but that might have disastrous consequences in the long term?" For some large companies, the certainty of short-term profit enormously outweighs the uncertainty of long-term disaster. In addition, we need to assess scientifically the risks of genetic modification. We need to ask about the Green Revolution, "Will monocultures ultimately degrade farming? Will the water run out?" As the century progresses, we must have the right answers to such questions. They all play a part in establishing food security.

6

DESTITUTE NATIONS

IN MOST SOCIETIES, there are certain types of thinking that are not politically correct. In a nunnery, some viewpoints can't be expressed. The USSR in the 1980s had an economy heading for total disaster, but to spell out the reality was as dangerous as expressing unacceptable viewpoints during the Spanish Inquisition. Sometimes the pressure to be politically correct prevents the facts from being discussed and understood.

It is politically correct in the West today to refer to the poor parts of the world as "developing nations," but the reality is that there are some nations that are not developing. Their GDP per head is steadily falling. They are in a cycle of steadily worsening poverty, disease, violence and social chaos. They are *destitute* nations. The UN sometimes refers to them as "failed nations." In order to discuss what is happening in the world, it is desirable to distinguish between "developing" nations and "destitute" nations. They are on very different tracks. As the world's population heads towards 8.9 billion, much of the increase will be in destitute nations unless drastic action is taken. It's startling to reflect, if you live in a rich Western city, that the number of people earning less than $1 per day exceeds the population of the First World.

THE SLOW DESCENT INTO DESTITUTION

I first visited Kenya in 1960. It had beautiful, highly organized farms and coffee plantations, and it was more prosperous than Singapore, which in those days was a poor, disorganized country full of mosquitoes and slums. Ten years later, because of Mau-Mau terrorists, many of the Kenya farm managers had left. By 1980, a significant number of the farms had reverted to jungle. Kenya was still a wonderful place to hire a four-wheel drive and explore, but it was steadily sinking into poverty. Its GDP was declining and its population was growing. By 1990, I couldn't rent a four-wheel-drive vehicle in Kenya because it wasn't safe. Tourists stayed in well-protected camps. By 2000, AIDS was rampant. Few children learned to read. Tractors were rusting in ditches. Kenya had terrible medical problems and little health care, and I was told it was unsafe to visit the beautiful areas except in highly secure tours. Kenya has a GDP per head of less than $1 per day, but it is not classified by the UN as a "Least Developed Country." Many countries in Africa are in even worse shape than Kenya.

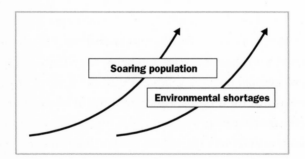

While the most advanced high-tech nations are becoming wealthier, the destitute nations are sliding into a situation from which it will be increasingly difficult to recover. When an area is very poor, it tends to make itself poorer by consuming what little is needed for future survival—the equivalent, in a general way, of eating the seed corn. In some parts of Africa, barefoot women walk for hours trying to find enough wood to build a cooking fire. They cut down trees, deforesting the land, which allows gales to blow away what is left of the topsoil, creating dust storms and deserts. As an area becomes destroyed in this way, it is difficult for it to recover.

Thomas Homer-Dixon, director of the Trudeau Centre for Peace and Conflict Studies at the University of Toronto, has done a great deal of research in the poorest parts of the world. His work has shown that there is a strong correlation between growing scarcities of water, cropland, forests and fish on the one hand and the spread of civil violence and dysfunctional government on the other.[1]

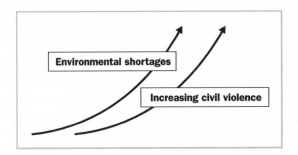

The converse is also true: Civil violence can impoverish a country. For years, Cambodia was a prosperous country with a French-influenced culture, but violence and political upheaval destroyed it in the late 1970s and left it impoverished for decades. Large countries like India and China use their government to manage the poorest parts of their economies, establishing education and moving industry to deprived areas. But poor countries, like Haiti, lack this government capability; they may be too poor to sustain adequate instruments of the state.

When Sierra Leone became independent in 1961, 60% of the country was primarily rain forest; now only 6% is. Throughout Africa, trees have been cut down, and very few have been planted to replace them. This has led to increased flooding and brought mosquitoes. Much of Africa has to deal with some form of malaria. In many areas where malaria is a killer, people sleep without mosquito nets. Many of these countries have stumbled into various forms of social violence, anarchy and war—a cycle difficult to escape.

Destitute nations have serious problems that will reinforce one another as these societies move towards the midcentury canyon. Many will have inadequate water and other resources for food production, excessive population growth, climatic disaster, extreme poverty, an absence of farmers and doctors, rampant malaria, diseases caused by bad sanitation, widespread AIDS and civil violence. Millions of people have become simply too poor to stay alive.

The situation in the poorest countries could be dramatically improved. Over the last three decades, however, leaders around the world have watched millions of people become poorer and more hungry but have done astonishingly little to improve the situation. In his 2005 book *The End of Poverty*, Jeffrey Sachs makes the case that we have the capability to banish poverty in our generation. He has produced detailed calculations to show how it could be done. There is no single model and no magic bullet, just a large array of difficult endeavours requiring a sustained effort on all fronts, including trade, humanitarian aid, education and governance.

SHANTYCITIES

One of the major momentum trends in relatively poor countries is the hordes of people who move from the countryside to the cities. Enormous numbers of Chinese are moving from rural to urban areas. By 2030, four-fifths of the population of Africa will live in cities. In 1950, there were only two cities in the world that had populations of more than 8 million people, London and New York. Now 30 cities exceed that. The migration to cities is driven by unemployment in rural areas and the perception (usually incorrect) that there are major opportunities in the cities.

Most governments have been unprepared for this migration. There is no appropriate infrastructure—inadequate sanitation, electricity, health facilities, schools and so on—for this influx. Shantytowns have grown up around cities and over time have become increasingly squalid. Hopelessly overcrowded, people live in corrugated-metal shacks, rusted shipping containers and junked cars. The streets become one long puddle when it rains, clogged with dead rats, mosquitoes and rubbish. People contract diseases associated with poor drinking water and sanitation, and many die from them. There are malnourished, potbellied children everywhere. When businesses are set up, they are usually illegal. Many shantytown dwellers refuse to pay taxes and are suspicious of outsiders. Often, local police dare not enter such areas, and the absence of law enforcement allows the levels of rape, violence and other crimes to go unaddressed. In some shantytowns, 40% of children are born HIV-positive, but murder is the leading cause of death for ten-year-olds.

The problem of shantytowns has worsened over the last 40 years because of the swift rise in population and rural poverty. In fact, these areas might be better called "shantycities." It is difficult to get a sense of their magnitude by viewing them from their edges. This sense hit me for the first time when I travelled over some of them in a helicopter to film them. Flying low, you could see the squalor of an utterly wrecked society; flying high, you could see the enormous scale of the destitution—shantycities stretching for miles and miles in all directions. I gazed down and wondered what these places would be like if there were not enough food. What would they be like if a new infectious plague infected them, as one surely will? The shantycities are incubators of disease.

The term *feral city* has come into use to describe cities devoid of law and order. Mogadishu is one. Most shantytowns are not the whole city, however. They are often massive areas of extreme poverty with their own boundaries, excluded from the police and other services of their country. They are largely isolated from their parent city but are not listed as separate entities. As a result, they disappear into an administrative blindspot.

Young people, even in the grimmest shantytowns, show surprising ingenuity in making the best of their conditions, but these efforts add nothing to net wealth. The same native ingenuity in an entrepreneurial culture would add greatly to net wealth.

Sometimes Japanese or German companies have moved into a Third

World town and set up a factory. After a few years, the people are making camcorders or microwave ovens. Protesters in the West complain that factory owners are taking advantage of cheap labour, but the factory transforms the town. The town soon has hotels, shops, better food, schools, perhaps a few scholarships. The birthrate drops. There is hope where no hope existed before—but this is an exception, and it's not likely to happen in most destitute shantycities.

One of the biggest problems in some of the poorest nations is the unspeakable level of government corruption. When famines occur and wealthy countries send food, much of it doesn't reach the people who need it because of inept or dishonest management. In Angola, a country of 33 million people, $4.3 billion of government money was unaccounted for in the five years 1997 to 2001.[2] Angola was a country where the average person's income was only 73 cents (US) per day, and women had 6.7 children on average. In 2001, the world supposedly had money-laundering controls in place, but $900 million still disappeared from the Angolan government coffers.

Many poor countries have, in effect, two economies side by side and barely communicating with each other. Economy A is like the West but less smoothly operating. Economy B is a shantytown economy, in desperate poverty, sometimes with Mafia-like control, often with social violence bordering on anarchy. Economy A has reasonable medical services and educates doctors and nurses. Economy B has rampant AIDS and many diseases caused by unsanitary conditions. The police of Economy A don't set foot in Economy B.

In a few places, we managed to get a film crew inside an Economy B shantycity. We were closely guarded by tough individuals, called "minders," who were strongly placed in the local power structure and could make sure we were not molested. We were not free to film whatever we wanted. One room I walked into had about a hundred babies and very young children all lying on their stomachs, unattended, effectively abandoned. Like most shantycity children, they were desperate to hug an adult, as though they had never been hugged.

In Economy A, on the other side of high razor-wire security fences, were supermarkets not much different from those in the West, supplied by local farming operations. The food provided had to be carefully packaged and perfect in appearance—supermarket chains demand perfection. A lettuce farm, for example, produced an assortment of lettuces in plastic wrapping for a supermarket with an internationally known name. More than a third of the lettuces in the packaging shed were not perfect; some had brown edges or minor blemishes. So huge quantities of imperfect lettuces were thrown into a dumpster. Much other food was similarly thrown away. Yet two miles away was a shantycity where people were hungry and destitute.

Nearby was the local head office of a wealthy global corporation that, because it wanted to demonstrate ecological correctness, used recycled carpets. Amazingly, the recycling of the carpets was done half a world away in the United States, with high shipping costs. It could have been done on the other side of the razor-wire fence, where people desperately needed jobs and income—but Economy A doesn't talk to Economy B.

It is sometimes said that country communities sinking into poverty could resort to subsistence farming by which they have existed for a thousand years. Why not let Africa be what it was before external forces disturbed it—but now with better medicines and seeds? The difficulty in doing this is the loss of country skills as rural communities migrate to cities. People then expect to get food from shops. The current generation is not being taught the farming methods that have been handed down for endless generations. By losing country skills, the community burns its bridges.

SANITATION

The 1980s was declared by the United Nations as the International Drinking Water Supply and Sanitation Decade, and governments and international agencies spent $134 billion to deal with drinking water and sanitation problems. Yet at the end of that decade, more than 1 billion people still lacked safe drinking water, and 1.8 billion lacked adequate sanitation. The gains in clean water supply and sanitation made during the 1980s were devoured by population growth in the 1990s. By the year 2000, 3 billion people were without sanitation and safe drinking water.

Six main diseases are associated with poor drinking water and sanitation—diarrhoea, schistosomiasis, trachoma and infestation with ascaris, guinea worm or hookworm. At any given time, up to half of humanity suffers one of these. According to the World Health Organization, they cause about 5 million deaths every year.

Experts on disease say that we humans are far more susceptible to new diseases now than ever before because cities throughout the tropics and subtropics are severely overcrowded, and the people there have poor nutrition and inadequate sanitation. If a mutant organism gets started, it meets conditions fertile enough to multiply fast. A new and lethal strain of flu starting in a densely packed shantycity could spread worldwide before protective vaccines could be manufactured.

LEAST-DEVELOPED COUNTRIES

The United Nations recognizes a category of nation it calls LDC (Least Developed Country). The criteria that cause a country to be classed as an LDC are a low GDP per head, a low Economic Diversification Index (measured by the share of the labour force in industry, the proportion of the GDP in manufacturing, the per capita commercial energy consumption and the level of exports) and a low Human Resources Weakness Criterion (based on life expectancy at birth, per capita calorie intake, school enrollment and adult literacy). The UN currently lists 49 Least Developed Countries—just under a quarter of the world's countries. They have 10.7% of the world's population but only 0.5% of its GDP. They have 0.3% of world trade—half the share they had two decades ago. From my own observation, having travelled in poor countries, I concluded there was not a good match between the Least Developed Countries and destitute nations. Some LDCs are reasonably stable and have hope for the future; some nations not on the LDC list are on the destitute, downward spiral.

The United Nations promotes an LDC into the "developing nation" category when it has a GDP per capita of $1,035 per year, averaged over three years—but 15 countries with a GDP per head of less than $2 per day, and some with less than 50 cents per head per day, are not categorized as LDCs, although most are destitute nations spiralling into worsening chaos. Some destitute nations are not on the LDC categorization because they were once colonies of wealthy nations, or part of the Soviet Union, and have a high enough Economic Diversification Index. Any remnants of a better-managed past are now gone, however.

There could be a precise categorization of a destitute nation in terms of its decline in GDP. For example, a destitute nation might be defined as one with a GDP per head that has declined on average by 1 percentage point or more per year for the last five years and is now less than $2 per day.

The following is a list of countries whose GDP per head declined on average more than 1% per year from 1980 to 1998.

NATION	AVERAGE ANNUAL DECLINE IN GDP PER HEAD (%)
Angola	9.5
Congo (Democratic Republic)	4.5
Nigeria	4.2
Zambia	3.6
Sierra Leone	3.1
Belarus	2.7
Algeria	2.3
Ivory Coast	2.2
Madagascar	2.2
Nicaragua	2.2
Niger	2.2
Slovak Republic	2.1
Cameroon	2.0
Jordan	1.5
Namibia	1.4
Estonia	1.0
Mali	1.0
Mozambique	1.0
Rwanda	1.0

We can see that there is a world of difference between developing nations such as Chile, Brazil, Malaysia and Thailand, where there is vigorous industry, good universities and young people who have hope for the future, and destitute nations such as Angola, Haiti, Estonia or Ivory Coast—where, by and large, young people have no hope. Perhaps the most important reason for distinguishing between destitute nations and developing nations is that, if humanity continues on its current patterns of behaviour as we approach the canyon—the midcentury crunch—much of the destitute world may not survive.

The right thing to do, without a shadow of doubt, is to keep that from happening, but as mentioned in the first chapter, we can look at the future in two ways: *What is the right thing to do?* and *What is the most likely thing to happen?* If efforts to transform the destitute nations remain feeble, then the massive forces of the approaching canyon will overwhelm the ability to make rescue operations succeed.

DEAD CAPITAL

The Peruvian economist Hernando de Soto asks the question, "Why does capitalism thrive only in the West?" Many Third World nations have adopted the inventions of the major capitalist nations, from the business card to the computer, but they haven't been able to make capitalism work as it does in the West. De Soto concluded that this is because there are fundamental flaws in Third World systems. Correcting them presents leverage factors of immense power.[3] De Soto tells us that the capital of poor countries, including his own country, is held in defective forms, and this does immense damage.

In the 19th century, an American might have had a sack of money hidden in a mattress. De Soto calls this "dead capital" because it couldn't be used as collateral or put to use by the banking system as it would be in the West today. Such money could only be used once. In today's society, the same capital can be put to use multiple times. In the First World, people have mortgages that allow them to own a home. Banks lend money secured by mortgages; so, in effect, the same money is spent multiple times. The total of the loans greatly exceeds the capital on which it is based. The home may be used as business collateral. Businesses depend on borrowing, and successful businesses have many shareholders. People setting up businesses can take risks because of laws about limited liability and bankruptcy protection. There are sophisticated methods of financing involving derivatives or futures markets. These are vitally important mechanisms enabling money to do more than its face value would suggest.

Westerners take land titles for granted so completely that they hardly ever think about them. In the Third World and ex-communist worlds, the majority of homes are held without a title—it's often almost impossible to get a title. Most homes are outside the legal system. If a home has no deed or title, its owner can't get a mortgage on it, and it can't be used for business collateral. Possessions that are not recorded can't be turned into capital that has other uses. Similarly, many businesses are outside the legal system. They don't have limited liability and can't obtain funding from banks or finance organizations. They are cut off from the lifeblood of a modern society.

The inability to raise capital keeps poor countries poor. Individuals are prevented from starting businesses that could become valuable. The magnitude of this problem is huge. Hernando de Soto's researchers found that 80% of all real estate in Latin America was held outside the law. In Egypt, 92% of city dwellers and 83% of people in the countryside live in dead-capital homes. The total value of Egypt's dead-capital real estate is about 30 times the value of all shares on the Cairo Stock Exchange.

Contrary to popular belief in the West, the poor in poor countries often have substantial savings. De Soto's team did extensive work trying to estimate how much, and his conclusions are astonishing. He estimates their savings at *40 times all the foreign aid received throughout the world since 1945*. In Egypt, for example, the poor have accumulated savings worth 55 times as much as the sum of all the direct foreign investment ever recorded there, including that received for the Aswan Dam and the Suez Canal. De Soto calculated that, if the United States were to increase its foreign aid to the level recommended by the United Nations, it would take it 150 years to transfer to the world's poor people *resources equal to those that they already possess*. These substantial savings are dead capital. They might be cash in a mattress. Much larger, they are a house without a deed or land without a title. The house may have been built slowly, brick by brick, until it is a substantial home, often with farm buildings, but it is outside the legal system. Such capital doesn't have the multiplier role that capital has in the West. As a result, the economies of such countries are seriously undercapitalized.

In principle this situation is correctable. Local government should record the ownership of all property and should issue deeds and titles, as in the West. When this is done, homes become much more valuable. Illegal businesses could be flushed out and made legal in such a way that people can invest in them. A legal business is likely to make more money than an illegal one. It's necessary to reorganize the property system so that people can legalize their homes, secure mortgages and use their property as collateral if they want to. People need to be educated about the multiple uses of capital.

Politicians in poor and ex-communist countries don't seem to know how to change the situation, but de Soto says the answer is in the history books, particularly the journey of America from a nation of gunslingers to one of venture capitalists. The Third World needs quick-to-implement standards of incorporation for all businesses, deeds for all real estate and the removal of extralegal activities. Their extralegal ownership needs to be incorporated into recorded ownership so that the finance system can make capital go to work. Then there can be mortgages, home insurance, use of collateral, open share ownership, stock options, use of futures markets and all the modern means of extracting more value from existing capital. Banks can help people to set up shops, orchards, workshops, corporations with limited liability and ingenious entrepreneurship.

Is it practical to achieve this change in the Third World? Certainly. De Soto has an organization bringing about this change in more than 30 countries. A similar transformation occurred long ago in First World countries, but took a hundred years or more, and the Third World needs to correct its problem quickly. In one way, it's much easier now than it was a century ago: We know the procedure. There are appropriate computer software and accurate surveying techniques, and the legal documents needed are commonplace. In another way, it's more difficult because there are enormous extralegal communities in each nation with rules of their own invention. Collectively, they are a force that has to be handled with care.

The financial benefits from achieving this would be immense, but the difficulty lies in doing this without hurting vast numbers of people. The political problems in making the change to active-capital countries are likely to be great.

Unfortunately, establishing a title to your home or land in the majority of Third World countries is such a nightmarish process that most people don't start. De Soto's researchers set out to buy and establish a title to small homes in various countries and recorded what happened.

In Lima, to obtain legal authorization to build a house on state-owned land took 6 years and 11 months. Then, obtaining title to the land took much longer.[4] In Egypt, to acquire and legally register a small lot on state-owned land takes from 5 to 15 years, and obtaining permission to build a home on former agricultural land takes 6 to 11 years. That is why so many Egyptians who build homes build them illegally. If an existing homeowner tried to make his home legal, he might end up having it demolished, and the law says he could serve up to ten years in prison. In Haiti, to settle legally on government land, a citizen can lease it for five years and then buy it. To obtain the lease takes two years; to convert the lease to purchase takes 12 years—a total of 19 years. The situation is similar elsewhere in the Third World.

The ability to legally own or occupy a home was defined as a fundamental aspect of human rights in the UN's Universal Declaration of Human Rights of 1948. Many forms of human rights violation are visible, but the crippling administrative procedures that prevent home ownership are largely invisible and misunderstood in the West.

De Soto's researchers produced charts showing the steps a person had to take to obtain legal rights to a home or a one-man business. In Peru, obtaining legal authorization to build a house on state-owned land took 207 steps in 52 government offices—and then another 728 bureaucratic steps required by the Municipality of Lima.[5] By any standard, this is utterly insane. A system could be designed with appropriate databases so that the process could take ten minutes on the Internet. It might take a few years to create the system and populate the database with the information needed, but it would be a worthwhile investment.

If Western company managers skilled at business redesign were confronted with an organization that operated like a Third World bureaucracy, they would know how to transform it. They might set a goal that the time taken to obtain a title and a deed to a property never exceed two weeks. To transform a nation's bureaucracy, though, would have major political repercussions. Most of the proponents of the old methods would have to be removed. The change would occur only if it were driven by a determined head of state and the will of the people. In most poor countries, such leadership is missing. In 1992, Libyan leader Muammar al-Qaddafi was reported to have incinerated all Libya's land titles.[6]

HOW TO CRIPPLE BUSINESS

Today, if you open a one-man cake shop in Cairo and want to make it legal, it will probably take you over 500 days to complete all the government paperwork. Because of this, most cake-shop owners find government administrative procedures so burdensome that they set up their businesses outside the law.

According to the International Labour Organization, 85% of all new jobs in Latin America and the Caribbean since 1990 have been created in the extralegal sector. In Zambia, 90% of the work force is employed illegally. To most of the world's poor, only less-than-legal arrangements are available if they want to build a home, open a shop or manufacture a product to sell. Government procedures that prevent people from participating in the processes of active capital do immense damage. They constitute a totally unacceptable form of de facto apartheid.

The administrative cesspits of the poor world simply prevent ordinary people from having a decent chance. They have to be cleaned up. Some nations, such as Singapore and Taiwan, have managed this effectively. Singapore's GDP per head today is about $21,000. If it had kept a legal and administrative system like Kenya's or Nigeria's, its GDP per capita could be closer to $1,000.

The long-term benefits of converting dead capital to active capital are immense. In a typical Third World country, it would create enormously more wealth than the foreign aid given by rich countries. Not only would people be able to afford nicer homes because they could get a mortgage, but society in general would grow more affluent. Helping such countries reengineer their systems would be far more beneficial than giving financial aid to the present cesspits.

ESCAPE FROM DESTITUTION

For a very poor nation to have hope for the future, a long staircase has to be climbed. Some nations have demonstrated that it can be done. In 1960, South Korea and Ghana had almost identical economic profiles in terms of per capita GDP, relative importance of their primary manufacturing and service sectors, nature of their exports and amounts of foreign aid. American officials at the time predicted that abject poverty would continue in South Korea. Its GDP per head was $400 (in today's prices). Two decades later, South Korea had become an industrial giant, due to hard work, discipline and education. Within only one generation, it had become a world-class industrial power with massive exports. Ghana, since then, has slowly declined, and its people now earn 84 cents per day on average.

The "Four Tigers" of Asia—Singapore, Hong Kong, Taiwan and South Korea—focused on exports, joint ventures, encouragement of foreign investment and a high savings rate. They encouraged students to get degrees at American and European universities and then return home. They emphasized that young people should become engineers and entrepreneurs, not lawyers or soldiers. These "tigers" have fairly small populations. Now some huge nations want to be "tigers," too. They want to spread the success to their huge populations.

To climb the staircase, several factors are necessary. First, there needs to be a serious focus on education. A high literacy rate is essential. The percentage of the public gaining advanced degrees needs to be steadily increased. The more skilled workers a nation has, the more it can evolve from cheap-labour sweatshops (including white-collar sweatshops) to work that pays higher salaries.

While some developing nations are climbing the staircase, others are stumbling—for example, much of South America. Most destitute nations are failing to reach even the first rung of the staircase. I visited Southern Rhodesia when it became Zimbabwe in 1980. It was a very attractive country with great hope for its future. People said, "The new government will give us free education for our children and free health care." On a recent visit there, people told me, "It's impossible for us to get our children to learn to read or to learn arithmetic, and we have no access to health care." Its GDP per head has dropped from $2,000 to $260 and continues to fall fast. The few remaining skilled farmers and doctors are leaving if they can. The rule of law has broken down.

An essential challenge of the 21st century is to implement the millennial goals set by the United Nations at the start of the century: to eradicate extreme poverty and to make education available to everyone. It's vital to build food security so that the fear of famine is gone. Less well understood, but a vital aspect of building a decent world, is the need to eliminate dead capital—to cast aside the corrupt administration and bureaucracy that keeps capital dead. By the end of the century, it's possible that extreme poverty will have been eliminated almost everywhere. The question is, will we have famines and Rwanda-like slaughter on a massive scale before the mechanisms of decency are finally put to use?

7

CLIMATE CATASTROPHE

P RIME MINISTER Tony Blair's chief scientific adviser, Sir David King, told Blair that climate change represents a greater danger to our future than terrorism. If James Lovelock's view is correct, and we reach a state in which we are harming the basic control mechanisms that regulate the planet, then climate catastrophe would eclipse any other problems.

The 21st century will have to cope with such climate change. It may be gentle, or it may be Gaia reaching a tipping point—at which it becomes out of control. There are plenty of things we can do to avoid catastrophic change if we do them soon. But we are not doing them, and the cost of apathy could be staggering.

I interviewed US senator John McCain and was surprised that his view on climate change was radically different from that of the Bush administration. He said, "Climate change is real. It's taking place as we speak. The race is on. I'm very, very concerned. I don't think America is going to be like the movie *The Day After Tomorrow*, but I do believe that we are going to experience serious consequences. The question is: Is it going to be too late for us to reverse them? Will we take action to reduce the greenhouse gases, which are affecting our climate, or is it going to be too late? Will we leave our children and grandchildren an environment that is far different from the one that I enjoyed growing up in?"

The Earth has natural cycles of warming and cooling and is high on a natural warming cycle at present. The sun is a little closer than normal. Lovelock states that it is bad luck that this hot period coincides with a time when we are pumping vast amounts of greenhouse gases into the atmosphere.

An uncertainty of how much of the Earth's warming is caused by human activities and how much by nature gives the coal and petroleum industries a reason to argue with authorities who say we must lessen carbon emissions. It also gives politicians a reason for avoiding politically difficult action.

Almost all climate scientists now agree that human-generated greenhouse gases are the main culprit that will cause temperature rise in the next few decades. In my interview with Lester Brown, the global environmentalist, he stated emphatically, "There's consensus in the scientific community, without question, on the greenhouse effect and the effect of rising carbon dioxide levels on the ocean temperature. But we don't seem to be able as a civilization to respond to it. As the effects of climate change become more apparent, our children and grandchildren may disown us. They'll wonder how we could have been so incredibly shortsighted."

At present, humankind is pumping far more carbon dioxide into the atmosphere than nature can absorb. Michael McElroy, who directs Harvard's Center for the Environment, maintains that, in the next few decades, if we create the atmospheric concentrations of carbon dioxide that are expected, we will create an environment that the Earth has not been in for 30, 40 or 50 million years.[1] We know this by analysing air bubbles trapped many millions of years ago in the Antarctic ice.

Even a minor difference in temperature between the oceans and the air above them causes the air to move and can produce severe weather patterns. It created Katrina and the other severe hurricanes of 2005. Katrina grew from being a tropical storm to being a Category 5 hurricane in a single day because the warmth of the sea made it spin faster than normal. Recently, we have seen much unusual weather. Bangladesh had floods with 83% of the land surface under 2 metres of water. In Guam in 1997, there were record-breaking winds of 368 kilometres per hour. Because the force exerted by wind is proportional to the square of the velocity, those 368 k/h winds were six times the force of a 150 k/h hurricane.

Australia is one of the world's top exporters of bulk farm foods, but recently it has had severe droughts that, in some places, have cut farm incomes in half. Scientists believe that this is caused by a vortex of winds around the Antarctic, which is accelerating.[2] Both the North and South Poles are centres of systems of strong swirling winds. The South polar area is cooler, while Australia itself is becoming slightly warmer. This causes the wind system to spin faster, dragging the winter and spring rain from Australia into the ocean south of Australia. Because of this, Australia may be facing a period of worsening droughts, severe fires and low crop yields.

EARTH'S GREENHOUSE

Life on Earth is totally dependent on the greenhouse effect.

Several naturally occurring gases in the atmosphere—such as carbon dioxide, nitrous oxide and methane—act like the glass in a greenhouse and trap some of the heat from the sun as it is radiated back from the surface of the Earth. If these gases were not there, our planet would be too cold for life as we know it.

In the 20th century, we created these gases artificially with a massive use of carbon-based fuels—mainly coal and oil. In 1990, humankind artificially pumped about 16 billion metric tons of carbon dioxide into the atmosphere. By 1999, that figure had risen to 25 billion. In the last few decades, the concentration of carbon dioxide in the atmosphere increased by 25%. Tropical rain forests absorb large amounts of carbon dioxide, but we are cutting them down on a grand scale. The director general of the

United Nations Environment Program gloomily stated that it would "take a miracle" to save the world's remaining tropical forests.

The artificial increase in greenhouse gases is causing a slow, small rise in the Earth's temperature. If today's energy technology is not changed, there may be an average temperature rise of 5 degrees Celsius in the next 50 years. That may not sound like much, but if we said a river was 5 feet deep, you may find, as you go across, that some parts might be 2 feet deep and others 20 feet deep. The same with temperature increase—some places would experience little change, but others may be 20 degrees Celsius hotter. In summer, there is now open water instead of deep ice at the North Pole. In parts of Alaska, houses and buildings lean at strange angles because the permafrost on which they were built is thawing. In 2000, a ship sailed through the Northwest Passage, which had been considered impassable before then. Climate models predict that, by 2070, Glacier National Park will have no glaciers.

One indication that we need to take global warming seriously comes from the insurance industry. The insurance industry is preparing itself for "mega-catastrophes," including storms that could do more than $30 billion worth of damage. Such "mega-cats" are so large that no one insurance company could handle them, and there is major activity to share the risk among many "reinsurance" companies.

In the 1950s, the damages from natural catastrophes cost about $4 billion per year. This had risen to about $40 billion per year by the end of the century. In 2002, the damage was about $55 billion, and in 2003 it was $60 billion.[3] The UN Environment Program estimated that the amount would be $150 billion by 2010. The world's largest insurance company, Munich Reinsurance, said the losses would reach $300 billion a year within a few decades, and Britain's biggest insurance company said that unchecked climate change could bankrupt the global economy by 2065.[4] Long before this possibility becomes reality, advanced countries will have taken protective action. Today it's not difficult to create houses or office buildings that withstand winds of 160 miles per hour—even when large walls of glass are used. If winds reach 200 miles per hour, glass and building materials will become appropriately stronger, but this won't protect the billions of people who live in shacks and shantytowns.

NEW ENERGY CONSUMERS

Today the United States has 4.5% of the world's population, but generates 23% of its carbon dioxide emissions. The United States has, by far, the highest carbon dioxide emissions *per person* of any country (except Singapore). Still today, more than half its electricity comes from coal—the worst fuel for CO_2 emissions. The United States generates 18 times as much carbon dioxide per person as India and 100 times as much as most of Africa. When estimates are made of how much carbon dioxide the Third World will produce when it develops, the figures are alarming. Singapore is one of the most impressive examples of a country going from poverty to prosperity, but in about three decades, its carbon dioxide emissions went from 1 metric ton to 22 metric tons per person. India is now developing rapidly and could easily go from 1.1 metric tons per person to 12. China's economic surge could take it to 14 metric tons per person in three decades. China and India, the world's two most populous nations, plan to satisfy their exploding demand for energy by tapping vast reserves of coal. China has plans to build 600 coal-fired power stations in the near future. It sits on a disproportionately large share of the world's coal, and it's exceptionally dirty coal.

Until recently, most of the carbon dioxide emissions have come from about a billion people in the advanced world. In the future, most emissions will come from about 4 billion people in the developing world. The future well-being of everybody depends on rich nations helping developing nations move as fast as possible to the next era of energy, with India and China understanding that clean-energy products could represent a massive export market for them.

BEYOND KYOTO

1997, the year of the Kyoto agreement, will be remembered as the time when many nations of the world began to confront the problem of global warming. The agreement called for industrial countries to cut their aggregate emissions of carbon-based gases (to 5.2% below 1990 levels) by 2012. It was specified in the Kyoto Protocol that the terms would not come into force until the agreement was ratified by the 55 nations that were jointly responsible for 55% of the world's carbon emissions. Seven years after its acceptance, that had not happened. One hundred and sixteen countries signed the agreement, but the United States would not.

Although there were many strong words about the Kyoto Protocol, the Kyoto reductions would be hopelessly inadequate. *Much* greater reductions are needed if the climate is to be stabilized. The protocol now appears more like a cosmetic attempt to hide the political embarrassment of global warming than a serious attempt to stop climate change. Many scientists and policymakers believe it impossible to avoid *a further doubling* of carbon dioxide emissions. The reason is that "the world's economic and political systems cannot depart from business-as-usual rapidly enough."[5]

In 2003, the British government announced a plan to cut carbon emissions by 60% (far beyond the Kyoto number) by 2050. This was the amount that scientists calculated was necessary to stabilize atmospheric carbon dioxide. Tony Blair and Sweden's prime minister, Göran Persson, jointly urged the European Union to adopt the same goal.[6] It may be relatively easy to achieve this goal by 2050, because by then multiple noncarbon energy technologies will have matured. What is of greater concern is to cut down carbon emissions in the near future. In 2002, Germany announced that it would cut its greenhouse gas emissions by 40% by 2020, and it proposed a 30% cut throughout Europe by that date.[7] Some authorities have estimated that the world's use of carbon-based fuels needs to be cut by 70%—and cut quickly—to prevent major disruption to the Earth's climate. That kind of cut would be disastrous for the coal and oil industries.

CLIMATE SWITCHES

It's usually assumed that global warming will be a slow, gradual process, but certain climate changes in the past have occurred with catastrophic suddenness. Elaborate supercomputer models have been built to simulate the Earth's climate.[8] These models show that the climate is much more vulnerable to human activities than is generally thought and that the climate will often change with large swings rather than gradual moves from one stage to another.[9]

Many complex processes are stable up to a point. If we push them gently from their equilibrium state, they return to that state, but if we push them too far, their stability breaks down. This is referred to as being *metastable*—a condition in which stability is maintained only within a certain range of values. Taken beyond those values, conditions become unstable. A cyclist freewheeling down a steep hill is metastable. If you pushed the bike gently from its course, it would wobble, but a rider could bring it back under control. You might push it harder and watch it recover again. If you push it too hard, however, despite the rider's efforts, it will wobble out of control, career down the hill and crash. Various factors that effect climate change are like this. Up to a point, they can be perturbed and restore themselves if the source of the damage is removed, but when disturbed by more than a certain amount, a major transition occurs. A concern with the Earth is whether our activities will cause a major irreversible transition.

Scientists' models show that Gaia is almost certainly metastable. It can be disturbed in diverse ways but returns to its basic condition. If it is pushed beyond the range from which it can recover, however, then it swings to a fundamentally different condition. The forces of Gaia keep themselves in balance and regulate the planet, giving us a breathable atmosphere and the comfortable temperatures that we know. But if we upset the balance, we would cause catastrophe on a grand scale. We don't know where the limits are. We don't know how hard we have to push the cyclist to make her crash.

It is possible—if we seriously cut carbon emissions in 30 years' time—the Earth will recover from its spell of global warming. There are serious indications, however, that we are close to an irreversible tipping point. Many scientists now believe that there is evidence that the Earth will react to human pollution at a faster rate than they had expected.

Greenland is larger than Mexico and covered in ice about 2 miles deep. Scientists' models considered this to be a solid block of ice that would take a thousand years to melt, but new satellite evidence shows that the melting ice at the surface forms lakes that trickle down crevasses to the bottom of the ice, creating rivers of water underneath the ice. As a result, the ice slides more easily towards the ocean. James Hansen, a climate modeller with NASA, has commented that, once the ice sheet starts to disintegrate, progression can reach a tipping point beyond which breakup is explosively rapid. He and Gaia scientists such as James Lovelock give many examples of processes that reach a tipping point.

In his book *The Atmospheric Environment: The Effects of Human Activity*, Michael McElroy says that "climate switches" can result from global warming.[10] In the tropics, there are three cycles in which huge columns of warm moisture-laden air rise—over South-east Asia, over Brazil and over Africa. As a column reaches the cool air of high altitudes, it releases rain, the air descends over the ocean, and the cycle begins again. The Hadley Centre model suggests that, as the Earth's atmosphere warms, the three columns of air could coalesce into one enormous column. This would cause intense rain and severe flooding in one area, somewhere around Indonesia, and severe drought in many other areas.

A climate switch will be expensive for America, but it could be disastrous for many of the world's poorest countries. A large section of Africa could lose much of its crop-growing capability, for instance.

THE OCEAN CONVEYOR

There is a thousand times more heat in the ocean than there is in the atmosphere. An ocean current, called the Ocean Conveyor, circulates slowly around the globe, a massive conveyor belt that transports heat, rising in some areas and sinking in others. It's extremely slow compared with the circulation of the atmosphere. The Ocean Conveyor is rather like a giant flywheel, making the Earth's climate relatively stable, but it does have an Achilles' heel.

The Ocean Conveyor travels north up the Atlantic from the equator, becoming the Gulf Stream, which makes Bermuda surprisingly warm. As much water as 75 Amazon Rivers is carried by the Gulf Stream. As it meets the landmass of the northern United States, it swings sharply east with much turbulence. It then travels across the North Atlantic towards Britain. This transport of heat makes Britain and Western Europe's temperature 5 to 10 degrees Celsius higher than it would be without it.

The Gulf Stream is metastable. England's Midlands and Labrador are at the same latitude, but in England we don't have ice and dogsleds. Greenland is a huge country, however, covered in mountain ice that is 2 kilometres thick, and this ice has started to melt vigorously. Fresh water floats on top of salt water, but warm water floats on top of cold water. If too much cold fresh water joined the Ocean Conveyor, its surface would become cold. The large volume of salty water would sink and flow slowly to the South Atlantic, no longer available to warm the British Isles and Western Europe; they would, instead, be chilled by the cold, fresh water from the melted Greenland ice. The Ocean Conveyor is robust—except in the North Atlantic.

The history of the Conveyor in the North Atlantic can be recounted from the study of deep-sea sediments and from cores drilled from the ice. On several occasions in the past, the current of the Gulf Stream was replaced with an icy current, and in the lush, green climate of Ireland, most of the plants died. Ice spread across England and Western Europe. Scandinavia became tundra. The 16th-century paintings of Pieter Brueghel show frozen landscapes, and Dutch masters painted elegant people skating on Holland's canals. A rapid change in ocean circulation occurred 8,200 years ago, causing an extreme cooling period to start very suddenly, wiping out the woolly mammoths, while drought spread throughout the American West and Africa.

The possibility of a North Atlantic disruption of this global Conveyor is being intensely studied by the Woods Hole Oceanographic Institution in Massachusetts, which employs over a thousand scientists. These scientists are deploying large numbers of robotic sensors in the ocean and in the melting ice.[11] The models of climate change give no certainty of exactly what is going to happen or when it will happen. The Woods Hole scientists cannot say when the North Atlantic current will be disrupted or, in detail, what the consequences will be. The director of the institution, Robert Gagosian, posits that such a disruption would not be like the rheostat in your house; it would be more like a light switch—either on or off. In past events, a disruption of the current has occurred in as short a time as 10 years. In the event 8,000 years ago, the Conveyor went below the surface in three years.[12] According to models, if it happens in the 21st century, it would take longer than that, but the consequences would be momentous. We are fooling with a system of massive forces.

CATASTROPHE FIRST

Because there is no certainty, the oil and coal industry can choose to ignore the possibilities and carry on with business as normal. But if we wait until the evidence of catastrophic climate change is conclusive, it will be too late to stop it. The forces provoking climate change cannot themselves be changed quickly.

The possibility of climate change represents what Ross Gelbspan describes as a titanic clash of interests that pits the ability of the planet to support our lifestyle against the profitability of big oil and big coal—two of the largest commercial enterprises in history. Gelbspan documents how he believes the coal and oil industries have recruited dubious scientists to damage the credibility of the climate scientists, creating false science in the same way that the tobacco industry found bogus scientists to "prove" that nicotine was nonaddictive.[13] He argues that the study of global warming has become so visible that profit-driven falsification of science constitutes "a clear crime against humanity."[14]

As with the other troubles we describe, there may be a catastrophe-first syndrome with global warming. Unlike some of the other catastrophes, however, this one may creep up on us slowly, leaving much scope for argument until it is too late. Scientists with their extensive instrumentation will have warning that the Ocean Conveyor is beginning to change its position in the North Atlantic, but by then, it will be too late to stop it. Once the Ocean Conveyor starts to change, the huge momentum of the forces at work will carry it forward irreversibly.

Today we are driving without a clear map, close to a cliff edge, at steadily accelerating speed. The Kyoto Protocol and other efforts amount to only a marginal slowing of the speed. We need to not only slow down but reverse direction. If the public could see the cliff edge, they would clamour for action, but the cliff's not visible. Ecological problems like a lake becoming green and stinking are clearly visible, but progression towards disrupting the Ocean Conveyor is not, and politicians can comfortably do nothing. Somehow or other, we need government leadership to convince the public that action is needed before the momentum towards catastrophe becomes too great.

The US Department of Defense commissioned a study on the consequences of the Ocean Conveyor's changing its position and worked out three scenarios in detail. In the worst, Britain and Holland have such frigid temperatures that many people leave. Much of Sweden turns to tundra. The Riviera becomes like the Maine coast. As this is happening and Western Europe turns icy, central Africa becomes hotter, and rainfall rates decline. Many parts of the world experience severe droughts. There are mass migrations of people and wars caused by shortages of water and food. Despite this scenario, presented by its own Department of Defense, the US government took no action.

A NEW ERA OF ENERGY

One of the worst culprits pumping greenhouse gases into the atmosphere is the car. This can be lessened in two ways: One is to redesign the petroleum car so that it uses much less petroleum and produces less exhaust. Much more effective is to switch from petroleum to hydrogen-based fuel, which doesn't generate carbon dioxide. "Concept cars" using both of these approaches are being driven and indicate that a massive change is coming in the car industry. The sooner the better—because much of the damage done by global warming is cumulative.

After the oil shock of the 1970s, American cars were redesigned for fuel efficiency. Between 1973 and 1986, the average new US-made car halved its gas consumption from 13 to 27 miles per US gallon. Then gas prices declined, and fuel efficiency ceased to be a significant sales incentive. What became profitable and popular were SUVs (sports utility vehicles) with high fuel consumption. Comedians joked that the smaller the woman, the larger the truck she drove.

In 2003, Toyota started selling hybrid cars that combined a gasoline engine with an electric drive. When you touch the brakes, your car loses momentum. This is a loss of energy, but much of that energy can be saved by allowing the momentum of the car to charge a battery. The battery gives back the energy when you accelerate again. This flywheel effect saves fuel, especially when driving in start-stop city traffic. One Toyota hybrid car achieved around 80 miles per US gallon, depending on driving conditions, but other hybrid cars were less efficient. The Toyota hybrid engine is used in Nissan cars also.

A revolutionary change will occur when alternate energy sources become cheaper than oil. This will be brought about by improved technology, mass production and marketing. Particularly important, we need to get rid of the huge government subsidies for oil and coal. There are massive subsidies every year both for oil exploration and to support the coal industry. The US government spends billions of taxpayer dollars to militarily defend access to oil in the Middle East. The true cost of carbon-based fuels includes the costs of pollution, military protection of oil assets and the costs of abnormal storm damage that is related to global warming. The medical expenses for coal-related sickness and lost workdays are enormous. When these costs are included in the accounting, the cost of power from wind generators and solar panels could be less than the cost of power from coal and oil. If that happens, there will be mass production of alternate energy sources, and that will bring down their costs substantially.

FUEL CELLS

A technology critical to our future is the fuel cell.

A fuel cell is similar to a battery in that it produces electricity by using chemicals—but it doesn't run down like a battery because the chemicals are supplied continuously. The basic chemical is hydrogen. Today fuel cells are expensive. They will be improved and eventually mass-produced in vast numbers and will be cheaper than they are now. The fuel they use will be packaged conveniently and made easily available. They will be used in both cars and the home. Fuel cells may become popular for the home before they are common in cars.

Fuel cells for cars will be lighter than a battery of the same power. A battery produces only about 1% of the energy that the same weight of hydrogen produces. Weight is important because a battery-powered car expends much of its energy just carrying the battery. It can't go very far or very fast. To do so would require a very heavy battery. Prototype fuel-cell cars have been built that could travel a thousand miles without refuelling.

Fuel cells will power machines of all sizes, from small to very large. Military applications range from power for night-vision goggles to power for submarines. They run at efficiencies considerably better than those obtained in even the largest fuel-burning engines and power stations.

To generate electricity from hydrogen, a fuel cell employs a catalyst to induce the hydrogen to separate into protons and electrons. In a common type of fuel cell, an electric field causes the protons (which have a positive charge) to pass through a thin membrane, while the electrons (which have a negative charge) are attracted in the other direction. After protons pass through the membrane, they combine with oxygen molecules from air to form water. The electrons are channelled into an electric current, which is the power output. Like a battery, the fuel cell has no mechanical moving parts.

An electric current passed through water can split it into hydrogen and oxygen; a fuel cell does the opposite. Hydrogen and oxygen join together to make water and electricity. In the future, internal combustion engines with their constant explosions will seem very crude. A fuel cell doesn't cause explosions or generate high pressure. It's simple and silent.

Although fuel cells are simple in principle, they have been difficult to develop. The chemical reactions do not take place readily unless special materials are used for the catalyst and membrane. Progress has depended on making these materials reliable and cheap enough.

Eventually cars may be powered by refuelable cartridges that generate hydrogen. Hydrogen is one of the most abundant atoms in the universe. Two of the three atoms in a water molecule are hydrogen. Hydrogen fuel will be produced by electrolysis of water, as described above.

"Concept" cars powered by fuel cells represent the most fundamental change in car design for decades. Instead of a heavy central engine, they have an electric motor on each wheel. There are no axles, transmission, gearbox, differential, starter motor or alternator. The parts of the car that require expensive maintenance are gone. General Motors demonstrated an elegant mass-producible chassis of a hydrogen car. It is flat, with four wheels, like a huge skateboard. Many different types of vehicles could be built on it, ranging from high-performance sports cars to big-family SUVs. The concept cars built on this chassis are roomier than today's cars. Many believe they are more attractive and more comfortable to drive than current cars. They produce no pollution and are relatively inexpensive to maintain.

Using ultrastrong composites instead of steel, a car could be made three times lighter, but no more vulnerable in a crash. Aerodynamically, it could be made two or three times more slippery so that it moved "less like a tank and more like an airplane."[15] This enables the redesign of tyres so that less energy is lost where the rubber meets the road. As the car becomes light and aerodynamic, the weight savings snowball. Components become smaller, and many of them—like power steering or power brakes—can be omitted.

At first, fuel-cell cars will be more expensive than conventional cars. In early production runs, fuel-cell cars might cost $2,000 or so more than gas-powered cars. To fall in price below gas cars, the membrane and catalyst of the fuel cell need to be improved. Then, with mass production, prices will drop substantially.

Several types of fuel system are possible for fuel-cell cars. Much research is going into finding the best way to pack a large amount of hydrogen into a small rechargeable cartridge. There are many options. It's possible that we'll fuel our cars with cartridges we buy at supermarkets. To encourage you to buy a fuel-cell car, part of the deal may be that cartridge refills will be delivered free to your home when you send an Internet request. The same fuel cells will power both the car and the home. Today's fuel stations will slowly become part of the romantic past. Fuel station operators and oil companies, of course, have a different view. Some believe that gasoline may be the source of hydrogen in the short term and that the hydrogen could be extracted from gasoline at the garage rather than in the car. The car is then lighter and more efficient.

Hydrogen can be produced by using coal or oil as a power source—that defeats the purpose because it generates greenhouse gases. It can also be produced by solar and wind generators. When the output of a large wind generator has nowhere to go, it can be charging hydrogen cartridges. Hydrogen can be extracted in many other ways that are environmentally benign. It can be extracted from methanol gas, which comes from decaying organic material. Brazil grows biomass crops for the purpose of extracting methanol for car fuel.

A small step beyond that is to extract hydrogen directly from biomass crops. Biomass fuel is one of the important options of the future, and some inventors have schemes for producing vast quantities of hydrogen organically. When the hydrogen economy is mature, society will generate few greenhouse gases, although coal and oil will fade from the scene slowly.

In recent years, breakthroughs in fuel-cell technology have made a "hydrogen economy" seem practical. A home may use rooftop solar panels to both heat its swimming pool water and generate electricity. Small wind generators like those on boats may be used in homes to warm water or generate hydrogen by electrolysis. City skyscrapers will have glass walls facing the sun that are made of photovoltaic material so that windows can generate electricity as well as let in heat.

There will be many opponents of an alternate-energy economy. It will play havoc with OPEC, and the petroleum industry will go through major changes. Oil companies today advertise themselves as *energy companies* that sell future forms of energy, but because virtually all of their profits still come from oil, they are determined to hang on to it. Governments could change this by insisting that the huge hidden costs of oil and coal be accounted for.

The United States imports over a billion dollars' worth of oil every week, and this has a major impact on the economy. It would be better to spend this money on the hydrogen economy and make America a leading exporter of hydrogen cars and fuel cells. The United States needs to be self-sufficient in energy rather than being hostage to the Middle East.

SOLAR POWER

In Tibet, many homes use homemade parabolic solar reflectors, 5 feet wide and lined with silver paper salvaged from chocolate bars. These focus the sunlight onto a large iron cooking pot or kettle and keep it hot for most of the day.

It is often said that solar energy can provide only a very small part of the energy needs of an industrial nation, but an enormous amount of energy from the sun reaches the Earth, and we use almost none of it except in agriculture. The sunlight energy reaching the United States each day is several thousand times more than all the electricity generated by all the power stations in the country.

The *Solar Living Sourcebook* gives details of solar-energy products that you can buy today. One of the best for generating electricity from sunlight is the Astropower 120-watt photovoltaic panel. The technical editor of the *Sourcebook* calculated how much electricity would be generated by large arrays of such panels. He concluded that all the electricity used in the United States could be generated by covering the Nellis Air Force Base and surrounding Nevada Test Site with arrays of Astropower panels.[16] The panels come with a 20-year manufacturer's warranty.

Based on measured real-world performance, he concluded that an Astropower panel typically produces 107 watts. He estimated the small loss of converting this DC output to AC. The panels would have fixed mounts facing south, with an angle of tilt of 38 degrees, with space for maintenance access. This would allow 1,858,560 panels per square mile. Records going back to 1961 show that this area of Nevada averages 6.1 hours of sunshine per day. Thus, the square mile would produce 425 million kilowatt-hours of electricity per year. The Nellis Air Force Base and Nevada Test Site cover a large area, but one only half that size would be needed to generate as much electricity as all the power stations in the United States.

Nobody's suggesting that this is a good idea, but it does point out the amount of sunlight available. Today's solar panels are small, designed to fit on somebody's roof. For every job making these small solar panels, there are four jobs needed to install them and service them. If, instead, solar panels were designed to cover large areas, they could be much less expensive and more efficient. Large-field solar generation is planned in China, where it won't have to compete with the subsidies of the oil industry.

Electricity generated from large-field solar systems could become cheaper than electricity generated from oil. When that breakpoint is passed, solar panels will be mass-produced in vast quantities and their cost will become much lower.

WIND

In the 1970s, few people took windmills seriously. A few wind farms were built because a tax break was provided. The first ones used wind machines with three blades on a tall pole. These generated around 100 kilowatts. Today's towers are higher, and the blades are longer. The generator captures more wind, and the biggest generator produces 5 megawatts. A tower can be in a farmer's field with the farmer farming the land below the blades while he earns money from the power generated. Originally, wind cost about 40 cents per kilowatt hour. Today, new-technology wind farms in high-wind places cost 3 cents per kilowatt-hour. The cost varies because the winds vary, but in a good location, energy from wind is cheaper than energy from coal or oil. In one study conducted by the US Department of Energy, it was concluded that three states—Texas, Kansas and North Dakota—have the wind-power potential to provide most of the electricity needed by the whole country.

Large wind generators can have a very long lifetime. A major cost is the tower. Once built, however, it lasts for ever. The blades last for many years. An anti-wind story has it that wind generators kill birds, but modern generators' blades move surprisingly slowly, and birds easily avoid them. Components of the generator do wear out, but they are now designed for easy replacement. The one-time capital cost is high; the annual maintenance cost is low.

Some large wind projects are being planned. A wind company, Winergy, plans to generate 9,000 megawatts from a network of wind farms stretching along the Atlantic coast. A European consulting firm has calculated that, by 2020, Europe could be getting *all* its residential electricity from wind if European governments get serious about developing offshore wind resources, especially those in the North Sea.[17] Germany leads the world in generating power from wind. It generates 16,000 megawatts today—the equivalent of about 12 large power stations. Schleswig-Holstein, its northernmost state, generates about a third of its electricity from wind.

Solar panels work only when the sun shines, and wind generators work only when the wind blows. The power generated needs to be stored in some way. Solar panels or wind generators could charge batteries or generate heat for later use. Particularly important, they could generate hydrogen from water and charge the cartridges that are used in fuel cells. The main objection to wind generators comes from rich people who don't like their appearance, and that objection is widespread.

MEGAWATTS AND NEGAWATTS

The profits of power companies are regulated. Generally, the more electricity they sell, the more money they are allowed to make. At a time when it is vital to lower greenhouse emissions, many power companies are motivated to do the opposite and have plans to build big power stations.

In California, the power regulatory authority is more enlightened than most. It calculated how much cheaper it was to save megawatts than to create new megawatts. If every household in California replaced four incandescent lightbulbs (average 100 watts) with fluorescent lightbulbs (average 27 watts) burning five hours per day, it would save 22 million kilowatt hours per day—enough to shut down 17 power stations. Even more efficient are the new white light-emitting diodes (WLEDs) that enable us to see at night. If every household in California replaced one average-flow showerhead with an energy-saving showerhead, it would be equivalent to closing another 15 plants. If every household installed a solar-power hot-water heater, it could shut down another 67 power stations,[18] and if the state paid for the lightbulbs, showerheads and solar panels, it would cost much less than building one new power station. Around the First World, an extraordinary amount of energy could be saved. It is ironic that almost nobody is doing so—especially at a time when the Earth's control system is heading towards an unstable condition.

The public utility commission of California invented the term *negawatts* (*negative watts*) to refer to electricity saved. There are many ways in which electricity can be saved. A typical saving scheme might cost one cent per kilowatt-hour saved, while to build and operate a new power station might cost 10 cents per kilowatt-hour—it often costs much less money to satisfy demand by *saving electricity* than by building new power stations.

The California public utility commission set out to reform the regulatory process by devising ways to reward power companies for lowering electric consumption rather than increasing it. This was referred to as *the negawatt revolution*. The commission worked out formulas for letting utilities keep as extra profit part of whatever they saved their customers. In this way, an era of wanton waste of power could evolve into an era of saving power.

Domestic refrigerators vary greatly in their energy efficiency. The best are eight times more energy-efficient than the worst. There is no correlation between energy efficiency and price. Most other household appliances have a similar absence of correlation between energy efficiency and price. Electric utilities tried giving rebates (say, $50) to customers who bought energy-efficient refrigerators. They discovered that it was far more effective to give rebates to retailers who sell them—the inefficient refrigerators then quickly vanish from the shops. Still more effective was to give the rebates to the manufacturer, whose profit is small compared to the retail price, and because the rebate to manufacturers adds directly to their profits, the rebate can be smaller. Similar rebates can be used on many other types of resource-wasting industrial equipment.

The Pacific Gas and Electric Company serves northern California and was the largest investor-owned utility in the United States. Around 1980, it planned to build 10 to 20 new nuclear power stations every few miles along the coastline of the entire state. Then the rules of the game changed, and the power stations were not built. The power company decided to meet demand with negawatts rather than new power stations. It stated that it intended to meet most of its new power needs from more efficient use of power by its customers. This was the "best buy."

The "second-best buy" was privately bid renewable resources, such as wind or water power. Coal and nuclear power stations, which were once considered the only practical options, became regarded as so costly and environmentally unfriendly that new ones shouldn't be built.

Regulations in California have demonstrated that providing incentives to save power instead of constantly expanding its consumption can have a massive positive effect.

NUCLEAR POWER

To most of the public today, because of concerns about safety, it is outrageous to suggest that nuclear power be a substantial part of the world's future energy supply. Technology can change rapidly, however, and sometimes needs to be looked at with a fresh eye.

Traditional nuclear power stations certainly have potential problems that are very serious. First, certain types of failure could be very dangerous. The reality of the Chernobyl disaster was far worse than people realized at the time. Second, there are major difficulties in disposing of radioactive waste. Decommissioning a traditional nuclear power station is very expensive. Third, and just as serious, there is concern that traditional nuclear power facilities inadvertently could be means for nations or terrorists to obtain the fuel needed to make atomic bombs.

Much of the power-generation industry is private enterprise, and it has to make a profit, but the nuclear power industry has often found itself entangled with nuclear weapons. In most countries, it has been run or controlled by government and characterized by outrageously false accounting and public misinformation. In 1954, a massively publicized piece of misinformation came from the chairman of the US Atomic Energy Commission, who commented that electricity generated by nuclear means would be too cheap to meter.[19] Five decades later, the cost of nuclear electricity was double that of electricity generated from oil or coal.

In the 1973 oil crisis, oil prices quadrupled, and President Nixon later called for the building of 1,000 nuclear reactors by the year 2000—part of a programme called Project Independence—but all nuclear power plants ordered in the US since 1973 were cancelled, and no new orders were placed after 1978. The public often assumes that this was the result of the accident at Three Mile Island, but that accident occurred in 1979. The real reason utilities did not build nuclear power plants was that they were too expensive, partly because of new environmental regulations and associated lawsuits.

In 1988, Margaret Thatcher's government set out to privatize the electricity-generating industry, including nuclear power. Britain's civil service had assured Thatcher that the nuclear power industry could be operated profitably. When private corporations did their own calculations, however, they concluded that there was no way nuclear energy could make a profit. In a desperate attempt to make it profitable, the British government proposed grossly dishonest accounting rules. They planned to prolong to 135 years the accounting lifetime of nuclear power stations—in other words, they could be depreciated over an outrageously long period of time.[20] Also, the government would do the very costly operation of processing the nuclear waste, but at the taxpayers' expense.[21] CEOs in other industries have gone to prison for such schemes.

In 1989, the British government abandoned its attempt to privatize nuclear power. In 1994, it was calculated that decommissioning a nuclear power station and cleaning up the site would cost more than $15 billion.[22]

The Chernobyl catastrophe happened in 1986, and its true effects were hidden by the USSR. In January 1993, the Chernobyl Committee of the Russian government stated that, of those who took part in the clean-up of the Chernobyl site, 7,000 died during the seven years following the disaster.[23] At the following Davos World Economic Forum, the president of Ukraine, Leonid Kravchuk, stated that 11 million people had been affected by the Chernobyl accident and that the accident cost $55 billion in medical aid.[24]

FOURTH-GENERATION NUCLEAR POWER

Given this history of nuclear power, how could a person possibly advocate that nuclear power stations could be important to our future?

The answer is that there has been a fundamental change in the technology. The nuclear power stations in existence are *second-generation*. Those that can be built today are *fourth-generation*. There is an enormous difference between the second generation of aeroplanes in the 1930s and today's Boeing 777. Second-generation nuclear power stations represent the state of the art in the 1960s and 1970s—the technology of Three Mile Island and Chernobyl. Fourth-generation nuclear power stations have been fundamentally reinvented and specifically designed to avoid the problems described above.

Fourth-generation nuclear power stations need to satisfy the following four criteria:

1. It must be technically *impossible* for them to have a runaway chain reaction. No accident, failure or human carelessness could produce mass radiation. No matter what mistakes the operators make, the power station is inherently safe.

2. The nuclear power industry must be entirely divorced from the nuclear weapons industry. It must be *impossible* to use their fuel to make atomic weapons.

3. The spent fuel must be easy to dispose of and must not leave radioactive problems for future generations. The large radioactive rods that are difficult to dispose of today would not exist. No uranium could possibly be in contact with the atmosphere or the environment.

4. The nuclear power stations must generate electricity at lower cost than with coal or oil power (when all factors are considered, including the eventual decommissioning of the power station).

Designs for fourth-generation reactors have been developed in various countries. Particularly interesting is the pebble-bed reactor, in which uranium fuel is entirely enclosed in a spherical casing like a tiny ball bearing, 0.03 inches in diameter (0.75 millimetres). These spheres are ultrastrong and have a four-layer shell that can withstand very high pressures and temperatures. The ball bearing cannot be crushed, corroded or melted. There is no possibility of uranium dust spreading.

A traditional nuclear power station has a containment dome of metal up to 100 feet across and around 9 inches thick. The intent is that, if something goes wrong, any radioactive substances will be confined within the containment dome. In a sense, the containment zone of the pebble-bed reactor is the equivalent of the four-layer shell less than a millimetre across.

When the process begins in a fourth-generation nuclear reactor, neutrons pass through the wall of this shell. A neutron may cause a uranium atom to break apart into two atoms of lesser atomic weight and, in so doing, release heat. This fission releases two or three neutrons that fly out of the shell and enter other shells. Thus, there is a chain reaction, but the uranium is always contained inside the shells of the ball bearings.

In a pebble-bed reactor, about 15,000 of these ball bearings exist in a fuel ball—the "pebble"—which looks like a billiard ball but is slightly larger. Each pebble generates about 500 watts of heat (as much as a small electric kettle) when the reactor is in full operation.

This whole reaction takes place in an underground steel pressure vessel, about 20 feet in diameter and about 65 feet high. Inside this vessel are 310,000 fuel billiard balls, as well as 130,000 graphite billiard balls that moderate the reaction. The reactor is continuously refuelled, with new balls being added at the top and spent balls being removed at the bottom. Each ball passes through the reactor about 10 times during its life. This continuous refuelling eliminates the long periods of downtime that are necessary with traditional reactors.

The fuel balls are used to heat helium gas that enters at the top of the reactor, passes among them and leaves at 900 degrees Celsius. Its thermal expansion is transformed into rotational motion to generate electricity, and the gas is recycled back into the generator. The outlet temperature of

900 degrees Celsius is much higher than the limit of 320 degrees Celsius for conventional water-cooled nuclear reactors, and that allows greater electricity-generating efficiency. The helium is used both as the gas that turns the turbine and as the coolant, which eliminates much of the equipment and expense of conventional nuclear reactors.

With today's nuclear power stations, disposing of the large radioactive fuel rods is a nightmare. With pebble-bed reactors, the uranium remains sealed inside the very strong ball-bearing-like silicon carbide shells. It cannot leach out into the area where the waste is stored. The tiny shells are designed to last for a million years. They reside inside the billiard-ball-like casing, which gives further very strong protection. When the billiard balls are spent, they are set into a further solid material that can be disposed of easily.

To be inside a traditional nuclear power station that generates between 1,500 and 3,800 megawatts is a massive, thundering, awesome experience, like being in a science-fiction movie. By contrast, a pebble-bed reactor generates 100 to 200 megawatts—enough to power a small town—and is about the size of a 40-foot shipping container. A larger facility can have multiple such reactors operated from a common control room. There is a tendency in the power industry to move from giant power stations to smaller distributed power units. This avoids the cost of massive electricity grids spanning long distances.

A small prototype pebble-bed reactor was built in Beijing in 2004.[25] A particularly important characteristic of a fourth-generation nuclear reactor is that it is designed to be "meltdown proof." In any imaginable accident scenario, it shuts itself down—as was demonstrated to journalists in a brilliant publicity exercise.

When the reactor was in full operation, the doors were locked, and the journalists were told they couldn't leave the room. Engineers had told them that the pebble-bed reactor is "walk-away safe"—meaning that the control staff can just walk away if something goes wrong. Without any warning to the journalists, the staff were told to leave, and the doors were locked behind them again. It was then announced that the coolant system was being disabled.

The nervous captive audience watched the red number displays showing the reactor temperature steadily climb. It rose to about 1,600 degrees Celsius. Then, with no human intervention, it began to fall, the reactor slowly cooling down by itself because the tiny fuel ball bearings are designed so that neutron production slows if the temperature rises. As the atoms of uranium get hot, they spread apart, lowering the probability that an incoming neutron will strike a nucleus, and the chain reaction slows down. If the coolant system fails in a conventional nuclear power plant, the fuel rods would overheat. If a conventional nuclear reactor has an emergency, humans have to decide what to do very quickly. With the pebble-bed reactor, *no* human intervention is needed to prevent a crisis.

It is essential that future nuclear power be totally isolated from the possibility of atomic bomb production (unlike in the past). The pebble-bed reactor uses fuel that is only 9% uranium 235 and 91% uranium 238. With such fuel, it is impossible to make atomic bombs—they need uranium that is highly enriched (or plutonium). An industry that makes, distributes and disposes of fuel pebbles worldwide could be entirely separate from the industry that runs the new nuclear power stations. The production and disposal of pebbles would be tightly controlled. Extensive use of fourth-generation nuclear power would be incomparably safer than allowing the public to drive cars.

POTENTIAL IN DEVELOPING COUNTRIES

A vast quantity of new power generation will be needed in the developing world. Nowhere is new power more urgently needed than in China, whose economy will grow so rapidly. Its railways can't deliver coal fast enough for its filth-belching power plants, and there are frequent blackouts. As we have stressed, it is highly desirable that China's long-term plans for coal power are avoided because they would have a devastating effect on the world's climate. China has massive medical problems caused by its extreme pollution. It will use solar and wind power, but its booming economy will need far more energy than that. Oil and gas are in short supply in China. A hydrogen economy with mass production of fuel cells makes sense, but not if the hydrogen has to be created by burning coal.

Chinese scientists have estimated that China will need 300,000 megawatts of nuclear power by 2050. To create this at low cost, standardized mass-produced design is needed. Nuclear power stations of the past have been custom built, at great expense. In the future, modular reactors should be churned out like Model-T Fords. The pebble-bed reactor is designed to be modular—easy to ship and easy to construct.

Pebble-bed reactors can also produce hydrogen for fuel cells. Creating hydrogen from water can be done much more efficiently at high temperatures than at low temperatures. The temperature of the gas in a gas-cooled nuclear reactor is ideal. So these reactors can be designed to produce the fuel for a hydrogen economy. Fuel cells will be a critical part of future energy, particularly in the developing world. To power the vast number of cars the world will have, there is no viable alternative to hydrogen, but producing the hydrogen needs to be done without creating greenhouse gases.

If cars operate with fuel cells, and the main forms of power are solar, wind, hydro, fuel cell and fourth-generation nuclear, then the energy needs of the West and the future needs of China and India can be met without planetary damage. Pebble-bed reactors will drop in cost substantially if they achieve large sales. Countries such as China and India could have a thriving industry selling clean energy technology to developing nations.

Coal-fired power stations cause lethal pollution. In addition to the carbon dioxide, power stations pump into the atmosphere other gases and fine particles that cause medical problems. In 2000, a study by the nonprofit Clean Air Task Force reported that tens of thousands of people die prematurely in the United States every year as a result of power-plant pollution.[26] The study found that fine-particle pollution from US power plants cuts short the lives of nearly 24,000 people each year, including 2,800 from lung cancer. On average, these people lose 14 years of their life. Hundreds of thousands of Americans suffer each year from asthma, cardiac problems and respiratory issues associated with inhaling fine particles from power plants. These illnesses result in tens of thousands of emergency room visits and hospitalizations each year. Power-plant pollution is also responsible for 38,200 nonfatal heart attacks per year.

In masterpieces of warped PR, the US coal industry continues to state that coal is the cleanest and safest form of energy.

DREAMS OF FUSION

In the 1970s, one of the great hopes for humankind's energy was fusion. Fusion is the process that takes place in the sun—a reaction between the nuclei of hydrogen atoms—and it releases immense energy. Hydrogen is one of the commonest elements on Earth. If power stations had a controlled fusion reaction, humankind could have abundant energy without generating greenhouse gases or pollutants that cause medical problems.

For several decades, the governments of the United States, Russia, Europe and Japan have spent a fortune trying to make this grand idea practical. A fusion reaction takes place in plasma, a gas so hot that molecules don't exist in it, but the fusion of hydrogen nuclei can occur, as it does in the sun, releasing endless energy. The problem is that the plasma is so hot that ordinary materials cannot contain it.

There are two main approaches to building a fusion reactor. One attempts to contain the burning plasma within an intense magnetic field shaped like a doughnut, in a device called a tokamak. The other, at the Lawrence Livermore National Laboratory, bombards a tiny pellet of hydrogen (deuterium/tritium) with many intensely powerful lasers. Both approaches require massive-scale engineering and major design breakthroughs before a fusion machine is able to generate more power than it consumes.

There is now an unprecedented international collaboration to build a tokamak fusion generator, called ITER ("way" in Latin). Scientists and engineers from China, Europe, Japan, Korea, Russia and the United States have joined forces. The burning plasma, as hot as the sun, will be confined inside a doughnut-shaped magnetic bottle by superconducting magnetic coils. Fusion reactions take place when the plasma is hot enough, dense enough and contained long enough for the atomic nuclei in the plasma to start fusing together. At a temperature of at least 100 million degrees Celsius, the hydrogen nuclei will combine to produce helium, neutrons and enough energy to generate electricity and sustain an ongoing reaction.

The intent is that ITER will become the first fusion device to produce thermal energy at the level of an electricity-producing power station. ITER will be much bigger than the largest existing tokamak machines, and its expected fusion performance will be many times greater.

The laser fusion machine is perhaps even more dramatic than the tokamak. Today it uses 192 of the most powerful lasers ever built, each 400 feet long. These fire with a timing accuracy of less than a billionth of a second, at a target a fraction of a millimetre across. Eventually the process will need even more lasers, and they must be far more powerful. They now fire about twice a day and take hours to cool; in the future, they'll have to fire continuously at a stream of targets.

Fusion enthusiasts reluctantly admit that neither tokamak fusion nor laser fusion will be good enough to make power stations technically feasible for many decades—not in time to stop the burning of carbon fuels causing catastrophic climate changes. Large government subsidies are going into giant fusion schemes, perhaps because they won't compete with today's oil or coal industry in the near future.

A new type of hot fusion has been invented, and its development is being kept so quiet its owners asked me to remove a section on it from this book. While the US government continues to spend vast amounts of money on a technology that won't work for many decades, a start-up company has set out to build a small fusion reactor that might be shipped around the world in competition with the pebble-bed reactor.

THE ENERGY TRANSITION

Half the population of the world today has no electricity. Rather than building an expensive power grid with power stations belching out pollutants, developing countries can have relatively small power stations and use of alternative energy. The cost of replacing today's coal power stations with nuclear power would be high. If a worldwide effort were started to replace them with fourth-generation reactors, it would require two 200-megawatt-sized reactors being built every day for four decades—at a cost of about $100 billion per year. Such an industry could be profitable and would lessen the climate damage. In two decades or so, the products and services of noncarbon-based energy (including cars) will be valued at trillions of dollars per year. Major countries will jostle to have a leading competitive position in the new-car and power-generation industries. The race for new energy products will become as intense as competition in the computer industry was. It makes sense to strive for all the noncarbon solutions—negawatts, green buildings, energy efficiency, eco-affluence, wind, solar, fuel cells, biofuels and pebble-bed reactors. Which will give the highest leverage factor? Almost certainly, eco-affluence—the spread of lifestyles that are affluent and enjoyable but do no harm to the environment—and electricity wherever possible. It's easier to change lifestyle than to change the oil industry.

The tragedy of the present time is that there is government commitment to and aggressive lobbying for the wrong solutions. There are massive subsidies for planet-wrecking technology. The American automobile industry is on welfare. It's appropriate to worry about China and its plans for dirty coal power. The Chinese government is well aware of the problems. Sir Crispin Tickell, a top British environmental adviser to governments, comments, "I find that those at the top of the Chinese government understand the implications of environmental issues better than those in almost any other government in the world."[27] The Chinese government has decreed that as much as 10% of that vast nation's energy must come from renewable sources by 2010. China won't have effete objections to the aesthetics of wind generators or false subsidies that block the development of fuel cells.

By midcentury, Tony Blair's goal of a 60% reduction in the use of carbon-based fuels may have been reached. The problem is what happens *before* then. How powerfully will the carbon-based fuel industry resist the 21C Transition? How much irreversible climate change will occur? How devastating will its effects be on the poorest countries? Much global warming is already set to happen and is probably irreversible, but there is still a window of opportunity for action to keep climate change from becoming a catastrophe of unprecedented scale. We need to act now, not two decades in the future, because the window is steadily closing. Incredibly, the time for action with the least cost and the biggest payoff has already gone by. One giant oil company ran full-page ads in leading newspapers saying exactly the opposite—that there is a covert duty to burn oil faster than ever in order to exploit the oil company's window of opportunity available until global warming slams it shut.

By the second half of the 21st century, we'll reach an era of abundant energy without pollution. The many-decade effort to create fusion power will probably have paid off (I suspect with a route different from ITER). Fuel cells will become inexpensive and will be mass-produced in great quantities. There will be widespread use of large-field solar generators and 10-megawatt wind generators. Third World countries may have a massive use of fuel-cell motorbikes (which exist today) and light three-wheel vehicles rather than cars.

Abundant energy will enable us to create fresh water from seawater. This will enable desert cities to bloom spectacularly, as Abu Dhabi has done. Abundant energy and water will be critical to grand designs for civilization. At the same time, buildings and cities will have become "green" and machines will have become energy-efficient.

The change to noncarbon energy will have occurred later than it should have. As a consequence, the world will have had to cope with intense storms, droughts, heat waves, climate change and disruption of farming practices. Coastal cities will have seawalls and dykes, and many buildings will be designed to withstand Category 7 hurricanes. In the second half of the 21st century, humanity will be concerned with how to live well on a planet that *has* been severely damaged by climate change.

If anyone ought to be able to speak definitively on this subject, it's Rajendra Pachauri, the chief of the UN Intergovernmental Panel on Climate Change, which has over 2,000 top climate scientists. When asked if nations would adapt to climate change, he explained that in some cases we've gone beyond the ability to adapt. "Where ecosystems are threatened and social systems have been strained, you might actually have disruption in human existence. Take the case of sub-Saharan Africa. There are places where it's going to be very, very difficult for people to construct any kind of social system that gives them sustenance. They'll probably have to move to other locations."[28]

Far more dangerous, the Earth's control mechanism (Gaia) may reach a tipping point of relentless positive feedback and heat up regardless of what humankind does. A sign that humanity is in deep trouble will be when the weaker trees in tropical forests start to die because the temperature is too hot for them. Once that happens, the remaining forest will absorb less carbon dioxide from the atmosphere, so global warming will increase. Then more trees will die, still less carbon dioxide will be removed . . . and so on. The Earth will be at the start of a rapid change of state, after which most of its surface will be too hot to be inhabitable, and air-conditioning will make it worse.

We'll then have a changed planet. We'll create artificial environments with habitats for survival, but humanity will be able to survive only in much smaller numbers. Engineers will have to invent various types of grand-scale engineering that can lessen the sunlight reaching the Earth or counteract the changes in climate in other ways.

James Lovelock warns that by the end of the century, massive numbers of humans may die.[29] Given the extreme consequences of reaching a Gaia tipping point, it is insane to allow the growth in carbon emissions to continue.

8

INVISIBLE MAYHEM

THE DAMAGE TO the rain forests or the Aral Sea is photographable and dramatic, and abnormal weather patterns associated with global warming look good in film special effects. There are other forms of damage to our environment, however, that don't get such exposure. They are out of sight and out of mind yet insidiously harmful.

Since the Second World War , the chemical industry has created tens of thousands of valuable new substances—plastics, fertilizers, paints, insecticides, food additives and so on. Chemical engineers had no idea that a small number of these chemicals would do damage to humans. Some of them interfere with cells in the human body and cause subtle malfunctions. In the worst cases they cause cancer, birth defects, brain damage and problems in children for which the cause is not obvious.

We should have reached the end of the Wild West era in the use of chemicals (although we may be entering the Wild West era with other technologies). The traditional view has been: charge ahead, market the product, enforce trade secret laws; the environment is large enough to absorb any damage. Corporations have had the attitude, "If you can't line up the dead bodies, don't interfere with us." In the absence of hard evidence of widespread harm, they insisted they had the right to proceed.

An important part of the meaning of the 21st century is that we must understand with scientific precision which artificial substances interfere with natural systems. If the interference is harmful, it needs to be stopped. As new chemicals are created, there needs to be a clear way to prevent damage to nature and, particularly, damage to humans. We are shrouding ourselves in such an increasingly artificial world we have to create strong protection mechanisms.

DISRUPTION OF HORMONES

Our bodies, and those of other creatures, work by means of three communication-and-control systems, each exquisitely complex and subtle. First, our nervous system, which is rather like a giant computer. It has communications paths—our nerves connect our nerve endings to our brain and interconnect the huge number of neurons in our brain so that we have thoughts, memories and conditioned reflexes. Second, our endocrine system, which broadcasts messages throughout our body by means of chemicals. Third, our immune system, which has the job of keeping us healthy by learning to detect harmful germs and viruses and attacking them. These three systems work in entirely different ways but are intricately interconnected.

This chapter is about the endocrine system and the man-made chemicals that accidentally interfere with it. The endocrine system uses two basic components: hormones and receptors. Hormones are chemicals that carry signals to other parts of the body. They are released in very tiny quantities by our glands. They direct growth, reproduction, neurological development and behaviour. Hormones tell bears when to hibernate, tell salmon when to return to their spawning grounds and cause women to menstruate every 28 days or so. They profoundly affect our immune system. Hormones have a very important role to play during sexual attraction, mating and pregnancy. As an embryo develops, the minutest trace of hormones in the womb affects how the baby evolves.

Hormones are received by molecules called receptors. The body has hundreds of different kinds of receptors, each one designed to receive a particular kind of hormone. The hormone and its receptor have a made-for-each-other attraction. When a receptor encounters a hormone that it is intended to receive, it grabs hold of this molecule. The two powerfully embrace each other in a process known as "binding." This is a "lock-and-key" relationship; the receptor is the lock and the hormone is the key. Once joined, the hormone molecule and its receptor move into the cell's nucleus and trigger the production of particular proteins that initiate the biological activity associated with the hormone.

Unfortunately, a small number of otherwise valuable chemicals get into our body and appear to be a hormone. They inadvertently unlock the receptor in cells of the body so that binding takes place. These artificial chemicals mimic a hormone and induce a response like that that would normally be triggered by a natural hormone. They are hormone impostors. The body's chemical messages are highly complex, and when synthetic chemicals interfere with these messages, severe problems can occur.

Our body receives hormones from pollen and other substances in nature. These hormones usually don't cause a problem because the body evolved defence mechanisms long ago. Such hormones are part of the natural world, and the body is able to break them down and excrete them, but it can't break down man-made compounds. Unlike nature's compounds, these unwanted man-made substances accumulate in the body.

Natural evolution has not equipped us to deal with these new artificial substances—it would take natural evolution millions of years to adapt to them. Where we understand that an artificial substance is causing a problem, we need to stop it from happening. The correct action is to stop selling that substance.

With today's technology, any chemical can be tested to determine whether it interferes with the human endocrine system. It is relatively easy to detect such chemicals in our atmosphere, food and water and in other substances we use. Any chemical that interferes with our endocrine system should be banned, but some companies make a profit from them. The consequences of putting lead products into petrol as antiknock additives, decades ago, are well documented—tens of millions of Americans suffered brain damage, and their IQs were permanently diminished by exposure to lead in the atmosphere. Outrageously, filling-stations in many countries still offer customers a choice of leaded petrol. Endless birth defects, too, are caused by chemicals that should be totally banned.

POPS

The term POPs (persistent organic pollutants) refers to substances that are not part of nature (pollutants), that interfere with organic systems and that are "persistent," meaning that they don't dissolve in water and are not degraded by physical, chemical or biological processes. They do dissolve in fat and do accumulate in the fatty tissues of humans and other creatures. They can *disrupt the endocrine system* because it mistakes them for hormones. Because the body doesn't decompose them as it would decompose and flush out natural substances, they persist and hang around for a very long time. Their fat-seeking nature causes them to migrate to body fats, and this results in their becoming more concentrated as they pass to successively higher levels of the food chain. Once POPs are released into the environment, they sooner or later find their way into mothers (of humans or other creatures), where some pass through the placenta to the developing embryo or foetus. They also reach nursing infants via their mothers' breast milk.

Since POPs are from man-made chemicals, they didn't exist in the early decades of the 20th century. Even in very low concentrations, they can harm humans and other creatures because they are mistaken for the chemical messengers in the body. They are particularly harmful to a foetus in its early stages of development.

Documented injuries from POPs include: (1) tumours and cancers; (2) gross birth defects; (3) feminization of males and masculinization of females; (4) compromised immune systems; (5) behavioural abnormalities; (6) reproductive failure; (7) abnormally functioning thyroids; (8) other hormone system dysfunctions.

DURING PREGNANCY

The mechanisms of pregnancy in animals evolved a very long time ago and are remarkably similar in different animals. Regardless of whether the creature is a whale, a bat, a wallaby or a human, hormones regulate the development of the foetus in essentially the same way. We can study hormones in pregnant animals and most of the results of the study apply to humans. Hormone-disrupting chemicals have broadly the same effect on the growth of the embryos and foetuses of laboratory rats as they do on those of humans.

After conception, the fertilized egg divides into two cells, then four, then eight and so on until a solid embryo is formed. After four weeks, the human embryo is about four centimetres long, but its sex is not yet determined. From the fourth to the eighth week, the embryo is highly vulnerable to disease viruses, such as measles, and to certain drugs. At this time, even the minutest amounts of hormone-disrupting chemicals can play havoc with the development of the embryo. After the eighth week, the foetus is on its way to being well formed, and it is less vulnerable.

Nature learned, long ago, that it must protect the embryo and the fragile foetus very carefully. The foetus is surrounded by something rather like a permeable pouch we put food in—the "placenta." Toxic substances can't pass through the placenta, but nutrients from the mother do. The placental barrier works very well, but it is not quite perfect. Minute traces of hormone impostors *can* leak through the barrier and affect the tiny embryo when it is most vulnerable.

Every pregnant woman today has artificial chemicals in her body, some of which are hormone impostors. These are transferred to the foetus. She also has measurable concentrations of hormone impostors in her milk, and these are also transferred.

Because hormones trigger important events before birth, including key steps in sexual development, hormone-disrupting chemicals are a particular hazard to the unborn. Research has identified all current chemicals that falsely mimic hormones and can interfere with the development of the foetus. There can be no sensible excuse for not banning them.

TESTICLES AND SPERM COUNT

Niels Skakkebaek, a paediatric doctor in Denmark, had built up a strong reputation for his studies of testicular cancer. This had been a rare disease in Denmark, but by 1990, nearly one man in a hundred had it. Skakkebaek was seeing many boys with malformed penises. Some penises had their opening on the underside of the penis rather than on the tip. A study in 1984 of 2,000 Danish schoolboys found that 7% of them had one or both testicles lodged inside their bodies. Skakkebaek read about similar findings in studies of the alligators in the Florida Everglades. He wondered what Florida alligators and Danish schoolboys could have in common.

In his research, Skakkebaek examined foetuses that had been aborted. He found the precursor to testicular cancer cells. He has hypothesized that testicular cancer might be caused by some prenatal event, which didn't become a serious illness until many years later—after puberty. Just as a female foetus could be set up for vaginal deformities or cancer later in life, so can a male foetus be set up for testicular deformities or cancer.

There were reports about men having a declining sperm count, but Skakkebaek was highly sceptical about them—until 1991, when he started to worry about the sperm of Danish men because the sperm banks in Denmark were having difficulty establishing a core of donors. To his alarm, he found that 84% of Danish men had sperm quality under the low standards set by the World Health Organization, although the men themselves seemed to be normal in every other respect. He then set up a team to study results elsewhere. The team reviewed 61 scientific studies involving 15,000 men from 20 countries and found that average male sperm counts had dropped 45% between 1940 and 1990.[1]

Skakkebaek concluded that the average sperm count, which was 113 million per millilitre in 1940, had fallen to 66 million in 50 years. In addition, the average volume of semen ejaculated was 25% less. There was a three-fold increase in men whose sperm count was at or below a level at which they would have difficulty siring a child.

Pierre Jouannet, director of the Centre d'Étude et de Conservation des Oeufs et du Sperme Humains in Paris, was sceptical about Skakkebaek's results. Jouannet had data on 1,351 healthy men in Paris who had donated sperm to a sperm bank maintained by a hospital, starting in 1973. All had fathered at least one child and, therefore, were of proven fertility. So he analysed them, expecting to refute Skakkebaek's studies. To his astonishment, he found that the average sperm counts in this group had dropped steadily at 2.1% per year for the past 20 years.[2]

Then sperm-count studies became highly fashionable. In 1992, Elisabeth Carlsen analysed 62 separate sperm-count studies and concluded that sperm count among men throughout the industrialized world had declined by about 50% in the previous 50 years.[3]

Such studies were criticized on the grounds that a man's sperm varies depending on when he last ejaculated. Because of this, men were sought who had been abstinent for a given time. Among the Paris group was a subgroup of 382 men in a narrow age range (twenty-eight to thirty-seven years) who had all reported a period of abstinence. This subgroup showed a clear decline in sperm count from 101 million per millilitre in 1973 to 50 million per millilitre in 1992—a reduction by half.

Not only was the number of sperm declining, the quality was declining also. The Parisian study showed a significant increase in the proportion of sperm unable to swim and the proportion of misshapen sperm. Much of the damage to sperm can be seen through a microscope. Some sperm have two tails; some have two heads. Some have no heads. Some of the tiny tadpole-like sperm can't swim properly. Instead of a strong swimming motion, some have no motion; others have frenetic hyperactivity.

Both quantity and quality of sperm correlate inversely with age. Young people today have both less sperm and more damaged sperm than their parents' generation. Some men in their fifties had an interesting line with women: "Don't go out with a man below 30, because I have more sperm, and mine aren't deformed!"

Skakkebaek began to realize that the decline in sperm count was part of a larger set of problems with male reproductive organs—undescended testicles, penises with misplaced holes and high rates of testicular cancer. Such abnormalities have doubled in frequency during the past 30 years in many parts of the world.[4]

DERAILED SEXUAL DEVELOPMENT

Wildlife researchers have been documenting cases of derailed sexual development in fish, birds and other animals—impaired fertility, strange mating behaviour, both sexes in one creature, alligators with tiny penises and females nesting with females. Such abnormalities have been found in many species, including terns and gulls, harbour seals, bald eagles, beluga whales, lake trout, panthers, alligators and turtles. They have been found in many parts of the planet. Such abnormalities were investigated in detail and appear to have been caused by exposure to man-made chemicals that disrupt the endocrine system. Most of the damage is probably done in the early stages of pregnancy.

In humans and other mammals, a growing embryo always starts off female. Schoolchildren are taught that the sex of a child is determined by whether the sperm that won the race had an X or a Y chromosome. If Y, an elaborate sequence of events has to occur in order to start the creation of a male foetus. Hormones arriving at the appropriate receptors trigger these events.

You might imagine an automated production line where the standard model produced is a female. However, some parts are designed so that they could become male. If, in the seventh week of pregnancy, the Y chromosome delivers the message that a male is to be produced, a long set of actions takes place. Hormonal messengers go forth with instructions about how to make a male. First, the male testicles are developed. The tiny penis is made, followed by the scrotal sack, the sperm delivery system, the prostate gland, the genital skin, the parts of the male body and the beginnings of the male brain. Hormones are sent out to trigger the disappearance of the female options. Other hormones make the testicles descend.

The development of the male depends on getting an elaborate

sequence of hormonal messages at the right time. Until recently, this pro-grammed procedure rarely went wrong. Then the chemical industry pro-duced vast numbers of new synthetic chemicals. The majority of them caused no problems, but a few could be mistaken for hormones. If one of these false messages disrupted the elaborate ballet of forming a male, there would be serious lifelong consequences when the new child grew up.

SEXUAL CONFUSION

In the early 1990s, researchers at Brunel University noticed that male fish in a river near London had become hermaphrodites—exhibiting the sexual characteristics of both males and females in the same fish.[5] The researchers then explored the rivers of England and found hermaphrodite fish in many places. Sewage was being carefully processed in England and the hermaphrodite fish were downstream of sewage treatment plants.[6]

The researchers placed trout in cages downstream from sewage treatment plants, and after a few weeks, the males began to have elevated levels of a protein called vitellogenin in their blood.[7] Vitellogenin is the protein responsible for making egg yolks in female fish. Little, if any, vitel-logenin is normally found in the blood of male fish. Something coming out of sewage treatment plants was having an oestrogenic effect on the fish. Every sewage treatment plant in England had the same effect.

The researchers tested a few common industrial chemicals to see if they could stimulate the production of vitellogenin in male trout under laboratory conditions. They found several chemicals that did so. The more of these chemicals the male fish were exposed to, the more they had female characteristics.

Wildlife toxicologist Michael Fry then demonstrated that derailed sex-ual development in bird colonies was caused by chemicals that disrupt hor-mones. Though the male birds in these affected colonies looked normal, their sex organs showed severe sexual confusion. Some males had the egg-laying canal normally found in females. Similar studies were done with dif-ferent creatures in different countries, which showed that even a very slight exposure to a hormone-disrupting chemical can have dramatic and permanent effects on offspring at a certain critical stage during pregnancy. It can change everything from sperm count to mating behaviour and occurs across a wide range of species including mice, dogs, amphibians, cattle, sheep, monkeys and songbirds. When false hormone messages reach the developing foetus and cause it to have both male and female characteristics, the resulting creature is referred to as "sexually confused."

The mechanisms of the early stages of pregnancy are similar in creatures and humans, and there is good reason to believe that much of the same sexual confusion that is being found in birds, bees and bears also applies to humans. Discussion has arisen whether gay behaviour in fish and animals has similar causes to that in humans, but this is very hard to investigate scientifically. Tracing cause and effect in humans is difficult because controlled experiments are unethical. A researcher can feed hormone-disrupting chemicals to guinea pigs but not to humans. One research team found that 42% of women exposed to DES (a synthetic oestrogen) in the womb had a lifelong bisexual orientation, but such a team can't experiment with women as it can with laboratory mice.[8] With humans, more importance may be attached to being politically correct than scientifically correct.

PCBS

PCBs (polychlorinated biphenyls) are a family of synthetic chemicals, introduced in 1929 as electrical insulators. After the Second World War, chemical engineers found more and more uses for them—as lubricants, cutting oils, liquid seals and hydraulic fluids. They became ingredients in paints, varnishes, preservatives, pesticides and, eventually, carbonless copy paper. PCBs were nonflammable, apparently nontoxic, and stable. They were highly profitable for the chemical industry.

By chance, PCBs happened to interfere with the endocrine system. When it became clear that this was happening, scientists in many places studied PCBs. They found them everywhere—in soil, air and water; in lakes, rivers and ponds; in fish, birds and animals. In 1967, the manufacture

of PCBs was banned in the United States and then banned in other countries, but because PCBs are highly stable, they remained in the environment. More than 3 billion pounds of them had been manufactured.

PCBs are everywhere—in penguins in Antarctica, in the monsoon
rains of India, in humpback whales near Boston and in the fish served in
smart Manhattan restaurants. PCBs are being found in the bodies of
Eskimo children in the most isolated villages on Earth. They are found in
the milk of nursing mothers and in the sperm of teenagers worldwide.

How do PCBs travel so far? In their liquid forms, they get into waste-
water that flows to the oceans. On rubbish tips, they get into smoke, their
molecules attaching to dust particles that are blown in the wind. They
attach to organic matter—the leaves of plants or algae. Fleas, insects, zooplankton and small shrimplike creatures ingest them. Birds eat the insects.
Fish eat the zooplankton. The birds and fish migrate long distances. Ocean
currents carry zooplankton and small fish to the Arctic or Antarctic, where
they are eaten by larger fish, which in turn are eaten by polar bears. The
PCB molecule might reside in the fat of a fish for many years before the
fish is eaten. PCBs can go on journeys that take decades.

INCREASING CONCENTRATIONS

We commented that the body breaks down and flushes out nature's compounds but cannot always flush out man-made compounds. Evolution
hasn't taught it how to deal with man-made chemicals. As organic compounds such as PCBs steadily accumulate in the fatty tissues of living
organisms and are passed from the bodies of creatures low on the food
chain to creatures higher on the food chain, their concentration can
increase by factors of many thousands.

Tiny zooplankton are at the bottom of the food chain. Even zooplankton have some PCB molecules in their bodies. An anchovy eats a vast
amount of zooplankton. The natural hormones in the zooplankton break
down in the anchovy's body, but PCBs and other man-made molecules
do not, and they steadily accumulate in the anchovy. A cod eats vast
numbers of anchovies and other fish with PCB in them. The PCB steadily

accumulates in the fatty tissues of the cod, and a cod lives for decades. Some cod have been found with PCB concentrations of 48 million times those in the surrounding water.

A seal eats hundreds of fish. Again, the PCB molecules from all these fish accumulate in the seal's fatty tissues. PCB levels in seals are eight times greater than in cod.[9] Still higher concentrations are found in the Inuit villagers who eat the seals. This process is referred to as "biomagnification." Natural chemicals break down and are flushed out of creatures' bodies. Man-made chemicals such as PCBs that do not break down are referred to as "persistent" chemicals—they accumulate in the body and stay there.

BROUGHTON ISLAND

After Rachel Carson's book *Silent Spring* became popular, it was clear that DDT was causing serious problems. In 1964, a chemist at the University of Stockholm was measuring DDT levels in human blood and found a mysterious chemical contaminant wherever he looked. He found it in wildlife samples, in the sea, in his wife's hair and in his baby daughter. It took two years to identify the pollutant as PCBs.

By accident, they have the appearance of a hormone and can be accepted by hormone receptors in the endocrine system. They interfere with the development of the embryo in animals, fish, birds and humans. They have caused gross birth deformities, mental damage, deformed penises and vaginas, decreased fertility, testicular and breast cancers, and they have damaged the immune system. Their production has been banned, but they are still in the environment.

You might ask in a dinner party conversation: "Where on the planet do you think people have the highest concentration of PCBs in their body, or other chemicals that create hormone damage?" The dinner guests guess Manhattan or Calcutta, or perhaps an area of China where the visibility through industrial smog is 5 yards. In fact, the highest concentration of PCBs exists in the people on an island about as isolated and pristine as you can get.[10]

Broughton Island is west of Greenland, near Baffin Island, thousands of miles away from sources of pollution. The islanders live by hunting and fishing. Broughton Island became written about in popular magazines, and people on neighbouring islands talked about Broughton Islanders as the PCB people, as though they had a deadly disease. Marriage to them was discouraged. Worried that Broughton Island was getting bad press, the islanders, in a masterpiece of PR, changed its name to Qikiqtarjuaq— which the press could neither remember nor pronounce.

How could this beautiful, utterly isolated place, so apparently free from pollution, have won the prize for PCB contamination?

The Broughton Islanders fish and hunt, but catch creatures at the end of a food chain that has given them high concentrations of man-made chemicals that nature doesn't break down. Inuit from Greenland on one side of the Atlantic to Arctic Quebec on the other, have seven times as much PCB in their bodies as people living in industrialized parts of Canada. Greenlanders have more than 70 times as much of the pesticide HCB (hexachlorobenzene) in their bodies as industrialized Canadians.

Twenty years after PCBs were banned, they were found in the bodies of polar bear cubs in the remotest areas of the Arctic. Six years after that a study found female polar bears with both male and female sex organs. Hermaphrodite bears exist in all the Arctic areas.

BRAIN DAMAGE

The damage to the sexual development of the embryo, the birth deformities, the cancers and compromised immune systems tell a grim story. Another aspect of endocrine interference may be even more alarming— damage to our brain.

In 1995, 18 prominent scientists, including brain, neurological and behavioural researchers, met in Erice, Sicily, and shared evidence that endocrine-disrupting chemicals, at levels found in the environment and in humans, threaten brain development. In the first months of pregnancy, the developing brain is very sensitive to this chemical disruption, and permanent damage can be caused. It shows up as reduced intelligence, learning disabilities, attention-deficit problems and intolerance to stress—which are not evident when the baby is born.

The scientists who met in Erice produced the Erice Statement of 1995, summarizing what was known, what was suspected and what research was needed.[11] The statement said that the group was *certain* that exposure to endocrine-disrupting chemicals during pregnancy can lead to profound and irreversible abnormalities in brain development. This can be expressed as reduced intellectual capacity and reduced social adaptability. It can impair motor function, spatial perception, learning, memory, auditory development, fine motor coordination, balance and attention span. In severe cases, mental retardation may result.

The Erice scientists make the point that a small shift in brain capability in a population can have profound economic and social consequences. An across-the-board drop in IQ or increase in learning disability can be very expensive for society. The medical costs associated with endocrine-system damage are potentially huge, and they don't even take into account the damage to human lives and happiness.

Many regulations now invoke the "Precautionary Principle." According to the Principle, if two conditions exist, (1) scientific uncertainty and (2) a reasonable suspicion of harm, we should proceed with caution. If we aren't sure what we're doing, we should proceed slowly and carefully. The Principle says that decision-makers have a general duty to take preventive action in order to avoid harm even if scientific certainty has not yet been established.

The Precautionary Principle shifts the burden of proof and insists that those responsible for a product must vouch for its harmlessness and be held responsible if damage occurs. Corporations introducing a chemical or proposing an activity should prove that what they intend to do will not cause undue harm to human health or the ecosystem. This is quite different from the traditional approach—where, if a product causes damage, the victims and their advocates generally have the difficult task of proving that the product was responsible.

The Precautionary Principle, and treaties and conventions that reflect it, are ushering in an era in which gross damage to the environment is less likely. But the principle has to be applied with care. There are numerous human activities that could raise threats of harm to human health or the environment. Ill-drafted laws could open the way for a witch hunt by any group that has an interest in attacking some product or activity. Guidelines and precedents must govern the application of the Principle.

Parts of the chemical industry see some of their profits going up in smoke and have desperately tried to refute the Erice and other such statements—rather like the tobacco industry saying that its scientists had proved that there was no link between smoking and lung cancer.

THE STOCKHOLM CONVENTION

In May 2001, a convention in Stockholm established a legally binding treaty that mandates the elimination of 12 of the worst POPs. The convention stated that other chemicals need to be added to this list—and that all POPs that interfere with the reproductive system and development of the foetus should be banned. It is outrageous to allow such damage because some companies make a profit from it and make political contributions. Part of the meaning of the 21st century is that we establish the capability to protect nature from man-made technology.

The Stockholm treaty supports the "Polluter Pays" principle, under which the producer, exporting company and/or exporting country is responsible for the clean-up and destruction of obsolete POP stockpiles.

At the Stockholm Convention, 104 nations indicated their willingness to support the treaty. The United States was not one of them.

PART TWO

TECHNOLOGIES OF SORCERY

9

GENETICALLY MODIFIED HUMANS?

B ARONESS SUSAN GREENFIELD, director of the Royal Institute, lives in Michael Faraday's old flat. She bent my ear: "We're at a very special time—we—this generation. All of us alive at this moment are at a very exciting and very responsible time, where we have to decide, as a society, to use these technologies and harness them to create a better world."

As the century progresses towards its midcentury canyon, with a soaring population, declining amounts of arable soil and usable water, a stressed environment, increasing civil violence and cheap mass-destruction weapons, the extreme poverty in some parts of the world will contrast with extreme development of technology in other parts.

It's common to say that, whatever else changes, human nature doesn't change. That notion is about to end. In the 21st century, we'll have diverse ways to enhance the nature of the human being. Some capabilities will be highly controversial, and they'll make this century different from any other. Technology will enable diverse forms of environmentally benign affluence. It will present us with our most interesting question of all: How can we build civilizations appropriate for our time? While wealthy countries could build much more interesting civilizations, in the most destitute countries civilization is shattered to pieces. While the technology now on our radar screen can greatly enhance our lives, it can also plunge us into deeper trouble unless we manage technology better.

With 20th-century technology, there was a huge gulf between nature's systems and man-made systems—an uncrossable chasm between the living and the nonliving. At the start of the 21st century, we're beginning to see developments that combine living and nonliving systems. Our new knowledge of nature's genomes makes possible new forms of medicine and farming. Computer technology has started to produce "artificial life."

I called this part of the book Technologies of Sorcery because we're becoming like the sorcerer's apprentice, having started something that we can barely control. In the legend of the sorcerer's apprentice, the apprentice knows the magic is dangerous, but he plays with it anyway when his master is away. He can't resist it. In that story, there was a sorcerer and only one apprentice. Now we are all apprentices.

This chapter and the ones following describe some of the technologies that will become inherently difficult to control but are, nonetheless, powerful enablers of advanced civilization. This raises profound questions about what we want a future civilization to be like. For the first time on Earth, a species is capable of either directing its own evolution or destroying itself.

TAMPERING WITH LIFE

A technology that raises the image of "playing God" is the capability to tamper with life. More than cloning sheep and creating new plants, we are becoming capable of modifying the human creature itself. We are rapidly improving our understanding of genes, what they accomplish and how to modify them. The ability to modify human genes and to pass new genes on to future generations will present humanity with some of the most controversial choices it has ever faced.

Genetic modification (GM) may follow a track something like the growth of computing. In its early days, computing was done in only a few major research centres. It was crude and very expensive. Slowly and tentatively a few corporations made a business out of it. The press and the public were generally negative about computing, equating it more to George Orwell's *1984* than to an optimistic future. Over time, computing became easier to master, more powerful and less expensive. The computers in research labs and government facilities gave way to widely marketed mainframes, then minicomputers, then personal computers and then microcomputers. In less than a generation, every schoolchild began to learn about computers.

Genetic modification, in its early days, was similarly crude and expensive, and it was in the hands of a few large corporations. The press and public were generally negative, with Europe imposing a ban on GM crops. Like computing, genetic modification will become easier, more powerful and less expensive. Eventually plant growers all over the world will buy GM seeds. Many farmers will increase their profits by growing GM crops, and forestry will depend on fast-growing GM trees. Chinese fish farming in freshwater ponds will have GM fish. Some food in people's cupboards will be organic, and some genetically modified. Every living thing will have its genome sequenced. Just as computing became a hobby for many people, many home plant growers who try to create beautiful new flowers with cross pollination will do the same thing with GM techniques because they are so much faster.

Today's gene therapy is very primitive, but it is clear from our experiments modifying animal genes that modifying human genes will become practical. Our growing understanding of the genome will tell us when it could be desirable.

A number of terrible human diseases are caused by defective genes. Huntington's disease is totally horrifying. Its victims develop increasing nervous disorders, jerking limbs, muscular seizures and deep depression. They become completely demented. It takes 15 to 25 years to run its course, and there is no cure. The end is terrible. Huntington's runs in families and is caused by a single bad gene. It's a dominant mutation, not a recessive mutation, and the person's children are doomed. We can easily test people early in life to see whether they have that bad gene. From the coding, we can predict, with remarkable accuracy, the age at which the first symptoms of madness will start—usually between 40 and 50 years of age, by which time the person with this gene has probably already passed it on to his or her children.

The cells that cause Huntington's disease are in the brain, and we can't replace brain cells as we do other cells because most don't subdivide or replace themselves. At least now it will be possible to stop a person's children from having Huntington's disease. Huntington's disease could be and should be eradicated from the planet.

But if we start to modify our children's genes to save them from deadly disease, does it end there? If we could cure a child who has a defective memory, few people would say that is bad. There is a very thin line between tackling dreadful diseases and trying to make improvements in our children. The ethical questions about crossing that line need to be clarified.

At present in our knowledge of gene modification, we play like a chimpanzee at a grand piano. In a few decades, we might become like Rachmaninoff. The potential benefits of gene modification are large, but we need to proceed with the caution of a test pilot and with the very deepest respect for nature. We need comprehensive data collection and the best scholarship possible to make these complex and far-reaching decisions.

We interviewed the maverick scientist Craig Venter, who beat the US government in mapping the human genome. Describing how he was trying to artificially create life, he said, "We're trying to actually make the first single-cell organism by reproducing the chromosome and seeing if it will result in a living cell. I think that's going to become one of the biggest changes in science this century if that becomes possible. We can set up robots to actually build life forms and put in different sets of genes to understand empirically what they do and which ones are required for life and which aren't."

A DIGITAL MAP OF BIOLOGY

The entire set of genes in a creature is referred to as its genome. The 21st century began with the complete mapping of the human genome—an extraordinary achievement. Scientists can now map the genome of mice, microbes, Christmas trees, orchids and just about anything else. Your genome contains 3.2 billion letters and defines what you are.

Matt Ridley, author of *Genome: The Autobiography of a Species in 23 Chapters* and former editor of *The Economist* magazine, describes the human DNA molecule as being like a book with 23 chapters—one for each of our 23 pairs of chromosomes, which make up the gene material (macromolecules) found in the nuclei of cells.[1] Each chapter is divided into sections—genes. You have about 30,000 genes. A typical gene has about 100,000 letters (called "nucleotides" or "bases," each of which can be one of four combinations). If printed, the letters of your DNA would require a stack of paper 200 feet high. The letters of an average-length gene would fill a manual of 30 pages. It is amazing to realize that the entire 23-chapter book is coiled up in the DNA double-helix molecule in every cell in your body. So you walk around with 10 trillion copies of it.

Every living thing on the planet is built with DNA—fish, giraffes, roses, microscopic parasites, giant redwood trees and the dinosaurs before they were extinguished. A DNA molecule is like a computer disk on which any life form could be encoded. If we detect microscopic life in the sea of Jupiter's moon Europa, it will be interesting to find whether it is also built from DNA. It seems almost certain that life on planets trillions of miles away around distant stars is utterly different and not built with DNA—but we don't know.

The ability to map the genome of living things gives us an extraordinary world to explore. Most of the code sequences of our own genome evolved a very long time ago in creatures very different from humans. Geneticists do many experiments with fruit flies because they breed quickly and have a short life so that it's easy to study genetic mutations. Amazingly, some of the text in fruit-fly DNA also exists in human DNA. For example, a gene used in making eyes in fruit flies is similar to a gene used in making human eyes. It appears that fruit flies, humans and other animals descended from a common ancestor and that certain gene mechanisms from hundreds of millions of years ago worked so well that nature kept them while radically different creatures evolved. We are an accumulation of the results of untold millions of Darwinian experiments, almost all of them occurring long before man existed.

The DNA of a chimpanzee is remarkably similar to the DNA of a human. The letters of the chimpanzee DNA and human DNA are about 98.4% the same. A chimpanzee's muscles, stomach, eyes, liver, ears and so on work in about the same way as those of a human. Every known chemical in the chimpanzee brain is also present in the human brain.

Now that we can map, digitally, the letters in the DNA of creatures past and present, we can use computers to explore the pathways of ancient evolution.

AN EXTRAORDINARY TOOL KIT

Scientists have developed a set of tools for understanding and manipulating genes.

Because the blueprint for our nature is digital, we can edit it as though we were using word-processing software. We can snip out sequences of letters, modify the text, overwrite words and, if we wish, add sequences from other creatures.

To cut and paste genes we need scissors and glue. The scissors, glue and many other chemical tools are referred to as "enzymes." There are many types of enzymes; enzymes do all the work that is done in a cell. A glue is an enzyme called "ligase." It connects loose sequences of DNA whenever it encounters them. The scissors are called "restriction enzymes." Their purpose in nature is to defeat viruses by cutting up their genes, but they can only cut where they encounter a particular sequence of letters. It is like a word-processing function that can cut text only if it finds the word *dog*. Gene engineers have hundreds of restriction enzymes, each of which finds a specific sequence of DNA letters and cuts there.

Gene specialists often want to search for a particular sequence of letters—for example, a sequence indicating that you might become a victim of Huntington's disease. Word-processing software could easily conduct a similar type of search. The way we do it with genes is by means of a gene chip. A gene chip lays down a long sequence of DNA letters in physical grooves on a silicon chip. When bodily fluids are washed over it, it can detect DNA sequences that are identical to what is on the chip. Therefore, we can make a Huntington's disease chip that will detect the bad gene if we smear our saliva over it.

A human has about 10 trillion cells. If scientists want to genetically change a person, they have to insert a new gene into every relevant cell. This seemed impossible in the 1960s, but in the 1970s scientists found out how to do it by using a retrovirus. A retrovirus is a virus that contains genetic code, in the form of RNA, and can insert the code into our chromosomes, doing an RNA-to-DNA transcription. Genetic engineers use special retroviruses in a controlled way to spread modified genes throughout the body. It travels to vast numbers of cells in the body with a message saying, "Make a copy of this gene and stitch it into your chromosomes." When the cells that are modified divide, they replicate the inserted gene.

Uncontrolled retroviruses are responsible for the spread of viral infections, including at least one type of cancer. Some retroviruses destroy the cells they alter, as with HIV (the virus that causes AIDS).

The mechanisms for cutting and splicing genes, along with the mechanisms for inserting modified genes into existing cells, give us an increasing capability to redesign nature. It feels a little like the time around 1950 when we had invented programmable computers; we realized that, in principle, we could make a machine do almost anything we wanted. There are immense technical problems to be solved in gene splicing, just as there were with early computers. With computers, they were solved, and 40 years later, creative people had Macintoshes on their desks. The home gardener 40 years from now will be creating beautiful genetically modified flowers in his computerized home hydroponic environment. Dare we ask what the section on designer babies will say in *Cosmopolitan* magazine?

MODIFYING FUTURE GENERATIONS

Not surprisingly, there is great nervousness about modifying human beings. In one way, however, modifying a human can be less risky than modifying a plant. Human gene modifications can be done so that they affect only the one person. A genetically modified human, unlike a plant, doesn't produce pollen that blows in the wind. The risk is contained to that one person.

There are two ways in which the genes of a human (or any creature) can be modified. They are referred to as germline modification and somatic-cell modification. Germline modifications are passed on to our future generations; somatic-cell modifications are not.

Sperm cells in a man and egg cells in a woman are called germ cells. If the DNA of these cells is modified and an embryo is formed, every cell in the resulting baby contains the modification. All other types of cell are called somatic cells—the working cells of the body. When the DNA of somatic cells is modified, the change is not passed on. All modifications of human genes so far have been somatic-cell modifications, but it will probably become much easier to make germline modifications than somatic-cell modifications.

There are risks inherent in gene therapy. The Law of Unintended Consequences is always with us. When DES (diethylstilboestrol) was being prescribed to pregnant women on a massive scale in the 1950s, nobody predicted that it would cause cancer and severe problems in unborn children. So it's unwise to tamper with human genes frivolously. Some people argue that human DNA should be sacred and untouchable. Even if one accepts this, there is still a strong case for repairing damaged DNA (and much of our DNA is damaged).

A key question about gene therapy is, "Do the benefits to the patients outweigh the risks?" Gene therapy has had formidable problems so far, but as it progresses the risks will decline and the potential benefits will increase. The techniques for modifying genes will become more standard, more reliable and less expensive.

To pass modified genes on to future generations, we must alter sperm or egg cells, or alter an embryo—in other words, practise germline modification. If we modify an embryo, when the baby grows up, it will pass the changes on to *its* children. Scientists have done this with mice and monkeys but not yet with humans (at the time of writing). Doctors would need to be very sure that they knew what they were doing before they made human modifications.

Germline modification is much simpler than somatic modification because it can be done with an embryo, and when a baby grows from the embryo, all of its cells contain the change. This avoids the difficulty of having to modify 10 trillion cells in a grown person. Germline modification has been made illegal in some countries, but not others.

Just as a plane crash is not usually caused by one failure but by the coincidence of multiple failures, most human diseases are associated with multiple defects. Some of these are genetic defects, and some are not. Some genetic defects are dangerous, and it is desirable to correct them. If you understood your genetic defects, you might want to avoid passing them on to your children. If you survived a nightmarish flight in a plane with a faulty hydraulic system, you wouldn't want your children to fly in it. When genetic repair becomes safe and affordable, it will lead to a medical tool kit with many applications.

Germ cell manipulation might be used to remove an inheritable disease. People in future societies will become familiar with their own genes and gene defects. Just as today you may have a printout of a comprehensive blood test, in the not too distant future, you will have a computer disk highlighting aspects of your genetic makeup. It will point out areas where you need to take precautions.

IN FERTILITY CLINICS

The first so-called test-tube baby was born in England in 1978. Her name is Louise Brown and she has lived a healthy life. There was a hysterical outcry when she was born and a demand that test-tube babies be banned. Rather than being banned, the procedure evolved into a major industry. A million test-tube babies have been born. Numerous fertility clinics now employ IVF (in vitro fertilization). Such clinics use eggs and sperm from the would-be parents. They take an egg from their female client and fertilize it to form a cell, which is allowed to grow into an embryo that is transferred to the mother's womb so that a normal pregnancy can continue. This has become routine, safe and highly profitable.

Some IVF clinics have performed an extension of this procedure.

They similarly remove an egg, but fertilize it multiple times, allowing several cells to form with slightly different DNA. The technician examines the DNA of each of them and eliminates any that have Huntington's disease, cystic fibrosis, sickle-cell anaemia or other genetically based diseases. A carefully selected cell is used to grow an embryo, which is implanted in the woman. Would-be parents who have a severe inheritable genetic condition can avoid passing that condition on to their descendants. The royal families of Europe suffered from serious inbreeding, and many monarchs were outrageously mad. What would they have paid to clean up their germline?

Various companies have decided that it is good business to harvest and stockpile eggs for use in the future. A growing number of women sell or "donate" their eggs to such companies, and men donate sperm. On American college campuses, advertisements offer to pay as much as $50,000 to egg donors with high SAT scores or other desirable attributes. Some celebrities sell their eggs or sperm for a very high price, for future use.

This use of eggs can produce babies with selected genes without modifying genes. Also, in the not-too-distant future, the clinic may go a step further and introduce a new gene into the embryo. The embryo is allowed to grow a number of cells with the new gene. These cells are carefully examined and tested to make sure that the required genes are present. The embryo is then transferred to the mother's womb so that a normal pregnancy can continue. The new genes will be in every cell of the resulting baby. This has been done extensively with animals, but not yet, as far as is known, with humans. Sooner or later, it probably will be done with humans, perhaps in secret until it becomes an accepted procedure.

There are still many difficulties with gene modification. Many abnormalities have occurred in genetic work with animals. Geneticists are steadily learning how to avoid the problems, and some believe that embryo modification soon will become a professional discipline, as safe as normal pregnancies. Infamous diseases that are caused by *one* faulty gene—for example, muscular dystrophy, cystic fibrosis and Huntington's disease—are early targets for gene therapy. Most defects are more complex because they relate to multiple genes.

Approximately 1% of people show remarkable natural resistance to infection by the AIDS virus. The genes that confer this resistance are being identified. It may become relatively easy to introduce the HIV-resistant genes into the embryo. This could prevent a person's children from being vulnerable to AIDS. Similar procedures might be developed that cover resistance to gene-related diseases—perhaps Alzheimer's, stroke, heart failure, diabetes and some forms of cancer. We have so much to learn before we know what is likely to work.

AN ARTIFICIAL CHROMOSOME

Germline modification is a serious step because the altered genes will be passed on to all future generations, and there may be unanticipated side effects. A more interesting option than germline modification could be to modify, say, your child, by giving him or her a new 24th chromosome. This form of genetic change doesn't change the original genome. The artificial chromosome would be placed in the embryo and would reside in every cell in the resulting person, along with his 23 pairs of natural chromosomes. Using a 24th chromosome in this way is not germline engineering because it is noninheritable. It wouldn't be passed on to its owner's children.

Furthermore, the person with a 24th chromosome in all the cells of his body need not use it. It would contain controlling code that would determine when and where particular genes are expressed. This gives the owner the means of switching the genes on and off. He may never switch them on if he so chooses. He can switch on genes he selects and switch them off if he doesn't like them.

An empty chromosome has been created that has no genes of its own but has "docking sites" where cassettes of designed genes can be inserted (by means of enzymes). When such a technology first comes into practice, there may be only a few gene-packs for specific diseases, but eventually there would be numerous gene-packs, each offering its own type of benefit.[2] Technicians in fertility clinics might be trained to insert the gene-packs. Eventually, there'll be a large library of gene-packs.

The ability to deactivate or update the artificial chromosome makes it less worrying than normal germline modification. Whatever procedures we use for modifying human genes ought to be reversible because surprises will occur. This is achievable with an artificial chromosome.

GREEN AND RED CHANGES

In my kitchen, I have a painting of a pig named Caesar, born in 1834, who weighed 1,948 pounds. Such enormous creatures were the result of selective pig breeding, which was a hobby of the British country squire in the 19th century. Natural gene selection has been more dramatic than high-tech gene modification (at least with pigs). The selective breeding of animals can be much more efficient today because we can map their genome and understand what parts of the gene produce desirable traits. We can find out which animals have those desirable gene sequences and interbreed them. Some companies are doing this to produce better beef cattle, for example. Although it depends on reading the creatures' genes, it is natural breeding rather than genetic engineering. A farmer could achieve the same result much faster by modifying the genes directly.

We might draw a line in the sand between creating genes that nature could have created and creating new types of genes. Let's call the former "green" genes and the latter "red" genes.

Caesar the pig has green genes. They were achieved by natural breeding. If they had been achieved by gene splicing, they would still be green. We could imitate horse breeding but do it fast and efficiently by gene splicing—without attempting to create something that nature couldn't create. Using gene slicing to take out bad genes, such as that for Huntington's disease, is green. So is the use of most gene therapy in medicine.

On the other side of the line, we create gene modifications that couldn't happen in nature—red genes. Scientists modifying plants have created plants that nature *couldn't* create, by combining genes from entities that could never interbreed. They might combine genes from a fish and a strawberry if they find that useful. Geneticists might find good reason to splice a gene sequence from a bird into human genes. This is crossing the line into red territory and creating something that nature wouldn't create.

There are already arguments for taking an ancient gene from a lizard and inserting it into a human chromosome. Because the results would be unpredictable, much experimentation would be needed. It is already being done with laboratory animals. Because chimpanzee DNA is so similar to human DNA, scientists experiment with chimpanzee enhancements. They might find some red ones that would be very valuable in humans, and there'll be an irresistible desire to try them out in humans.

One can imagine great media attention to a story alleging, inaccurately, that a chimpanzee has had its sexual performance improved. Videos of the chimp would be much more entertaining on TV than those of Dolly the sheep. There may be GM puppies that teenagers swoon over, but soon rebellious youths will say, "If chimpanzees can have their sexual prowess enhanced, why can't we?" Some red gene changes might be highly marketable.

Tay-Sachs disease, sickle-cell anaemia and other dreadful diseases are caused by inherited genes. In some societies, it may become regarded as immoral to have a child who is a dwarf or who has Down syndrome, because it could have been prevented. Some of our gene damage has been caused by artificial chemicals or radiation of our own creation. It seems especially wrong in these cases not to correct it if we know how to.

A gene region on chromosome number six has been associated with severe dyslexia. It may be an achievable journey to go from correcting dyslexia to creating humans who have a good chance of better learning ability. There will be safe gene changes that can increase longevity; will we resist them? A valid viewpoint may be that it is unethical to resist it. If a million people can live 10 years longer, and we prevent it, it is equivalent to killing a million people 10 years before their normal age of death.

Some countries have made genetic modification of humans illegal but have done so without understanding the key difference between red and green changes.

We may decide to enhance nature cautiously in numerous situations that are controlled and measured. Our genome mapping will tell us that certain genes give you perfect pitch, make you fat, make you worry-free or give you an addictive personality. If commercial organizations offer parents the possibility of improving their kids in some way, they will find no

shortage of customers. Gene modification may become a part of the consumer society, like breast enhancements or nose jobs. Some parents pay
heavily to give their children piano lessons or golf training; they would
pay for enhanced memory, height, or blue eyes, too.

It is easy to conjure up entirely false images of what germline change
might create. Many human attributes—such as intelligence, literacy, kind-
ness or athletic skill—are affected by many genes interacting in ways that
we don't understand yet. They are linked to enormously complex factors
of environment and nurture. Gene engineering won't create Einsteins,
Miltons or Tiger Woodses.

A long-term danger of human gene modification is that it could cause
a social class distinction. Wealthy or well-connected people may have
genetic enhancements that most people don't have. Future corporations
might do gene tests on prospective employees and pay for their brightest
people to have gene enhancements. There will probably be clubs for gene-
enhanced people, helping them to "improve" their children. As gene-rich
people travel around the world, they will meet and help other gene-
rich people. Gene-rich people may marry gene-rich people. There has
been much alarm expressed about this possible class distinction. In reality,
it may cause less differentiation than other forms of human enhancement—such as intense training enabling people to use advanced software
or do highly paid jobs.

The steady, and eventually massive, improvement in the capability of
human beings will be one of the momentum trends of the 21st century.
This will come because of better education, better nutrition, better jobs,
better tools, superintelligent computers, extreme-bandwidth evolution of
the Internet and the avoidance of mind-destroying drudgery. In some
countries it may come because of green genetic enhancement. If green
gene changes become common, however, will red gene changes be far
behind?

10

NANODELUGE

I N THIS BOOK we are concerned with events that change society in the most important ways—with the ocean tides rather than with the waves on the surface. If you can hitch your wagon to something that will accelerate relentlessly, year after year for many decades, you will see big changes. Computer power is one of the things that will accelerate relentlessly.

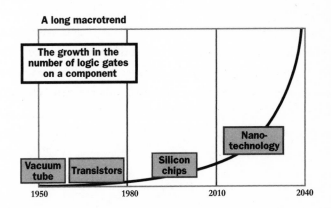

Most people associate the growth of computing with the 20th century, but the true computer revolution is yet to come—with ubiquitous sensors, nanotechnology, global data warehouses and totally pervasive access to networks of extreme bandwidth. The main reason the true computer revolution is ahead of us is that machines will become intelligent. Computers can be immensely more powerful than the human brain because their circuits are millions of times faster than the neurons and axons of the brain, and they can be designed to perform specific types of "thought" with great efficiency. Such computing will become an infrastructure that is everywhere, like the air we breathe, affecting almost every activity of humankind.

A popular view of our time is that when computers become intelligent, they'll be like humans. I think this is nonsense. They may talk to us with human faces on their screen and human voices, but deep under the covers their intelligence will be radically different from ours. There'll be diverse forms of such intelligence, some of which will be designed to constantly improve its own capability. As machine intelligence evolves, this will be fundamentally different from nature's evolution. Machines will be established to evolve with a specific goal in mind.

Freeman Dyson, one of the people who made the Institute for Advanced Study at Princeton legendary, used the phrase "infinite in all directions." It seems to apply (with just a touch of exaggeration) to 21st-century computing. Parallel computers will be built from an endless number of tiny computers working simultaneously. As well as being gathered into large machines, the simultaneous processors could be far away on global speed-of-light networks. The number of processors in one configuration will continually grow. Artificial-intelligence techniques will create new types of intelligence that can automatically learn, evolve and improve themselves. There may be some types of intelligent software that are designed to improve their own intelligence indefinitely.

The power of such computing can be applied to subject matter that itself seems infinite in all directions. The genomes of all nature's plants and creatures will be decoded and recorded. Every movement of every stock market in history will be available. The ecology of the planet and oceans will be instrumented. Genetic engineering has endless theoretical possibilities. Robotic devices will have no resemblance to the robots in movies, but as they become intelligent, they'll have a near-infinite number of applications. Nanotechnology transponders, so tiny that they are invisible, can, in effect, be made superintelligent by a wireless link connecting them to nearby chips, or to the near-infinite capability of a future Internet.

Ubiquitous machine intelligence that becomes increasingly powerful will be one of the enabling factors that will bring spectacular changes in civilization, but it also sends major alarm signals about whether we can control our technology. Part of the meaning of the 21st century will be learning to coexist with such technology.

THE CHESSBOARD

In the computer industry there's a well-worn story about a king in ancient times being advised by a consultant. The king asked the consultant how he wanted to be paid. The consultant suggested that a grain of rice be placed on the first square of a chessboard, two grains be placed on the second square, four on the third and so on, each time doubling the number of grains for the square. The king thought this was reasonable. To get his council to agree, he negotiated that in order to lessen the short-term drain on the treasury, the payment should take place every 18 months. One grain of rice now, two grains in 18 months, and so on. It would take almost a century to pay the full amount, which would go to the consultant's offspring. The king said at his press conference that the consultant was not asking much for his services.

For many years not much grain was needed. By the time the 21st square of the chessboard was reached, a sack of grain fulfilled the king's contract. The 33rd square needed four truckloads of grain. By the last third of the chessboard, things were wonderfully out of control. The 51st square needed a convoy of a million trucks, and the council had to substantially increase taxes. The 58th square needed far more trucks than there were on the planet. The last square will need 64 times that much. (Appendix 1 shows the chessboard of increasing computer power.)

The most famous prediction about the evolution of computers is somewhat similar. Gordon Moore of Intel, years before the moonshot or microprocessors, predicted that the number of transistors on a chip would double every year and a half. This was to prove a remarkably accurate prediction and became referred to as Moore's Law. When the first computer with transistors was marketed, in 1956, one component had one transistor. Ten doubling periods later (1971), the densest chips had a thousand transistors. After 20 doubling periods (1986), they had a million transistors. After 30 doubling periods (2001), a billion transistors. After 40 doubling periods, some components will have a trillion transistors, or their equivalent, and so on. By the last square of the chessboard, computer capability will have reached extraordinary levels.

THE PETACOMPUTER

The most famous computer in movie science fiction is probably HAL from Stanley Kubrick's *2001: A Space Odyssey*. Arthur C. Clarke, author of the book on which the film is based, tried to make the engineering as close to reality as he could. The nefarious HAL was a supercomputer with thousands of modules that operated in parallel, interconnected by very-high-speed channels. When the one surviving astronaut in the film was desperately trying to kill HAL, he was seen pulling out the separate parallel units as fast as he could. In 2001, IBM delivered a machine to the Sandia National Laboratories with hardware much like HAL's for a price of $85 million. While the hardware is similar to HAL's, the software has no resemblance. Instead of human-like intelligence, like HAL, it does bit-crunching simulation of nuclear explosions. Sandia says it will need ever-faster machines for this purpose.

The power of a computer is often quoted in terms of the number of floating-point operations it can execute in a second—abbreviated FLOPS (floating-point operations per second). The first massive vacuum-tube computers at the end of the Second World War (one in Britain and one in the United States) could manage around about 100 floating-point operations per second—the first grain of rice on the chessboard. At the start of this century, Sandia ordered a machine of 100 teraflops (100 trillion FLOPS).

(The prefixes *mega, giga* and *tera* mean a million, a billion and a trillion, respectively. The prefix *peta* means a thousand trillion, *exa* a million trillion and *zetta* a billion trillion—10^{21}.)

A machine with petaflops of power (a thousand trillion FLOPS) is referred to as a petacomputer. IBM completed its first petacomputer in 2005.[1] It has been estimated that the human brain has a processing power of about 100 petaflops. We'll have a machine of this power around 2015, but it is entirely wrong to conclude from this that such machines will be like people. They will not, in any sense, be close to the capability of the human brain, but they will have certain roles in science and commerce that enormously exceed human capabilities.

Such computers are needed for numerous military applications, including future antimissile defence systems. When advanced nuclear weapons are exploded in a simulated fashion *in the computer,* they can be experimented with to achieve successively better designs. *Physically* exploding nukes is crude and inflexible. Simulated explosion in supercomputers is used to explore tiny backpack weapons, weapons for small missiles, neutron bombs that wipe out the people in a city but not the buildings, as well as weapons for destroying incoming missiles.

How much will a 100-petaflops computer cost? If many millions are mass-produced, as with today's personal computers, the cost will be reasonable. A key to their future is the growth of million-unit applications such as the following:

- Mass-market games using virtual reality that is ever more realistic. Interactive virtual reality can burn up all the computer power the designers can throw at it.
- A large market for special-function robots as they become interestingly intelligent (very different from today). Robotics can burn up endless computer power.
- Military applications that give soldiers immense power without being near the dangers of the battlefield.

- Special effects in movies that are ever more exotic and realistic.
- Real-time language translation. Even though imperfect, this will lead to mass-market devices. It may become a service provided by telephone companies.
- Electronic commerce applications that consume massive computing power.
- The combination of complete DNA knowledge and supercomputing will set the stage for a new science of preventive medicine and the capability to create proteins, the functional structures that make biological creatures operate. Increasingly, we will acquire the capability to manipulate life.

By 2025 we'll have petacomputers everywhere, like today's microprocessors.

NANOTECHNOLOGY

Steam power enabled much that happened in the 19th century. Electricity enabled much that happened in the 20th. Nanotechnology will enable much that happens in this century.

A nanometre is a billionth of a metre—a very tiny distance—the distance your fingernail grows in a second. The smallest atom (hydrogen) is 0.1 nanometres in diameter; the largest naturally occurring atom (uranium) is 0.22 nanometres in diameter. Most molecules are less than a nanometre across. If you have very good eyesight, the smallest things you can see are about 10,000 nanometres across. When nanotechnology manufacturing is in high gear, the world will be flooded with devices so small that we can't see them.

Nanotechnology is concerned with the smallest usable items that mankind can make. It refers to items between 1 and 100 nanometres across. A human hair is 50,000 to 100,000 nanometres in diametre. The smallest components on today's silicon chips are about 30 nanometres across.

There has been a lot of hype about nanotechnology. For example, early visionaries described how factories might build tiny devices that are assembled atom by atom. This would be extremely difficult to do and may always remain impractical. It would also be a very slow process. It has been estimated that to build a few ounces of material atom by atom would take millions of years.[2] Instead it's necessary to invent highly automated fabrication processes that use *self-assembling* structures. With self-assembly, different nano-modules search for other modules that they can connect to as a key fits into a lock. This happens constantly in biology. Our body has trillions of cells, and biochemicals flood through our body until they find a cell that they can communicate with. A hormone has a key and searches for a lock that it fits. Similarly, in self-assembling processes nanomodules may flood through a mass of other modules until they find a module that they can connect to. Step by step they connect until a designed structure is built up.

Today, some inventors have a clean room in their house basement and play with nanotechnology or quantum entanglement. Some have a biological laboratory in which they can modify genes. They are able to share ideas with inventors on the other side of the planet. I depth-interviewed one such inventor, Bill Parker, in his clean room, both of us with face masks and elaborate garments to help keep the room clean. He demonstrated self-assembly of nanotechnology devices: "You pour in different components that connect to one another, swirl them around in a beaker, pull it out and now you have a billion sensors." Each device produced in this way is far too small for the human eye to see.

A carbon molecule may contain many atoms of carbon. Carbon has a special property: Its atoms can be linked to one another very tightly in different patterns. One pattern is diamond; others are soot or slippery graphite. The pattern could be like a tiny geodesic dome—an extremely strong structure, about a nanometre across, made of 60 interlocking carbon atoms (sometimes called buckyball molecules, after Buckminster Fuller). Scientists have recently learned how to make tubes of carbon atoms linked together in hexagons forming a tiny cylinder that is phenomenally strong because it is one single molecule. It has been demonstrated that such a tube can have a tensile strength more than 60 times that of high-grade steel. It is the strongest material ever made. In principle, the molecule could be very long. Carbon nanotubes have been woven like hemp into a rope so thin that you can't see it but strong enough to suspend a truck. Carbon nanotubes are possibly the best technology for building a home high-definition television screen 3 feet high and 9 feet wide. There are different types of carbon nanotubes, with different tightly locking configurations of carbon atoms. Work is progressing in many companies on mass fabrication techniques.

An electron can fly along a carbon nanotube with effectively no resistance so that the tube behaves like a superconductor. Carbon nanotubes have been constructed that act as a switch or logic structure such as an AND or OR instruction in a computer. Short nanotubes are about one-hundredth of the width of today's tiniest transistors. This means that future nanochips may have 10,000 times as many switches on a two-dimensional surface somewhat larger than today's chips. Carbon nanotubes do not generate the heat that is a problem with today's chips. Today's chips have a two-dimensional layout because of the need to dissipate heat. Nanotube modules will be three-dimensional, and this will massively increase the memory or logic built into one module. It will take many years to evolve the necessary mass-production technology. Nanoscale fabrication, superconductivity and three-dimensional structures will eventually make possible the building of immensely powerful supercomputers.

DIFFERENT PROPERTIES

Nanoscale devices are not only small but have different properties from larger devices. Chemical properties change when items are close to one molecule in size. Electrons move differently; so, electrical properties change. Ohm's Law, the fundamental law of electricity that we learn in school, doesn't apply to electrons moving in carbon nanotubes. Carbon nanotubes and other nanotechnology have radically different physical properties. Matter behaves differently in nano form from the way it does in bulk form. To understand it, multiple disciplines are needed. The different properties produce results that can't be produced in conventional ways. These can be used in the design, manufacturing and use of nanotech devices. At the time I write this, the biggest-selling nanotech device is a golf ball with a nanotech coating that reduces air friction. There are types of coating for buildings to which dirt and graffiti adhere only lightly and wash away in the rain. Nanotechnology is a field with immense scope for invention.

The basic properties of digital chips will change as we head into the nanotech era. The evolution of traditional chip production—championed, for example, by Intel—will compete with radically different design and fabrication methodologies, using self-assembly techniques. The latter—a total shift in manufacturing methods—will probably cause the chessboard to have many more than the 64 squares we described. Almost certainly, nanotechnology will still be growing in capability in midcentury and, perhaps, still by the end of the century.

Knowledge is represented by bits, which computers manipulate. There is a limit to how small *physical* things can be made, but bits have no inherent mass or size. They could be represented by modulations of light and can travel at the speed of light. Computers of the future will manipulate immense quantities of bits in overwhelmingly complex ways. There is effectively no limit to how large their warehouses of data can be. A biological cell contains many billions of bits, and an ant contains many billions of cells. The brain of an ant is the width of 10 of today's transistors. Nanotechnology will eventually take us to this level of miniaturization.

Even our best chips contain circuits that are fat and clumsy compared with a biological virus. A virus is a clever device that can use a cell to replicate itself. It can produce another copy of itself in about 20 minutes. In another 20 minutes, there are four copies; in an hour we have eight copies. At the end of 24 hours, if conditions were perfect, the virus could produce four thousand million trillion copies of itself. Still more clever, the virus can mutate into new forms. The vast majority of mutations are abandoned, but occasionally one is useful. The virus will, sooner or later, mutate into a form that is resistant to the antibiotics it encounters. A virus, then, seems like a diabolically clever machine. The size of this mutating machine is *less than a tenth of the diameter of a single transistor.*

We are nowhere near to competing with nature yet as far as the *size* of bits, but when we consider their *speed,* we beat nature hands down. The bits in our brain move through a tortuous tangle of nerves at an average speed of about 30 kilometres per hour. Bits move over optical circuits in computers or over long-distance fiberoptics at the speed of light—that is, 40 *million* times faster than the average speed of signals in our brain.

While computers are gaining power, programmers are not becoming much more productive. Computers will become a million times more powerful in the next 30 years, and by then, there will be billions of computers all over the world. So, how do we create programs for them? Programming high-speed computers in traditional programming languages is a painfully slow process. Trying to write large computer programs instruction by instruction is very slow and error-prone. The computer technology towards the end of the chessboard would be crippled if we had to program it by hand; it would be like pulling a Ferrari with a horse.

Fortunately, there are various ways to automate the production of program code. One is to set up a process whereby software evolves. We give it narrowly focused goals towards which it can progress incrementally. It tries endless mutations, assesses the results and selects the mutations that achieve the best results. Just as an Internet search engine can appear to be astonishingly fast, evolution—perhaps on many distributed computers— will eventually produce results with astonishing speed.

Self-improving software exists today in various forms, most of them fairly primitive. Some future software will be designed to improve itself over a long period, often with human guidance, so that it can develop into a resource far beyond anything that a team of humans could create in conventional programming.

A NEW TYPE OF INTELLIGENCE

We are seeing new forms of computer intelligence that are radically different from human intelligence and, in narrowly focused areas, immensely more powerful. It is time to give up the 20th-century notion that artificial intelligence will be like human intelligence.

Non-human-like intelligence will be one of the defining characteristics of the 21st century. I'll refer to it as NHL intelligence. It refers to a growing family of techniques that enable computers to evolve powerful forms of intelligent behaviour automatically.

When we create self-improving computer processes, interesting things can happen. The processes often go on improving until the results are so complex that we cannot follow their logic. They will race far beyond human capability. The growth of powerful NHL intelligence is one of the important momentum trends shaping our future.

There is a growing variety of techniques for producing NHL intelligence automatically. Computers can be made to recognize patterns that humans cannot recognize, "learn" behaviour that humans cannot learn, explore data too vast for human exploration, "breed" programs that humans cannot write, assemble logical reasoning too complex for humans, *evolve* behaviour that humans cannot design and exhibit emergent properties that humans can't anticipate. All this capability exists today—but in an early form, like motorcars in 1900.

Some of these mechanisms, once launched, can run largely under their own steam and can be designed to improve themselves automatically. They do so at electronic speed and can produce computerized "thought" that is entirely different from human thought. The vast and complex search engines of Google give us great capability. As the various techniques for creating these new forms of intelligence mature, they enable machines to do things that the human brain can't do.

Attempts to make machines like humans have repeatedly failed to deliver their promises. In 1965, one of the great pioneers of artificial intelligence, Herbert Simon, wrote, "Within 20 years, machines will be capable of doing any work a man can do." Ten years later, Marvin Minsky, cofounder of the MIT Computer Science and Artificial Intelligence Laboratory, apparently unfazed by any timetable slippage, wrote, "Within a generation, the problem of creating artificial intelligence will be substantially solved." After more than four decades of intense research with big budgets, the computer's ability to demonstrate human-like intelligence is still very limited. There are three reasons for this. First, massive computing power is needed—machine intelligence belongs firmly on the last third of the chessboard I discussed earlier. Second, it is too complex to create with conventional programming methods. We need techniques that enable us to

generate software, or processes that evolve automatically. Third, there has
been the mistaken assumption that machine intelligence will resemble
human intelligence.

Historians in the future will say that computing emerged from its basic
period when, instead of following human logic in a sequence of simple
steps, it *began to exhibit an intelligence of its own, fundamentally different from
human intelligence.* We are now seeing the first baby steps of this new intelligence.

When computer intelligence can improve its own capability automatically, machines will become more intelligent at a rapid rate until explosive
change occurs. The Internet will enormously amplify this chain-reaction
capability, as machines will transmit their capabilities to other machines.
Different chain reactions will occur in different subject areas.

A PARTNERSHIP OF UNEQUALS

The human mind, very different from computers, is a world of dreams,
imagination, chicanery, poetry, love, ingenuity, religion and storytelling.
Computers not only lack the poetry; they completely lack common sense.
The intricate, devious human mind is capable of all manner of thinking
that computers won't emulate in the foreseeable future. Life has an endless
variety of delicate experiences—the smell of home cooking, the joy of a
Mozart concerto, a subconscious reaction to birdsong in spring, a woman's
eyes unknowingly signalling that seduction might be possible, the appreciation of a clever idea, the reaction to a baby's stare, the shivers down the
spine from a great clarinetist. Computing, no matter how deeply intricate
and complex, is a barbarian world, devoid of these sensations.

The non-human-like thought of computers will have even deeper subtleties. As software forages through warehouses of data, machines can
learn (with electronic relentlessness) to detect patterns that humans cannot recognize and nuances that humans cannot appreciate. Our challenge
will be to harness such NHL capabilities so we can create better corporations, a world of preventive medicine, safety from terrorism, clean ecology,
new levels of creativity and extraordinary new forms of civilization.

Robots are not like the ones we see in the movies. This is hopelessly wrong. A robot may consist of swarms of nanotechnology transponders, too small to see, transmitting by wireless and controlled by a distant machine, perhaps detecting harmful pathogens in the air. There's little point in a robot looking like a human and clunking around in heavy boots. A robot may be a hydroponic farm meticulously directing the right nutrients to plants so that they taste perfect. A three-star Michelin chef might be a cooking unit in your kitchen that orders prepackaged fresh products from the hydroponic-farm robots.

What machines are good at is fundamentally different from what *we* are good at. Machines are completely different from biological creatures. This is one of the critical aspects of the future.

Once computers can improve their own capability automatically and transmit that capability to other machines, we'll have the prospect of machines becoming more capable at a formidable rate, but however brilliant they are, they may have no flicker of common sense. They can acquire a profound self-improving alien-thought that humans cannot comprehend, but they have no understanding of things that seem obvious to humans. They are good at deep, brilliant, unfathomable forms of computerized logic—but without the integrating capability to make sense of the world.

Human intelligence is very broad but relatively shallow, while machine intelligence is very narrow but can be miles deep. NHL-intelligent software will acquire a relentless self-improving capability for specific tasks, but it will have no understanding of many things that we take for granted. Some skilled humans will know how to put NHL intelligence to work—to help to make better investments, to build better defence systems, to stop an avian flu pandemic or to create a corporation that beats its competition.

Corporations will make computers imitate humans, perhaps very convincingly, to give operators or customers a feeling of comfort, but it will be pretence. Just because a machine behaves like a clever salesperson doesn't mean it has any intelligence. In the future, we'll interact with many different computers, but unlike today, they may have a pleasant face, voice and style of communicating. The design of these human-machine interfaces may be a highly competitive profession. The designer will try to give a computer application the nicest possible personality in the hope that it will have very high sales or earn high royalties. But, just because you will talk to a charming computer face on the screen when you make a restaurant booking, it doesn't mean it is real. Sometimes, the more intelligent a computer face seems, the less likely it is to have any real intelligence. The deep intelligence of computers will be non-human-like, doing profound processing that humans couldn't possibly do and often couldn't understand. It'll be quite separate from the cosmetic conversation face that people think is intelligent.

A NEW TYPE OF THINKING

NHL intelligence brings to humankind a radically new type of thinking. Most things programmed on computers today are human thought processes that are automated—the handling of invoices, production planning, engineering calculations and so on. Computers execute them unimaginably faster than humans, but they are still human thought processes. In the future, computers will be increasingly valuable for a different kind of "thought" that humans cannot do—"thought" that is quite alien to humans.

We thus have three types of thinking:

- human thought processes
- machine emulation of human thought processes
- NHL thought processes

When behaviour consists of processes executed one after another, we can understand it better by understanding each process. When behaviour consists of multiple processes happening simultaneously, with interactions among the processes, then we cannot understand it merely by understanding each process. The vast number of steps, degree of parallelism and self-modification of future computing will defy step-by-step human checking. Nevertheless, the process is logical. It is as precise as mathematics.

A baby can play with coloured blocks, happily learning how to build little towers with them. By trial and error, the baby steadily develops skills.

What the baby does naturally with gurgling delight is very hard for a machine to do. A machine can't do things that we find utterly trivial, but the machine can search through hundreds of millions of documents around the world on the Internet and find what papers are available on some aspect of, say, X-ray crystallography. While computers are often hopeless at things that humans do well, they can be great at things humans can't do.

Biology used to be a science that cultivated things in petri dishes and studied diverse species but had little need for number-crunching. Today, biology, like almost every other science, is drowning in data. A top genetics laboratory can easily produce billions of bytes per day. The amount of information stored in international repositories of gene sequences is doubling every month, and the doubling time is shrinking. The data deluge will be much larger when serious mapping of the brain becomes practical. Many challenges in biology, from gene analysis to drug discovery, have become challenges that need NHL-intelligence techniques to discover patterns and test hypotheses. America's National Institutes of Health (NIH) issued a report saying that there is "an alarming gap between the need for computation in biology and the skills and resources available to meet that need." This comment could be applied to almost all sciences. What astronomers need are not more expensive telescopes but the computing to make sense of what they already see.

NHL intelligence uses many types of computational processes that do not employ the equations of conventional mathematics. They can be applied to the messy deluge of data that the real world produces. As these techniques mature, they will fundamentally change our ability to monitor and manage the planet. The giant oceans can be brought back to life if every fishing boat has a GPS transponder and its activities fit into computer models of life in the oceans.

Powerful computers have led to a new approach to science in which, instead of trying to reduce everything to a mass of equations, we store the data and allow it to be explored with new computational tools.

A vital part of 21st-century education will be to develop the broadness, synthesis and, ultimately, wisdom of human thinking and link it to the deep, unfathomable, constantly evolving NHL intelligence. Humans should be made better at what humans are good at, and machines should be made better at what machines are good at. Because these are so different, it is not likely that humans will be entirely replaced by computers in the foreseeable future.

To make a great movie requires sensitivity to numerous subjects so subtle that they are difficult to express—the style of photography, acting that conveys conviction, the building of suspense, the development of complex human character, the palette of colours, the blend of story and music, convincing dialogue, beauty in lighting and the profoundly subtle interaction between actors. This sensitivity causes us to notice the chemistry on the screen, the best camera motion and the adaptation of a script to the visual medium. This type of intelligence will probably never be automated.

If we can't build an artificial Alfred Hitchcock, we also probably can't build a Tolstoy or a Richard Pryor. Computers have been programmed to write poetry, paint like Mondrian and compose music like Bach. There is such a difference, however, between Bach and computerized Bach, or Renoir and computerized Renoir, that a music lover would not want to listen to computer-composed Bach, and an art connoisseur would not want to live with computer-composed Renoirs. Human Monet forgers never produce works as emotionally satisfying as Monet. The most interesting humans will remain unique, imitable only in crude ways. When the wallpaper becomes more intelligent than we are, we'll spend our time becoming human in ways that machines can't imitate.

In the future we might say, "A computer should never direct films, but a human should never run a factory." An interesting challenge is to develop the best synergy between deep NHL intelligence and human intelligence. How can the human-NHL combination be designed to achieve the most valuable results?

Global networks employing NHL intelligence will become essential in our progression towards managing the planet better. Computer networks will help collect overwhelming amounts of data that are pulled together in vast data warehouses where the new techniques help to detect and prevent money laundering and drug trafficking. Vast bodies of data with NHL-intelligence tools may convert homeopathic medicine from what some view as a black art into an engineering-like discipline. New rules of corporate reporting made practical by computer will encourage powerful corporations to operate in ways that help solve the world's problems.

As we step from nature's world deeper into an artificial world, we like to think of artificial intelligence as being comfortably like human intelligence. We want to have a rope to the dock of familiarity. NHL intelligence severs that rope and casts us adrift in deeper waters. They will be much more interesting waters once we let go of old notions and learn to navigate them.

11

AUTOMATED EVOLUTION

I T IS AWESOME to reflect that evolution is in our hands now. To have a feel for the 21st-century journey, it is necessary to understand the increasing and ultimately breakneck speed of technological evolution. The 21st century is the first century when aspects of evolution can be automated. It seems that the grand plan of the universe is to progress (one way or another) towards a time when evolution can happen at great speed. Life on Earth had been evolving very slowly for a very long time until recently a fundamentally different type of change began. It seems as though evolution is destined to go through stages of primary, secondary and tertiary evolution.

- *Primary evolution* is the mutation, drift, mixture and natural selection of species—a glacially slow process. First, single-cell life came into existence. After about 3 billion years, life made the transition from single-cell to multicell organisms, which could evolve from ticks to dinosaurs and from monkeys to man. Eventually, a species evolved that was intelligent enough to write books and understand science. (Some schools teach that *intelligent design*—by God—was necessary to bring life into existence.)

- *Secondary evolution* refers to the process by which an intelligent species learns how to create its own form of evolution. It invents an artificial world of machines, chemical plants, software, computer networks, transport, manufacturing processes and so on. It learns how to manipulate DNA. On a planet with billions of people, many of them may be inventing such things and then improving what they invent. Different ideas occur to different people. There is a great diversity of evolutionary tracks.

- *Tertiary evolution* refers to something that is just beginning on Earth. An intelligent species learns to *automate evolution itself.* This artificial evolution may start with software and computer networks and progress rapidly to the evolution of procedures, control mechanisms and hardware. Secondary evolution happens immensely faster than primary evolution, but when tertiary evolution gets into high gear, things change with great speed. We have no idea what humankind will eventually produce on Earth, but in another thousand years (a mere blink of the eye on the timescale of evolution), the results will be so extraordinary that it is natural to wonder if we'll survive. The 21st century is probably the century that will determine that. If we are to survive, artificial evolution has to change from being unstable and reckless to being an intelligently directed process in which the risks are well understood.

It seems as though this three-stage evolution is inevitable. For a long time, primary evolution evolves biological species, as it has done on Earth. If primary evolution goes on for long enough (billions of years), it produces an intelligent species capable of thinking and communicating. Eventually, this species tries to work out how the world works and acquires an understanding of science. It creates cities, vehicles, automation and globalism. This secondary evolution gains strength until eventually (after a few hundred years, perhaps) it becomes an avalanche of self-improving capability. Eventually, the intelligent species learns to automate many aspects of evolution—tertiary evolution—that gains strength until its capability changes at great speed and produces great diversity. Once the tools exist, millions of entrepreneurs will start their own processes of tertiary evolution. Mummy says, "What's Johnny doing with the bedroom light on at three in the morning?" and Daddy says, "He's using his new tool kit to evolve an Internet game."

DARWIN'S ALGORITHM

Since Darwin published *The Origin of Species* in 1859, there has been controversy about whether the processes of evolution followed a strict algorithm. *Algorithm* is an ancient word, derived from the Latin *algorismus,* meaning a "repeatable procedure." An algorithm could be written in the sand with a stick. It could be a detailed cooking recipe or an instruction manual that is followed step by step. The term *algorithm* is used today to mean a precise procedure that is executed on a computer. For example, there are algorithms for calculating square roots or the taxes that a corporation must pay.

Could nature's immense complexity have been created by an algorithm? Darwin described how species evolve and are steadily refined by going through endless mutate-test-select, mutate-test-select cycles. In his phrase "Multiply. Vary. Let the strongest live and the weakest die," there is a succession of tiny steps, each step being mechanistic, with selection of the fittest giving slow and steady improvement of the species. If the sequence of tiny steps is followed for many millions of years, we get the grandeur of the rain forests, gorillas and screeching parrots.

The process of evolution itself has a strong random element to it. If, on Earth, we set the clock back a billion years and allowed a billion years of evolution to happen again, its random walk would produce very different results. You might have six legs and fur like a bear.

Since Darwin's time, there has been a vast amount of research on evolution. Our understanding of the fine details has steadily improved and, along with it, our awe at the ingenuity of nature. Sceptics have desperately tried to disprove Darwin. They have found one situation after another about which they would say, "You'll never explain this with a Godless algorithm," but over and over again, some form of Darwin-like algorithm has been shown to work, sometimes in subtle ways involving pollinating insects, sex, microorganisms or synergistic interaction among different species.

The philosopher Daniel Dennett has studied the subject in great detail and argues that no supernatural intervention is needed to explain evolution. The aspects of evolution that looked like miracles have turned out to be more interesting where we can explain how they worked. Fossil evidence demonstrates that long hauls of evolution were Darwinian, but science cannot prove definitively that no process other than Darwin's algorithm has been involved. Certain changes are difficult to explain, including the first transformation from chemicals to life. Some religious authorities argue that Divine intervention occurred at certain points, but most scientists argue against this, saying that if God was designing things He would have done a better job.

If nature's evolution, or most of it, was achieved with an algorithm, computers ought to be able to achieve similar forms of evolution. Work on automated evolution has started to show results. It will have "intelligent design" because intelligent humans will direct it and help it when it gets stuck.

AUTOMATED DARWINISM

At the southern tip of Africa there's a masterpiece of evolution. A floral kingdom exists that is called the "fynbos" (pronounced "fain-boss"). The mountains and valleys are covered with a massed spectacle of flowering plants different from those in other parts of the planet. There are 8,600 plant species, most of them flowering, in a self-sustaining ecosystem. Of these, 5,800 are endemic (meaning that they evolved in that place for millions of years). To put this in perspective, the British Isles cover a much larger area but have only 1,500 plant species, of which fewer than 20 are endemic.[1]

I sometimes walk in such places with companions who say, "Isn't it utterly awesome that God created this incredible beauty?" I tell them that I find a different aspect of it awesome—that this ecosystem of such complexity and beauty *evolved*. Every time we walk in the fynbos, different flowers are in bloom, and almost none of them existed 5 million years ago. This magnificent floral kingdom evolved with an enormous amount of trial and error, by means of a set of rules that Darwin tried to describe. What is awesome to me is that an infinite number of different ecologies *could* have evolved, using the same set of rules.

When Richard Dawkins, the Oxford biologist who writes books about evolution, started to program computers in the 1980s, an endlessly fascinating realization dawned on him. You could create Darwin-like rules in software, set them running and emulate evolution. You might, given a large amount of time and computer power, produce the equivalent of a floral kingdom in software.

Dawkins created a very simple program to achieve evolution of tree-like shapes on a computer screen, and he was amazed at its behaviour. "I never imagined it would evolve anything but tree-like shapes. I had hoped for weeping willows, poplars and cedars of Lebanon." Branching structures began to cross back over themselves, however, filling in areas until they congealed into a solid mass. They formed little bodies—and branches, looking like legs and wings, sprouted from those bodies. The drawings evolved, totally unexpectedly, into weird bugs and butterflies. Dawkins, far too excited to eat, sat at his screen discovering amazingly complex water-spiders, scorpions and other shapes. He wrote in 1987, "Nothing in my biologist's background, nothing in my 20 years of programming computers, and nothing in my wildest dreams, prepared me for what actually emerged on the screen."[2]

Dawkins couldn't sleep that night. He explored the startling world he had created and hunted for nonplant shapes in it, trying to compare the evolution he had studied and written about for many years with the unpredictable evolution in his computer. In the months ahead, he discovered "fairy shrimps, Aztec temples, Gothic church windows and aboriginal drawings of kangaroos." Dawkins scratched his head in astonishment at what his creations were doing.

He experimented with the use of "genes" that allowed rules to be stated controlling how the drawings evolve. A string of such genes (like a chromosome) provides a set of rules for how drawings are created and grown and can be used to automate aspects of the evolution. By modifying the genes Dawkins could experiment with the evolution of his creatures in faster and more interesting ways—he watched the *genes evolve rather than merely the creatures themselves.* A higher level of evolution then becomes possible. Genes can be selected, for example, that are champion evolvers.

Researchers experimenting with software evolution have found that it easily gets stuck. It comes to a dead end fairly quickly. Automated evolution will usually be done by (human) intelligent designers who will be alerted when the process gets stuck and who will quickly adjust the rules. Today's technology allows us to build ultraparallel supercomputers specially designed for artificial evolution. If we had enough processors, we could achieve evolution a billion times faster than evolution in nature. This implies that fynbos-like complexity could evolve in days rather than millions of years. An enormous amount of experimentation will be needed before we achieve this.

GOAL-DIRECTED EVOLUTION

The plants and creatures of nature are designed to evolve, adapt, learn and constantly improve. Much future technology will similarly be designed to evolve, adapt, learn and constantly improve *itself* (with, as in nature, many evolutionary dead ends). Because the processes of nature have evolved and improved over a very long time, they have become awesomely complex. Self-evolving technology will also become very complex.

Nature's evolution is a random trial-and-error process. Vast numbers of mutations occur, randomly, with no goal other than survival of the fittest. Technology evolution, on the other hand, has a goal set by humans.

Nature tries everything. It has no idea what will work and lets natural selection decide. Nature didn't set out to build a giraffe or hummingbird. These were playful products of nature's random shuffling of the cards. Evolutionists have desperately tried to show that the processes were in some way set up to achieve the ultimate in evolution—humankind. Such ideas haven't held water. Humankind evolved in a bumbling way and is certainly not the ultimate in evolution. It's just the current prizewinner.

If evolution is used as an engineering technique, there must be a precise and measurable goal for selecting the mutations that are kept. An engineer specifies the desired behaviour that he wants. He finds a measurement that indicates how well the evolving system is performing and some automatic means of assessing the results. After each cycle of evolution, the mutations that do the best are taken into the next cycle. As in nature, the process goes through thousands, perhaps millions, of mutate-test-select, mutate-test-select cycles until a good design is achieved.

Unlike nature's evolution, artificial evolution is set up with specific goals in mind. A horse breeder can breed racehorses more efficiently than nature can because the breeder has a goal. The genes of what are perceived to be the best horses are used for each new round of breeding. Automated evolution is similar. It examines the fitness of each variation with some goal in mind. It is set up to steadily improve the fitness by many, many mutate-test-select cycles. Unlike horse breeding, artificial evolution takes place at electronic speed, and its results can be much more interesting. A horse is just a horse.

The disciplines of automated evolution need powerful computers that go through vast numbers of mutations at very high speed and test how well they fit the progression to the target. Because such a large number of mutations are tried and tested, computers are used with many small fast processors operating in parallel. They process numerous mutations simultaneously and test them, rejecting almost all of them, as nature does, and keeping the promising few for ongoing evolution.

On Earth, life evolved for about 4 billion years before man was created. It took three centuries to evolve from the first crude flicker of the Industrial Revolution to today's industrial society. Now, with special-purpose supercomputers designed for high-speed evolution of designs, interesting results can be produced in days.

This raises many questions. Where should such evolution be used? What industries or disciplines could benefit from it most? Will certain corporations dominate certain markets by using automated evolution? Can software be set up to evolve with no human intervention? What would we want it to achieve?

The code that results from evolutionary processes is so different from conventional code that programmers often find it difficult or almost impossible to follow. It can be made reliable by allowing it to evolve with a very large number of test cases. As machine intelligence becomes deeper and lightning fast, it will evolve mechanisms we can't understand. Will we be able to control it? How do we set the most appropriate goals? Will machines develop goals of their own? Could they develop a high-level goal we wouldn't want? How do we put controls in place to prevent that?

SETTING INTELLIGENT GOALS

Where automated evolution is used as an engineering discipline, it is not random, like nature's evolution. Engineers have a measurable goal in mind and set up evolution so that it progresses towards this goal. They have to state precisely what they want the evolution to achieve. For example, we might try to create software for foreign-exchange trading. We set up procedures that are designed to steadily refine themselves by going through endless mutate-test-select cycles. Each mutation is tested against the precise goal: *Improve return on investment.*

Building a "black box" for mechanistic types of investing like foreign-exchange trading or arbitrage is a fascinating challenge. Many computer algorithms have been evolved for this. Some of them give an excellent return on investment for a time, but then market conditions change in some subtle way and the algorithm loses money. Sometimes the loss is sudden and spectacular. The algorithm needs to be evolved using a long market history, capturing, if possible, the diverse and subtle ways in which the behaviour constituting markets can change.

The Daimler-Chrysler company has used goal-directed evolution to refine the design of diesel engines. The aerospace industry has used it to refine jet engines. John Deere has used it for evolving factory schedules that allow efficient customization of products. It has much greater potential for evolving new types of software. There are limits on how much change can be made to a diesel engine, but software can evolve in unlimited ways. As the understanding of automated evolution matures, it will probably enable us to create many types of software and procedures that we can't create with conventional programming.

There are three main reasons why it makes sense to create software by automatic evolution. First, when it works well, automated evolution of software is much faster than traditional programming. Second, it's possible to evolve functions that we couldn't program with traditional languages. We can "breed" chips to recognize human faces, for example. Third, we can evolve functions of arbitrary complexity. Genobyte advertises that its evolvable hardware enables you to "create complex adaptive circuits *beyond human capability to design or debug,* including circuits exceeding best-known solutions to various design problems that has been already proven experimentally."[3]

EVOLUTION ENGINEERING

When a human baby learns, it is, in effect, wiring up its brain; it grows the nerve patterns between neurons and adjusts the synapses so that the influence of one neuron on another changes. Various researchers, most notably Rodney Brooks and his team at the MIT Computer Science and Artificial Intelligence Laboratory, found out how to do that with electronics. Hugo de Garis, former head of the Brain Builder Group at the ATR research labs in Kyoto, Japan, devised a genetic technique to grow a small module of brainpower—only a few hundred neurons. Some "brain" functions need many of these modules linked together. He wrote, "My neural circuits grow in billionths of a second. This is so fast that I can grow many of them, each with slightly different mutations and, hence, with slightly different abilities to perform some task that I give them. By programming a piece of hardware that measures how well the evolving circuit module performs, it is possible to evolve an elite circuit (the top-scoring circuit—the best of tens of thousands) in about a second." This remarkable speed allowed de Garis and his team to assemble tens of thousands of such circuits into overall architectures.

De Garis uses the term *evolution engineering*. He started to create an engineering discipline with two levels of professionals in a team: architects, who specify what modules are required and how they work together, and evolution engineers, who breed the modules. Eventually, there will be catalogues of off-the-shelf modules from which complex systems are assembled, and a relatively small proportion of the modules will be evolved uniquely for one project. De Garis's team created a powerful supercomputer for the breeding process, which he called a "Darwin Machine."

Natural selection has a problem well understood in computer science. Darwin's survival-of-the-fittest mechanism constantly provides localized improvement; it doesn't provide long-range vision. In a vast mountain range, Darwinian improvement can only climb uphill, and this can leave it stuck on a local peak. There is no mechanism for going downhill and crossing the valley to a higher mountain on the other side. Natural selection has no way to see the long-distance landscape and realize that it is stuck on a local peak, but the human brain and its science can see the landscape and plot a course towards a distant goal.

Long-range vision is a critically important capability, and the programs we build for automated evolution seem to need an intervention of human intelligence periodically. Humans can detect when an evolutionary process is stuck, can scan the long-distance landscape and apply human intelligence when it is needed.

DIFFERENCES FROM NATURE'S EVOLUTION

With the machines we envisage today, automated evolution may become a billion times faster than nature's evolution. Furthermore, it will be incomparably more efficient. Darwinian evolution is described as being random, purposeless, dumb and Godless.[4] Automated evolution is targeted, purposeful and intelligent and has humans directing it and changing its fitness functions on the basis of results. In Darwinian evolution, the algorithm stays the same. In automated evolution, researchers will be constantly looking for better techniques and better theory. The techniques of evolvability will themselves evolve.

A great potential advantage of creating systems by evolution is that, if the evolution goes on long enough, they can be very complex, as in nature. They can be far more complex than any systems that humans could design with traditional design approaches. This characteristic of evolutionary engineering has been referred to as "complexity independence"[5]—the evolutionary engineer doesn't care about the internal complexity of the system that is being evolved; he cares about only the score you get when you measure how well it performs. The internal complexity of the system being evolved can go well beyond what human engineers have the intellectual capacities to comprehend.

The following table summarizes the differences between living systems and automated evolution:

NATURE'S EVOLUTION	AUTOMATED EVOLUTION
A biological plant or creature	A process or software
Glacially slow evolution	Very fast evolution
Immense complexity	Low complexity today; eventually complexity exceeding that of nature
No long-term goal in mind	Long-term goals that are set
Random trial-and-error	Steered by an intelligent-design team
Survival of the fittest	Selection based on design criteria
The selected version can't be immediately replicated.	The selected version is replicated and spread through the Internet.

NATURE'S EVOLUTION	AUTOMATED EVOLUTION
Methods of evolution rarely change.	Methods of evolution evolve rapidly.
Each plant or creature that evolves is physically separate.	Evolving software can be interlinked on the Internet.

THE SINGULARITY

Most technologies that have a period of exponential growth eventually slow down the pace. However, if they are completely divorced from matter there may be nothing physical to make them slow down. This is because the manipulation of *matter* has inherent limits to growth, as does development that depends on human intelligence. Abstract self-evolving intelligence, however, may be able to race ahead with neither physical nor human dependency slowing it. Sooner or later, one of society's major concerns will be how we cope with machines much more intelligent than us.

NHL intelligence will become incomparably deeper and faster than human intelligence. This may occur in very narrow professional areas first and then in increasingly broad circles of activity.

Vernor Vinge, a mathematician and computer scientist, concluded that computing technology will feed on itself, becoming more and more intense, like the gravitational forces of a black hole. He borrowed the term *singularity* from the astrophysics of black holes.[6] In mathematics or physics, the term *singularity* implies a variable becoming infinite. Vinge used the term to refer to the curve of technology growth becoming almost vertical.

In 1923, Germany experienced singularity with money. The curve of inflation became almost vertical. Germans used wheelbarrows full of banknotes to pay for bread. On the peak day, they burned money rather than firewood because it was cheaper. Vinge and writers who support his view believe that computer singularity will be similar. The intelligence in one computer may grow by hundreds of times in a day.

When computers first become more capable than humans, it will seem like a good thing—corporations will make huge profits; the stock market may outdo the Internet boom; video-game special effects will be sensational. Enormous numbers of computers, connected to a speed-of-light Internet, will share in the bonanza, but the increase in machine intelligence will feed on itself, becoming more extreme, until there is no time left to think about how to control it. When you get too close to a black hole, it's too late for arguments about navigation.

There may be an explosion of intelligence far beyond anything that humans can either understand or control. Most writers about the Singularity spell it with a capital S and paint a dramatic image of runaway intelligence happening very suddenly, like the German inflation, with humans having no idea how to control it. We can't anticipate what events will be like after the Singularity. The Singularity spokesmen believe that it will cause an acute break in the continuity of human affairs.

The date of the Singularity, when human and machine intelligences form an alliance, is likely to happen around the time the world reaches mid-canyon—perhaps around the time the planet's population reaches its peak and the stress on the Earth's resources is at a maximum. Ray Kurzweil, author of *The Singularity Is Near: When Humans Transcend Biology*, estimates its date as 2045.[7]

Today, the US TeraGrid is the world's most powerful computer network. The computers connected to it are in five locations and have 20 trillion operations per second of tightly integrated supercomputer power and storage of over 1,000 trillion bytes of data, connected to a network that transmits 40 billion bits per second. It is designed to grow into a worldwide network with thousands of times more power than today. A 2045 version of today's TeraGrid might be called ZettaGrid. The prefix *zetta* means *1,000 billion billion*. A zettabyte is 1,000,000,000,000,000,000,000 bytes. The ZettaGrid would connect millions of zettaflops supercomputers, each with zettabytes of data. It is in such an environment, with automated evolution, that the Singularity will occur.

There used to be a view that computers were just machines. If we wanted to, we could pull their plug. In today's society, nothing could be further from reality. Industry has restructured itself to be totally dependent on networks, powerful computation and, eventually, robotic production. Without computers, there'd be no food in the shops. Every year, we become more inescapably enmeshed in a world of computation and more dependent on the machines.

A consequence of having vast numbers of ultra-intelligent computers and robots will be that most of the work in advanced countries will be done by machines. This will generate very high productivity and wealth, and society will need some way of distributing wealth to the unemployed. Many people will be looking for worthwhile occupations. Perhaps many will be volunteers in organizations trying to help the poorest parts of the world and to put the new intelligence to work to help eliminate poverty.

I don't really believe in the doomsday view of the Singularity—that one day it will be like the German inflation. One reason for doubting this is that we'll have decades of warning and years of time to prepare for it. It won't be like an unanticipated earthquake. There'll be hundreds of millions of large computers running organizations everywhere with undramatic software that won't change very quickly. Supercomputers will be doing things like modelling climate change and doing ever more spectacular special effects in movies and providing scientists with ever more power to understand the world. These different groups will have planned their own transitions for how they will use an exciting increase in computer power. It won't be one transition but numerous transitions, which occur in different areas at different times. There will be plenty of time to anticipate possible harmful effects of sudden increases in computer intelligence and to put measures in place to stop the harm.

Nevertheless, the overall effect will be a profound change in society.

OVERLAPPING REVOLUTIONS

The long evolution of computing is occurring in these four overlapping revolutions:

1. The growth of dumb computing
2. The growth of global computer networks
3. The growth of NHL intelligence
4. The Singularity

The first of these revolutions started with mainframes, moved to mini-computers, then moved to personal computers and is becoming widespread in very tiny machines. Large computers do large engineering calculations and exotic applications, like creating virtual reality of intense realism. We can observe the extraordinary impact of Moore's Law long before its crowning glory—the age of nanotechnology.

The second revolution gained strength as the Internet provided new types of interaction and services around the planet. It enabled major reinvention of business processes. Long-distance circuits of massive capacity came into use with fibre optics, and endless handheld devices could connect to them wirelessly.

Computers will become really interesting when they become intelligent. This is the third revolution. Because NHL intelligence operates at electronic speed, it will race far ahead of human capability, learning behaviour that humans cannot learn. Because computer intelligence and human intelligence are so different, we'll need a close synergy between the two.

The Singularity is an inevitable consequence of computer intelligence feeding on itself. Computers will become increasingly successful at imitating aspects of human intelligence, and this will help produce systems that enable humans to use deep NHL intelligence when it reaches the Singularity level.

The first revolution has been referred to as the overthrow of matter because it stores such a vast number of bits and logic in such a small space. The second revolution (fibre-optic networking and wireless) is the overthrow of distance. The third revolution (self-evolving non-human-like intelligence) is the overthrow of machine dependence on human programming. The fourth revolution is the overthrow of the supremacy of human intelligence.

Each of these revolutions has its own grand-scale chessboard but evolves at different rates. In the first revolution, a thousandfold improvement takes 15 years. In the second revolution, a thousandfold improvement takes 10 years or less. In the third revolution, once techniques of self-improvement become mature, a thousandfold improvement in NHL intelligence may occur in two or three years—perhaps faster. The fourth (Singularity) revolution will bring an extreme rate of change but with only a small fraction of humankind able to take advantage of it at first.

The brightest people, not only from the computer industry but among entrepreneurs everywhere who think about how technology might be put to use, will be anticipating with great excitement how they will employ the spectacular increases in machine intelligence. Before the Singularity happens, venture capitalists everywhere will be flooded with proposals about how to make a spectacular return on investment out of it. There will be feverish excitement about this, far exceeding the Dutch tulip mania. The world's stock markets will reach what is perhaps their most volatile period ever. The cleverest investors will be using "black boxes" that incorporate the most advanced machine intelligence, employed to maximize the return on investment from both market crashes and booms in the different markets worldwide. In attempts to maximize their profits, they will borrow as much money as they can from around the world and use the most ingenious techniques, such as global arbitrage, to multiply their returns, protecting their own money when things go wrong.

The Singularity may be a time when certain individuals make more money than anyone has ever made. There was a moment in the 1990s when Bill Gates's personal wealth was larger than the gross national product of Israel. In the year of the Singularity, there may be a brief moment when the Bill Gateses or George Soroses of the future have a personal net worth far exceeding that.

12

THE TRANSHUMAN
CONDITION

TRANSHUMANISM IS A MOVEMENT concerned with human enhancement. It explores any possible workable method for improving human beings, seeking to surpass current biological limitations, and it examines cultural and social changes that should accompany human enhancement.[1]

Usually, when there is talk about modifying humans, people think we might use genetic modification, just as we genetically modify plants. In reality, at least for the next few decades, the most effective forms of human modification will not be genetic. Our genes are so complex that it would be very difficult to create genetically modified people who are more intelligent or more capable. More workable forms of human enhancement are likely to come from prosthetics, nanotechnology, regenerative medicine, drugs that affect the brain and electronic devices that enhance human capability.

In an interview I had with Ray Kurzweil, a prolific inventor who created music synthesizers and machines that read books to blind people, he illustrated how we have the capability to reinvent just about every component of our body. Biological evolution, he said, has produced inefficient mechanisms. "It uses proteins that are made from linear sequences of amino acids, and that has a lot of limitations. When we figure out how these different systems work, we discover that we can re-engineer them to be thousands, sometimes millions, of times more powerful." Kurzweil expects that many biological parts will eventually be replaced with nonbiological systems that are much more capable.[2]

Many high-tech replacements of body parts will be stronger or more capable than the biological original and not subject to biological disease and degradation. The drug industry will evolve highly refined products. We will be able to substantially extend the human life span. Neuroscience will develop in extraordinary ways when we have the capability to map the human brain with increasing resolution until we can record the transmission of signals among its neurons and then attempt to rebuild what we record in electronics. We'll be able to connect certain types of electronics directly to our brain.

A major 21st-century debate will be: Should we, or should we not, use technology to fundamentally modify the human creature? The genome project is giving us a rapidly growing understanding of human genes, followed by the capability to make at least simple changes to them. The project to map the human brain and record its working in detail may lead to even greater controversies than the genetic modification. Should we replicate parts of our own brain using electronics? Should we enhance our own brain by connecting it directly to electronics? Kurzweil commented, in our interview, "We're not defined by our limitations. We're defined by the fact that we are the species that seeks to extend beyond its limitations." Transhumanism has strong opponents convinced that we should not modify humans. Some have religious, or crypto-religious, arguments. Some say that there are deep consequences that we don't understand. Some say that it would be a slippery slope that we should keep off because it will become steeper and more slippery. In reality, it seems likely that the slippery slope will take us to major enhancements of humans, not degradations, and that the path to improvements will be far more extraordinary than has been generally realized today.

A major argument is that our complex human nature, reflected profoundly in our best literature, is too valuable to mess about with. Bill McKibben, a philosophical Vermonter who wants his beautiful world to stay unchanged, says simply, *"Enough"*—we've had *enough* of disruptive technologies. Let's enjoy what nature gave us.[3]

Nick Bostrom, an Oxford philosopher, has a different view. He is concerned with how biological modifications can make lives more satisfying. He maintains that "having a faster car, warmer house or more art hanging on the walls doesn't really change the core of us, but if you could become longer-lived, happier, smarter, more richly emotionalized—that's much more profound." Transhumanists agree that there are serious dangers that should be studied carefully so that protective measures can be taken. They believe that we can make the potential benefits greatly outweigh the potential dangers, but the dangers cannot be known in detail until we make the journey.

It seems inevitable that major aspects of transhumanism will happen. Some aspects of it will be too enticing to resist. Most people will want better health and a longer life, and there are many ways to achieve this. Most want the best for their children and more fun in life. Some things on the transhumanists' agenda are short-term and inevitable, like refined psychological drugs; others are long-term and science-fiction-like, such as mapping our brain in detail and making parts of it operate in conjunction with a computer. By the time the Transition Generation ages, parts of the human race will be substantially enhanced, and the social consequences of this will be enormous.

ENHANCED PEOPLE

We can distinguish between enhancements that are desirable health improvements and enhancements that change the human creature. Leaving aside gene modification, there are now other means of changing humans, like subtle brain chemicals, which seem to sell profusely, and cleverer prosthetic devices. The US military has developed a wearable robotic suit, which detects the wearer's muscle movements and greatly amplifies them. The wearer can jump high, or pick up a 200-pound weight as though it were 10 pounds. He can carry a 500-pound weight on his back for a long time. There are myriad variations on robotic enhancement of our muscles.

Some of the most controversial changes will occur when we begin to map our brain mechanisms in detail and learn how to make major enhancements to our brain. Professional people will be changed, even without biological tinkering, because they will use wireless links to computers with capabilities immensely beyond those of today. A project might employ a thousand engineers linked on a fibre-optic Internet, each employing computer facilities approximately a thousand times more advanced than human intelligence.

The term *bioconvergence* is used to refer to technology in which biological and nonbiological processes are combined, as with the "bionic" people of science fiction. Nanobiology will emerge as a discipline combining nanotechnology and biology. A blood cell is about 7,400 nanometres in diameter. A substantial amount of nanotech gadgetry can be put into a capsule the size of a blood cell. Devices about this size will go through the bloodstream to destroy bacteria, viruses or sclerotic plaque and help keep us healthy. We don't have nanotechnology devices yet, but we'll have MEMS (mechanical and electro-mechanical systems) devices the size of a blood cell. There are already major conferences on biological MEMS. We can send such devices into our bloodstream for medical purposes. Such devices might release minute quantities of chemicals. They might go into the brain capillaries and release precise amounts of drugs that affect our brain. They will be designed to take cancer-killing chemicals to a cancerous cell and not release them elsewhere, making chemotherapy precise and targeted. Today's chemotherapy will look as crude as carpet-bombing campaigns.

ENHANCED SENSES

One way to enhance a human is to improve his or her senses. Some animals have much better eyes than we have. Some have much better ears. Our nose is primitive compared with that of a dog, but some laboratory noses have been created that are about as sensitive as that of a dog.

Deaf ears, like mine, can have tiny electronics inside them. Today, not only can they restore normal hearing, but they can also enhance hearing—for example, with a handheld controller, you could make them so highly directional that you could listen to a conversation at a nearby restaurant table. They could create artificial silence, as Bose sound-suppressing earphones do. They could inconspicuously play music from your iPod. You might go for a walk in the woods at night and be able to hear all the night creatures, some communicating at frequencies beyond the human range.

Both sight and hearing can be improved fairly easily. You might have glasses that enable you to see in the dark and detect infrared radiation to tell you where hidden creatures are.

When people walk in the woods of Vermont, they sometimes pass close to a bear but don't realize it. The bear is careful not to reveal itself. If the person is on a horse, the horse detects the bear. The horse bristles, and his ears prick up. A human could be given electronic senses as sensitive as those of the horse. Your hearing aid might alert you by pronouncing, "Bear—24 yards at 10 o'clock."

Machines have been demonstrated that can identify human emotions. A person might be equipped with a device that has this ability to recognize subtleties underlying emotions. You may get an electronic warning from your hearing aids, inaudible to anyone else, that the person with whom you are talking is lying, is emotional, or feels nervous about the way a business negotiation is going. You may receive a range of tones designed to help you with the human dialogue. Guests at a party may wear in-the-ear devices that signal which men (or women) are strongly attracted to them.

Some sensory improvements might enhance the senses that you were born with. Others may give you new types of senses that you weren't born with, such as the ability to detect ultraviolet, ultrasonic, X-ray or microwave radiation—the entire electromagnetic spectrum. They may work with radar, sonar, microscopic motion detectors, radiation that penetrates buildings, GPS sensors that know our exact geographic position, or wireless signals that link to other people. When nonhuman sensors detect anything of interest, they may send a message to our biological senses. Personal sensors may be used to protect us from infections, a new pandemic or the possibility of biological attack.

We can build a diversity of artificial senses. This raises an interesting question: Should a baby be equipped with artificial senses so that its brain learns to use them and finds them a normal part of life?

LIVING LONGER

In the next two decades, numerous improvements will occur in the prevention of disease. There'll be more effective means of preventing strokes, heart attacks and cancer. Preventive medicine will improve and will often be customized to an individual's genome. Scanning a person's DNA for harmful genes will become standard practice and inexpensive. Gene therapy will be used to prevent certain illnesses. Some people, determined to enjoy old age, will be careful to avoid cigarettes, drugs, AIDS and excessive alcohol. They may bicycle to work and use wireless-linked instruments to monitor their body. They'll eat vegetables and do everything their doctor or personal computer tells them to.

A type of human enhancement that appeals to most people is the ability to live longer. In the 20th century, life expectancy in the United States increased from 46.3 to 79.9 years for women and from 48.3 to 74.2 years for men, mainly because of nutrition, improving health care and reduction in infant mortality. We are now at the beginning of a more fundamental attack on ageing.

Cells, the building blocks for all tissues in the human body, appear to have been designed to live to about 70 years or so. Cells divide periodically, and this cell division plays a critical role in the normal growth, maintenance and repair of human tissue. Cells can divide only a limited number of times during their normal life span. This limit is called the Hayflick limit, after Leonard Hayflick, who discovered it. It sets a limit on how long we can live.

Each of our chromosomes has a cap on it, rather like the plastic end of a shoelace. This cap is vital because, if it didn't exist, chromosomes would stick to one another, forming a spaghetti-like mess. The cap consists of a sequence of six DNA letters, TTAGGG, repeated many times—TTAGGG TTAGGG TTAGGG TTAGGG. It is called a *telomere*.

Each time a cell divides, the chromosome in the cell loses one of its TTAGGG sequences at each end. Matt Ridley compares this to an infuriating copying machine that cuts off the top and bottom line each time a copy is made.[4] The telomere length at the start of life is different for different people. It varies from about 7,000 to 10,000 DNA letters. This aspect of longevity is inherited. Some people are from long-lived families.

Enzymes are chemicals that do the work in a cell. Recently an enzyme has been discovered that can add extra sequences of TTAGGG TTAGGG TTAGGG TTAGGG to the ends of the chromosome so that the cell can divide more often. This enzyme is called telomerase. Using telomerase, cells in laboratory experiments have been made to overcome the Hayflick limit. By doing this, laboratory creatures such as mice, fruit flies and others have been made to live longer—sometimes 50% longer—than their ancestors. Telomeres can allow cells to reproduce indefinitely.

Telomerase is not present in most normal cells and tissues of humans, but we can add it. This is one action that opens up the possibility of living much longer. Michael West describes how to make our cells "immortal," and is the CEO of a company, Advanced Cell Technology, that wants to enable us to be both old and healthy.[5]

While not present in normal cells, telomerase *is* present in tumours. It is what enables the tumour cells to subdivide indefinitely. Telomerase is used by all major types of cancers. It has been demonstrated that, if we can stop the telomerase activity, cancers will stop growing.

Geron, a company in Menlo Park, California, built a business centred on telomerase. It plans to do two things: to use telomerase to add lengths of TTAGGG letters to normal cells so that they have a long life and to deprive cancerous cells of telomerase so that they can't replicate. In the company's words, "We and our collaborators have demonstrated that the enzyme *telomerase,* when introduced into normal cells, is capable of restoring telomere length or resetting the 'clock,' thereby increasing the life span of cells without altering their normal function or causing them to become cancerous."[6] Geron hopes to play a critical role both in killing cancer and in making humans live longer.

Companies like Geron and Advanced Cell Technology are conducting research to identify *all* the genes involved in ageing, with the aim of eventually altering their function in the quest for a long life with good health.

NEGLIGIBLE SENESCENCE

A growing belief among researchers in gerontology is that ageing is a disease and that there are ways to cure it. One of the strongest advocates of this is Aubrey de Grey, a researcher in the genetics department at Cambridge University.

The term *senescence* refers to the progressive loss of physical robustness that happens when we age. It may be defined as *the probability of dying in the next year.* This can be measured statistically in a large community of people and is used mathematically by insurance actuaries to calculate the odds. The *probability of dying in the next year* becomes higher as we grow older.

De Grey points out that a radioactive material does not senesce. It decays, but the half-life of decay remains the same. He believes that we can use engineering to make the senescence in people negligible—in other words, as we grow older, the *probability of dying in the next year* would stay almost the same. He calls his research SENS (strategies for engineered negligible senescence).

To achieve negligible senescence, the biological killers that become more prevalent as we grow old have to be disabled. Of seven such killers identified by de Grey—all of which can be eliminated—one is the reduction in telomere length. Another is random mutations that change our DNA, some of which make our cells more likely to subdivide and some of which lead to cancer. Another is faulty degradation of cells—those that degrade as they should when they have done their job, which can be avoided by making cells better at breaking down. Also, as heart cells and cells of some other vital organs die off throughout life, there are ways in which they can be replaced. Some of the cures for senescence require the removal of damaged genes and the periodic introduction of healthy genes, which the body will multiply with its natural processes.

De Grey would like there to be intense research aimed at stopping the ravages of the seven biological killers. He regards them as diseases that can be eliminated. To some extent, the damage they do can be reversed. There may be other life-threatening problems not yet identified, some of which may appear only at an older age than today's life span.

Steadily, we will reduce senescence. The goal is to reduce it to a negligible level. That won't mean that we live for ever; your spouse's driving might kill you, and there are other randomly occurring causes of death, such as a sudden plague. A senescence-free life means that we retain our physical robustness and vitality. In other words, we stay, if not young, at least young-like. Some doctors disagree, saying that de Grey underestimates the complexity of the human creature and the subtle interactions among different aspects of our nature.

Researchers have set out to engineer a mouse with negligible senescence. They expect to achieve that within 10 years. It is easier to create an immortal mouse than an immortal human because you don't have to worry about safety. Gene therapies not allowed in humans can be done in laboratory animals. There is a prize for developing an immortal mouse—the "Methuselah Mouse." Once such a mouse becomes possible, the public will realize that ageing is not inevitable. There will then be massive funding for achieving long life in humans. The big question will be, "How quickly can we get from a Methuselah mouse to a Methuselah human?" De Grey thinks the human might come 10 years after the mouse and should be referred to as a "low-senescence human" because there may be other biological killers that don't reveal themselves until we are very old and that will require further research.

Surveys have been conducted asking people if they would like to live to be centarians. Surprisingly, most people say no. They say that they would become bored. My grandfather, when he was seventy, said there was nothing worth living for any more. (He died at ninety-seven.) I find that people who like the idea of living to an old age are exceptionally mentally active in some way. Perhaps they are involved in complex research or with a passion for art, music or intricate hobbies. Orchestra conductors go on to a ripe old age, forever perfecting their art. If joining the Methuselah club is expensive, it may be only certain types of people who join.

Of course, if many people around the planet live longer, it will have an impact on the population problem, but artificially induced longevity will occur mainly in affluent countries with low birthrates. Most countries with high birthrates have shockingly short life expectancies. The social factors that help increase longevity will be similar to those that make the birthrate go down.

Some young people reading this book will live to the end of the 21st century. Well before they become old, extraordinary changes will have happened. The ability to give them new cells and a young immune system will become common practice. The use of certain hormones will greatly improve their energy and vigour. This century will be the first in which we see youth in old people. Later in the century, some people will hope to live to a very old age. This is a change of fundamental consequences.

THE MAGIC OF STEM CELLS

A new area in medicine is being created, called regenerative medicine, concerned with the capability to regenerate old cells or worn-out body parts. The use of stem cells opens up extraordinary possibilities. The more we find out about how we might use stem cells, the more impressive their capability seems.

When a human sperm first fertilizes an egg, a stem cell is formed. This is a cell that is capable of *transforming into any type of cell in the human body*. It splits into two cells, then four, then eight and so on. Approximately four days after fertilization, a hollow sphere of cells forms—a "blastocyst." This has an outer layer of cells, which will eventually form the placenta that is used to protect the development of the foetus. Thus protected, the new cells become an embryo.

The stem cells in the blastocyst can be extracted and stored indefinitely at a low temperature. These cells have not yet specialized for any particular function, but they have the potential to develop into any of the approximately 200 different mature cell types in the human body—muscles, heart, brain cells, liver and so on. This remarkable property is referred to as being "pluripotent." These stem cells can be multiplied in laboratory dishes so that there are large quantities of them, and they are still pluripotent.

If embryonic stem cells are injected into a part of the body that has cell damage, they can replace the damaged cells, then multiply and repair the tissue. This capability makes stem cells very valuable in restoring parts of the body that are damaged or worn out. The study and use of stem cells will save enormous numbers of human lives.

There is a religious objection to using stem cells on the grounds that such use will stop a stem cell from growing into a human life. The pluripotent stem cells that are not yet part of the blastocyst cannot by themselves form a human being if placed in a uterus, however. Neither sperm nor preblastocyst stem cells could form a human being on their own. Because of this, most stem-cell scientists believe that there is no genuine religious argument against using preblastocyst stem cells.

Stem cells are sometimes obtained from fertility clinics, which often produce and freeze multiple embryos from one couple. After the woman is pregnant, they are often permitted to throw away the excess embryos.

Most researchers who appreciate the uses of pluripotent stem cells believe that their potential for saving lives vastly outweighs any arguments that they should not be used. All doctors are admonished by the Hippocratic Oath, to which they have sworn, to save a human life if they have any means of doing so.

Of two types of stem cells—embryonic stem cells and adult stem cells—the latter exist in a mature body and are in no way related to human birth; so, there is no controversy about using them. Using them is no more controversial than using blood cells or any other type of cell. Both types of stem cells are used to replace damaged tissue. It was once assumed that adult stem cells could be committed to producing only one specific cell line, but scientists have since found that they can be used to create a variety of cells. Instead of being pluripotent like embryonic stem cells, they are described as multipotent. These stem cells can be isolated and transferred into cell cultures in laboratory dishes, where they can be reproduced in large numbers and still maintain their stem-cell capabilities.

Researchers demonstrated the potential of such stem cells in mice. They used radiation to kill the blood cells in a group of mice and then injected them with blood stem cells from other mice. These stem cells replaced the irradiated blood cells as expected so that the mice could live. They also formed bone marrow cells, skin cells and cells for the liver, lungs and digestive system, a capability that could be vitally important after a terrorist attack with nuclear-radiation weapons. A bank of human blood stem cells should be kept for such an emergency, just as blood banks are kept in the cities today.

The blood from a woman's umbilical cord contains many stem cells. When she has a baby, large numbers of stems cells can be collected and stored. Today, all over the world, these stem cells are washed down the drain. Such cells could be grown into tissues that the woman may need later in life. Such cells have been stored in the New York Blood Center. Pablo Rubinstein, who initiated this, says that it should be normal practice to collect umbilical-cord stem cells whenever a woman has a baby.

A pig's heart is somewhat similar to a human heart. Experiments have been carried out giving pigs severe heart attacks, sometimes leaving the heart tissue so damaged that the heart will fail soon. Adult pig stem cells (not embryo stem cells) have then been injected into the pig. They find their way to the injured area and start to rebuild the heart tissue. In many cases, a pig that was near death made a full recovery and returned to a happy, squealing life.

This can almost certainly be done with humans after a near-lethal heart attack. Researchers say that stem cells should be stored, in liquid nitrogen vessels, at every hospital, ready for emergencies for victims of heart attacks. Many lives could be saved. Stem-cell research holds out the promise of a supply of replacement tissues for many of our worn-out body parts.

REJUVENATION

Regenerative medicine employs a variety of techniques to rebuild damaged tissues. A particularly exciting aspect of regenerative medicine is the possibility of rejuvenating an elderly person's immune system.

Our immune system is of immense complexity. Numerous pathogens and viruses attack us, and our immune system marshals troops to fight them. Without a good immune system, we'd be in deep trouble, like people with AIDS. The rejuvenation of cells can be done in such a way that it restores many components of the immune system. They may be regenerated independently, in different ways, perhaps introducing young cells with young immune capabilities into our bone marrow.

For example, at age seventy, many people start to experience a decline in eyesight, even if their eyeglasses are good. Michael West's company, Advanced Cell Technology, wants to bring products to the marketplace that will help rejuvenate your immune system. He searches for where such products will make money the soonest—the "low-hanging fruit." He thinks that rejuvenating immune system cells that help you to maintain your eyes may be one of his company's first profitable products. Rejuvenation of other parts of the immune system will follow. West summed up in my interview with him the developments in regenerative medicine that he has helped pioneer: "I see a world where we may make new, young, healthy cells to replace the old ones in the body. I imagine medicine one day giving us back cells and tissues identical to those we had when we were born."

Rebuilding our immune system will become a critical aspect of preventive medicine. It will transform 21st-century medicine. West and his colleagues have shown how to take old human cells with an old immune system and use stem cells to give them a new immune system. "Frankly, in my lifetime I never expected that at this point in history we would have already invented a way to literally take old human cells back in time to be essentially identical to cells that you and I were born with. That technology is now in our hands, and how we use it is going to be one of the major challenges for the coming century." Regenerative medicine has a long way to go before it fulfils West's dream. Some aspects of it may come into clinical practice soon, but its full potential may be 20 years away. Once it becomes common practice, it will be in immense demand. It will evolve at a rapid rate, probably for decades. Nobody knows how far regenerative medicine will go, or how long we might live, but it looks as though repeated rejuvenations might become common. The combination of stem-cell use, therapeutic cloning and returning old cells to a young state with a young immune system sets the stage for very big changes.

A grim thought is that technology might enable us to live to be 120, but we wouldn't want to spend the final 30 years of it in a wheelchair. To make life extension appealing, we must find multiple ways to stop our health from going downhill. A worthy goal would be to keep health on a plateau for as long as we can until we reach the cliff edge of death. A relatively sudden end would be much better than an intolerably long decline into painful senility.

It's appropriate to refer to a long *healthspan* rather than just a long life span. Our healthspan can be made much longer than it is today.

BRAIN-CHANGING CHEMICALS

In Shakespeare's plays are seen all the characteristics of human nature that we live with today—the greed, love, leadership, lechery, scheming, treachery, trust, heroism, murder, joy, oppression, seduction—described far better than in our textbooks. One would conclude that human nature hasn't changed for 400 years, which is why contemporary-setting versions of Shakespeare often work so well. But today, when Hamlet ponders, "To be or not to be?" we might say, "Oh, come on! Give him Prozac." The near future will have narrowly targeted psychological drugs that could change the behaviour of most of Shakespeare's characters.

There is great public concern about changing people's genes but surprisingly little concern about changing their brain chemistry. Pharmacies are full of tranquillizers, painkillers, antistress pills, antidepressants, pills to make you sleep, pills to make you alert and numerous others. Brain scientists are making large strides in understanding the brain and its chemistry. Our ability to control the levels of serotonin, dopamine, epinephrine and other chemicals in the brain enables us to control our feelings of happiness, self-esteem, aggression, nervousness, depression, fear, fearlessness and well-being, as well as euphoric behaviour, and thus much of what we describe as personality.

Today's psychotropic drugs are blunt instruments, and their means of delivery is crude. A diversity of better-targeted drugs is on its way, and many will be tailored to the individual. They will be designed for the timed release of chemicals so that they have a longer-lasting effect, and soon they will be delivered with instrumented capsules that know their location in the brain and can detect the chemicals in that environment.

There are many drugs for mental illnesses, but now the drug companies have a larger marketplace among people who are not mentally ill. Almost 30 million Americans have taken Prozac and its derivatives. Children who can't sit still in class are given Ritalin. Teenagers in nightclubs take a rapidly growing family of drugs that induce sexiness, euphoria, short-term amnesia, loosening of inhibitions and other rave behaviour. Ecstasy is used, inducing uninhibited behaviour at nightclubs, but it can have a beneficial effect; it increases social sensitivity and human bonding. It can induce deep reflections on human or family relations. New, more powerful drugs may focus on these good effects, delivering the good without the bad.

The brain consists of a vast number of neurons connected by a very complex nervous system consisting of axons. The axons behave rather like wires, transmitting signals among the neurons. Synapses, attached to a neuron, have the job of sending and receiving the messages that travel on the axon. If the neurons in our brain sent straightforward signals to other neurons, the brain would act like a computer, but it is more complex than that because there are many chemicals in the brain that affect the firing of synapses and, hence, the signals sent among the neurons. The chemical neurotransmitters that convey messages among neurons can be affected by the overall chemistry of the brain. Some chemicals affect a synapse before it sends a signal, and some affect it when it receives a signal.

For example, serotonin is a neurotransmitter chemical. People with a low level of serotonin tend to be depressed, have poor impulse control and exhibit uncontrolled aggression against inappropriate targets; people with a high level of serotonin in their brain tend to be happy and lose their feelings of anxiety and aggression. Violent criminals, arsonists and people who die by violent methods of suicide score low on measures of serotonin activity. Monkeys that have low sociability and high hostility have low serotonin levels. Prozac and similar drugs increase the serotonin levels in the brain in selective ways. They interfere with the reuptake of serotonin by synapses that transmit it, so that there is more available to transmit.

COMPUTERS IN NEED OF PILLS

Imagine you and your friends have personal computers that, as well as having today's circuits, are filled with strange chemicals. These chemicals can cause a computer to register feelings. Sometimes a person's computer seems depressed; sometimes it has feelings of fear, self-esteem, nervousness or happiness. It may have great enthusiasm for exploring the Internet and become euphoric about what it finds. Some of these computers are sleepy, some fall in love with their owners, some are aggressive and others have complicated religious feelings. Some behave in bad ways—they are disruptive, bad-tempered or inattentive; some seem addicted to chaos.

This is not a very satisfactory situation. Fortunately, scientists realize that they can control these computers by adjusting their chemicals. In a somewhat similar way, schoolteachers learn that they can control the behaviour of unruly children by recommending pills such as Ritalin. Army units can make soldiers more aggressive and help remove the fear of battle. A lonely woman can use Prozac to improve her feelings of self-esteem.

People who took Prozac to relieve feelings of depression discovered that it also made them feel good about themselves. Soon there will be many drugs that will have different effects on different personalities. The customer will be able to adjust the choice of drugs to characteristics of the individual. Peter D. Kramer, a psychiatrist from Brown University, used the memorable phrase *cosmetic pharmacology*.[7] Psychotropic drugs will be sold on television the same way that cosmetics are.

The drug industry has been busy testing drugs on normal people—people who are free of mental disorders. Such subjects were asked, for example, to perform a negotiating task in which they and a stranger approach a stressful problem. On medication, they were less negative and more collaborative, and they tended to succeed in their negotiations.[8] It seems clear that there can be pills to help people doing business negotiations. Many normal secondary-school kids are swallowing pills to get an edge when they take their exams. Stimulants can affect the results of scholastic aptitude tests. The days are not far away when teenagers will pop pills when going on a first date. There will be psychotropic products to lessen marital friction. Researchers using standard tests measuring "aggressiveness" and "irritability" have shown how drugs can lessen these.

In the University Hospital of Geneva, Olaf Blanke and his colleagues found that by stimulating a certain point in the brain, above the right ear and just inside the skull, they could produce out-of-body experiences.[9] Many occult and pseudo-religious experiences are probably caused by brain chemistry or stimulation. Shamans in primitive cultures were well aware of what plants they could use to bring about hallucinations.

Some drugs affect aggressiveness. A future boot camp for terrorists will enhance its psychological methods with drugs. Terrorists on a mission will use pills that produce uncontrolled aggressiveness. There'll be pills that give them feelings of euphoria, religious intensity, determination and absence of fear. Other drugs decrease feelings of aggression, hostility and irritability; others create feelings of self-esteem, happiness and tranquillity. Benzodiazepines can be used to reduce anxiety. Some drugs make you sleep, or lower the number of hours of sleep you need. Some psychotropic drugs will enhance memory and factual recall. Acetylcholine system enhancers will enable immigration officers to learn and recognize an amazing number of faces. Some drugs can improve a person's creativity, or allow him or her to work late at night in innovative ways. Some jobs at a computer screen need long periods of abnormally intense concentration, and drugs can help a person to concentrate longer and more intensely. Dopamine system enhancers can increase stamina and motivation, and other drugs enhance the capability to perform difficult tasks.

Having a full colonoscopy used to be unpleasant; now with an intra-venous drip it can be one of life's euphoric experiences. Some drugs generate intense pleasure for a short time. Some can turn a sourpuss into Mr Nice Guy. Some scientists believe that, as our understanding of brain chemistry progresses, we'll find that all of our emotions have chemical or electrochemical bases.

Eventually, the public may use antidepression pills as freely as they use aspirin today. They will pop happiness pills, sex pills and pills that make them fall in love.

ENHANCING OUR BRAIN CIRCUITS

It is very difficult for us to imagine the complexity of the brain. It has about a hundred billion neurons. This is roughly the same number as the number of stars in our galaxy. On the clearest starlit night, you can see only a very tiny fraction of these stars. Each neuron is connected to about a thousand other neurons. Imagine that the stars were communicating in this way, and you could see axonlike connections joining the stars light up for a thousandth of a second. The night sky would then be ablaze, shimmering in every direction with different and changing intensity.

A very important project in neuroscience will be to map the connections among the neurons in the brain. As scientists do this, they will learn to emulate them in computers. This is a very difficult task, but reading the human genome seemed so monumentally difficult in 1990 that some scientists thought it was impossible. The technology needed to photograph the brain in action has been steadily improving for two decades. As I write this, it is possible to photograph only broad areas of the brain, consisting of many thousands of neurons, with techniques like FMRI (functional magnetic resonance imaging), but the techniques for scanning the brain and making pictures or videos of it are rapidly improving. They are doubling in resolution every 18 months or so. It is thought that this will continue until it will be possible to create images of individual neurons and their synapses sending signals over axons to other neurons.

Today these neurons, synapses and axons can be photographed in a dead brain sliced into sections. The challenge is to do it in a live brain that is functioning, so that the signals among neurons can be recorded. A possibility is to use quantum entanglement, in which the spin (or other property) of a minute subatomic particle is exactly linked to the spin of a twin particle. The subatomic particle in the brain has its spin changed, and the corresponding spin of the twin particle changes in the recording equipment several feet from the brain. The subatomic particle in the brain is too small to interfere with the brain. Several different technologies will enable neuroscientists to take millions of recordings per second of the brain while it's performing its tasks and to steadily combine such recordings to create an integrated model of the brain. The result won't be a photograph; it will be a working model of the brain performing tasks.

This will lead to better understanding of human dysfunctions. What is schizophrenia? What is bipolar disease? Today we have vague psychological descriptions, but we'll actually see them and understand them neurologically. We'll begin to understand how learning, memory and reasoning work.[10]

Creating a detailed map of our brain wiring will not give us complete understanding of our brain. The chemicals in the brain are highly complex and have a large effect on our behaviour. Much research will be needed to understand how brain wiring and brain chemistry work together. As this understanding develops, neuroscience will become one of the most exciting sciences.

Neuroscience labs will have access to supercomputers with numerous models of different parts of the brain at work. These models may be created worldwide and gathered together to help form an integrated understanding of the functioning of the brain.

Our brain has some small areas that perform highly specialized activities. These may be early targets to study and perhaps to replicate with electronics. An electronic replica of them will operate much faster than the biological original, because wires that connect transistors are a million times faster than nerves that connect neurons. The cerebral cortex is the highly folded outer part of our brain where most of our thinking goes on. If we took it out of the skull and stretched it out, it would be a disk of meat about a metre across and two to three millimeters thick. Some of the nodules of specialized activity are a column through the cortex, about a millimetre in diameter—2 or 3 cubic millimetres in volume, containing a few hundred thousand neurons. These nodules are less complex than many of today's microchips.

As we learn how to replicate parts of our brain in electronics, we may want to directly couple the electronics to our biological brain. This may be done by putting an extremely fine wire into the brain. A preferable way may be to have transponders in the brain fluid, possibly about the size of a nanotechnology blood cell, which is much smaller than a neuron. The transponder transmits signals to a computer outside the skull and receives signals from it. The computer might be in your ear or attached to your neck. The transponder interface would be standardized so that many transponders could communicate with many computers.

We won't have to wait for the age of nanotechnology. An early, simple transponder could be built with MEMS (micro-electromechanical systems) technology, which has already been demonstrated in the bloodstream and the digestive system. A special coating will be used to keep our immune systems from trying to repel the intruder. Electrodes have been coated with molecules that adhere to brain cells. The initial uses of brain/electronics connections are for severe medical conditions: 30,000 epileptics are being treated by electronic stimulators connected to the vagus nerve, and recently, patients with severe depression have been treated with electronic stimulators in the brain, with surprising success.

Technological ideas advance on a broad front when vast numbers of young people can be involved in them and experiment. This happened with computer games and the Internet. Some authorities think it will happen with electronic brain enhancements once there are standard nanotechnology transponders. We may have a world in which college students everywhere are trying out new brain enhancements.

It may be years after first mapping the brain that we learn how to make brain-enhancement technology work well, but once it works, it probably will advance at a furious rate, with a great diversity of experimentation. Perhaps transponders in our brain will be wirelessly connected to modules that carry out brain functions in highly advanced ways. Corporations may pay for their best employees to have online brain connections and corresponding computing modules. Different functions may be allocated to different employees. As the technology advances, there may be thousands of such transponders in our brain fluid, with appropriately strong security protection.

UPLOADING THE MIND?

There seems to be nothing in the thinking part of our brain, the cerebral cortex, that couldn't be transferred to a machine. We would need to simulate the effects of brain chemicals. That would need a machine with far more components than today's machines, but we'll have those soon. Today there is controversy about whether we'll be able to replicate the *entire* human brain in a computer. Some scientists, such as Ray Kurzweil, say, "Of course we will." Others, like Susan Greenfield, a top brain researcher at Oxford, say that we seriously underestimate the complexity of brain chemistry.

Replicating our mind in electronics raises many questions to which we have no certain answers. If a human mind is replicated in a computer, would it have a soul? Cognitive scientist Steven Pinker tells us that the soul is merely the sum of all the information-processing activity of the brain. Most neuroscientists disagree, saying that the "soul" is not part of neuroscience. The "soul" is said to be immortal, and what neuroscience studies is definitely *not* immortal. It dies when the brain dies.

Would a silicon version of the human brain be conscious? Possibly *yes*. It is a popular view at MIT that when machines have sufficient complexity, they will acquire consciousness. The philosopher Daniel Dennett argues that consciousness is an emergent property of complexity, whether in the human brain or a sufficiently complex computer.[11]

Our mind in a computer would be a collection of bits, and a collection of bits doesn't wear out. The machinery it operates on might wear out, but the bits can quickly be moved to a new machine. In a sense, then, a brain that is digital is immortal. It might repeatedly be transferred to new hardware as better technology comes along. An uploaded human would be several trillion bits. Twenty years from now Circuit City will be selling backup storage units of that size, guaranteeing that they are cheaper than at competing shops.

A digital brain would not have Alzheimer's or other biological diseases. A collection of bits would have duplicate copies of itself in case its host machinery failed. It would be backed up, like computer software, and protected by firewalls. Duplicate copies might be stored in ultrasafe vaults.

As the 21st century progresses, there will be tight coupling of our brains and ever-faster computers. It may be online coupling or coupling in which the human interacts with elaborate displays, as today. It will replicate brain functions and also create new functions because they are much more efficient. A tightly coupled relationship between the human brain and computers seems inevitable within the lifetime of younger readers of this book (the Transition Generation). We can expect a massive stream of different enhancements of human capability.

In the history of technology, certain inventions stand out as landmarks that changed the future—for example, in the 18th century, the steam engine; in the 19th century, the telephone; in the 20th century, the computer. We can ask: In the 21st century, what invention will have the most effect on changing the future? I suspect it may be the creation of wireless links that connect our brains directly to external electronics, including global networks. This may be done with large numbers of nanotransponders in the brain fluid. Eventually, there will be many easy-to-use links between our biological brain and electronic devices. The first direct links between our nervous system and tiny computers have already been made. The brain/computer linkage will grow from tentative experiments to robust connections. There will someday be an immense diversity of types of brain/computer connections.

This form of brain enhancement will require the individual to go through a learning process in order to use it well, rather like learning to ride a bicycle. Like other training, this will establish the connections among the brain's neurons that enable the person to do new tasks. Once the learning has occurred in the brain, the person will become comfortable with the new faculties, like becoming comfortable with riding a bicycle. People will steadily become skilled at putting electronic brain appendages to good use. When we first connect electronics to the brain, the electronics may be a replica of some small part of the brain so that the brain can learn how to use it naturally. Less natural appendages will require more intense training.

Once transponders in the brain become standard and easy to use, many people, especially younger people, will want to try them. Many of the uses will be fun. Once people start using them, they will find that, as with the Internet, there is a whole world to explore. There may eventually be thousands of nanosize devices in a person's brain. As this technology matures, the brain will be connected to more complex applications.

It is sobering to reflect that this will be happening in high-tech parts of the world, at the same time that the poorest countries are facing the deprivations of water and food shortages and the most turbulent part of the canyon we described. The story we have to tell is one in which astonishingly different threads of a tapestry have to be woven together.

The coupling of human brainpower to the explosive evolution of computer power will bring an extraordinary change in society. The combination of human and non-human-like intelligence will create forms of intelligence far more powerful than anything we can imagine today. To quote Freeman Dyson's phrase again, the growth of such technologies seems "infinite in all directions."

After a decade or two of brain/computer connections maturing, the Singularity will happen. Many Singularity "experts" think that the only way we are going to be able to cope with it is to become much smarter by having direct brain/machine connections. The brains of many people will be online to Singularity capability (whatever that might become). When the Singularity is a decade away, many people will try to prepare themselves for it.

Transhumanism and the Singularity are destined to combine.

We commented at the start of chapter 2—"What Got Us into This Mess?"—how the tragedies of classical Greek theatre described man daring to reach out beyond reasonable limits in quest of some ideal. The theatre audience is aware of forces in the world powerful enough to topple even the most admirable of men. The sin of the hero of Greek tragedies is hubris, leading him to ignore warnings from the gods and thus invite catastrophe. We now have new stories for Greek tragedies, grander and stranger than those of ancient times.

PART THREE

THROUGH
THE CANYON

13

THE AWESOME MEANING
OF THIS CENTURY

So, what is the meaning of the 21st century?

Evolution on Earth has been in nature's hands. Now, suddenly, it is largely in human hands. The extreme slowness of nature-based evolution makes it almost unnoticeable alongside human-based evolution. As we *automate* some of the processes of evolution, the rates of change will become phenomenal. This change from nature-based evolution to human-based evolution is, by far, the largest change to occur since the first single-cell life appeared. Its consequences will be enormous.

One might speculate that such a change occurs on faraway planets when their creatures reach a high-enough level of intelligence. When it first happens on a planet, it is probably dangerous. The creatures that take evolution into their own hands have no experience in the game.

Now that we are in charge of evolution, we need to learn the rules. We need to be cautious, using our scientific know-how as responsibly as possible. The change to responsible, scientific management of our own evolution is perhaps the most critical aspect of the 21C Transition.

Nature's evolution experiments, constantly trying new things. We humans are a new type of experiment—a young trial species, still adolescent and playing with fire. Unlike migrating swallows or foraging ants, we are not biologically programmed to know what to do. Instead, *we are an experiment in free choice.* This gives us enormous potential. We will spur evolution of the technology and management capability to exercise that free choice on the grandest scale.

On Earth, nature hasn't played this game before. There hasn't been a species before that can set goals, invent technology or organize ambitious projects. We are nature's biggest experiment so far.

We are not masters of nature—we are a component of nature. We must have the deepest respect for what nature has taken 4 billion years to create. Nature's biodiversity is of staggering complexity, and we are immersed in this complexity. When we interfere with it, we damage it in subtle ways. Unless this is understood, we will live in an increasingly tattered environment. The environment should not be something that we manipulate, like landscaping, but something we understand and treat much more responsibly than today.

Britain's eminent "green" authority, James Lovelock, created the hypothesis that Earth is a self-regulating ecosystem, which he called Gaia. We can interfere with Gaia up to a point; it's highly resilient. If we interfere with it too much, we'll be in trouble. Lovelock says that if we don't stop pumping carbon into the atmosphere, "We'll suffer the pain soon to be inflicted on our outraged planet."[1] From the viewpoint of Gaia, it is possible that our urban sprawl, climate change, holes in the ozone layer and products that poison nature are like a form of cancer starting to metastasize. In the long run, adjusting itself on its very long timescale, Gaia will survive, even if humanity doesn't.

A vital part of the meaning of the 21st century is that we must not push the Earth's control mechanisms beyond the zone in which they are self-regulating and stable. We have already gone too far in harming the climate, the wetlands, the soil, the oceans and numerous smaller-scale ecosystems, so a vital part of this century is to correct the damage. This will be increasingly difficult to do as the Earth's population grows, consumerism races across China and India, water runs short and the stresses of the canyon worsen.

Nature is metastable (like the cyclist described earlier). Up to a point, when we perturb it, it recovers, but beyond that point, it wobbles out of control. A country pond may have clear water and healthy fish indefinitely, but if there is too much runoff into it from a farmer using fertilizers, the algae blooms and consumes the oxygen, and the pond becomes a green and stinking mess. The ecology of the Black Sea collapsed suddenly, its stench closing the fashionable Russian resorts. Then there were outbreaks of cholera and hepatitis.

Many scientists passionately believe it is their destiny to change nature, but we need to understand the ways in which nature is fragile and know when and how to be cautious. A naïve view from the past is that technology gives us mastery of nature. A more appropriate view is that it has given us a depleted planet and an artificial world increasingly dependent on technology. Advanced technology puts us in need of even more advanced technology. Rather than being masters, we are being swept away in the rapidly accelerating floodwaters of our own technology. The Faustian bargain may be that a magnificent lifestyle comes with deepening entrapment in our own, ever more complex inventions. For the most part, we enjoy this deepening trap because it gives us entertainment, anaesthetics, mobility, corporate profits and lifestyles that past kings couldn't have dreamed of.

Sooner or later, humanity has to learn how to control technology and avoid what is too dangerous, just as we must control a teenager if he is learning to drive a Ferrari. We are learning to drive something monstrously more powerful. This capability for control needs to exist before technologies become more dangerous than they are today.

A special part of the meaning of this century is to put into place controls to make sure that we don't accidentally destroy the utter magnificence of humankind's long-term future. We don't know the details of how to control future technology, but we know enough to believe that control is possible. It's a matter of good engineering with excellent safeguards and good management with strong discipline—all of which can be taught. This century, one hopes, will see a transition to a planet managed well enough to make its long-term survival likely. As man began to create science in the 17th century, he was unlocking a power that would grow in ways that he couldn't possibly understand then. Once unlocked, that power would grow over centuries until it could change the Earth, change biology, change civilization and change *Homo sapiens*. Humankind excitedly released the powers of science with no caution. We split the atom, created electronic intelligence, learned how to modify the genes of every living thing and are now learning how to modify ourselves. Science unlocks unimagined riches, but it also leads to forces that can destroy us. As the avalanche of technology gains speed and momentum, its capability for both good and ill becomes more powerful.

We live on a beautiful but totally isolated world. We won't find another world that can replace it. At the beginning of the 21st century, we are trashing this world, and we are gaining the power to destroy civilization. We have the intelligence to manage planet Earth well and, to a large extent, correct the damage we have done. We need to do so fairly quickly because the capability for destruction is growing fast. We need to make the planet work well with an excessive population.

By the end of the 21st century some technologies will have astonishing power; so, part of our learning how to manage our planet is learning how to live safely with these technologies and to build civilizations that live well with one another. Theoretical science today makes it clear that we are still only in the early stages of a much longer journey. Technology will grow in power and momentum for centuries beyond this one, taking us into areas that utterly defy our intuition and common sense—areas describable only with mathematics. We have no idea, for example, where quantum entanglement will lead. By the 23rd century, science will be unimaginably more powerful than today.

Sooner or later on this long journey, we have to learn how to control what we are doing, and this is the century in which it must happen. In this extraordinary 21st century, we must put the controls in place. This capability to manage what we are doing needs to be established before the game becomes too dangerous. Diverse types of controls must be established to enable Earth's civilizations to thrive.

The 20th century was the most violent century so far. It was like a teenage world swept with intense emotional traumas and violence, prior to adulthood. Because of mass-destruction weapons, the 21st century cannot withstand similar behaviour. It's the century when we have to achieve adulthood, and there are some early signs of this developing.

Improvement in behaviour is critical in the 21st century because, suddenly, we're all in the same totally isolated melting pot, amidst dreadful weapons. We'll have new wealth and capabilities that can greatly upgrade what humankind is all about, but we have to grow up pretty quickly. This is *childhood's end*.

The melting pot is global. In the long run, this is good if we all treat one another with decency and share democratic freedoms. Eventually, we'll understand the ramifications of our statement "All people are one people," but today, globalism has massive flaws. Large corporations have major bridges to countries where they can make a profit but are bypassing countries where they can't. Because of this, the disparity between rich and poor countries is enormous. Failed nations are not on the big corporate map of the world. Unless well-managed and well-funded action is taken, the worst poverty will become more extreme. The average cow in the European Union receives more than $2 per day from subsidies, but 3 billion people—47% of the world's population—live on an income lower than $2 per day,[2] a hopelessly inadequate amount to feed a family and give them the most basic education, and billions more people will be added to the poorest countries by midcentury. To make matters worse, in many of these countries, the water that is essential for growing food will be running out.

We may be trashing part of the Earth's environment, but billions of the Earth's people are being trashed in an even worse way. While globalism is spectacularly increasing the wealth of the richest countries, the poorest can barely feed themselves. I happen to be writing this in St Petersburg, where such extremes also arose, and there was violent revolution followed by seven decades of oppression and trauma.

Childhood's end is about much more than environmental correctness. It's about removing the reasons for future bloodshed and human horrors. It's about getting rid of the unspeakable poverty, disease and hunger that plague many on the planet. It's about learning what we should do with the extraordinarily transforming new technologies that we are unlocking. It should be understood that future science is both magnificent and dangerous and can open up counterintuitive worlds that are grandly different from what we can see and feel, and that these gifts of science should be global.

If we are to survive, we have to learn how to manage this situation. We need to put in place rules, protocols, methodologies, codes of behaviour, cultural facilities, means of governance, treaties and institutions of many types that will enable us to cooperate and thrive on planet Earth. If we can do that with whatever the 21st century throws at us, we'll probably be able to survive future centuries. If our 21st century world falls apart at the seams, civilization will be set back many centuries.

The 21st century, then, brings us the following challenges:

CHALLENGE 1: THE EARTH

The 21C Transition is a change from wrecking the planet to healing the planet. We need to stop actions that lead to climate change. We must heal the ozone layer, achieve sustainable water use, stop excessive cutting of forests, revitalize the soil and achieve food security. Such changes need to occur in the early part of the century, before the destruction goes too far.

When we stop polluting rivers, they clean themselves quickly; lakes recover more slowly. Tough measures are needed to preserve wetlands essential to marine life, so many of which are being destroyed by seaside property development.

A change in the capability to manage the Earth well is coming from the deployment of vast quantities of micro-instruments, which feed voluminous data to computer systems. Humanity is changing from being ignorant about the planet to having vast quantities of information linked to supercomputer models.

In the second half of the century, we will have learned how to live within nature's trust fund. One hopes that we learn this by science and good teaching. If not, we will learn from catastrophe-first patterns of events. The Earth's climate will change, and we will learn to live with the changes. Freeman Dyson reflectively commented, "The art of living is to make the best of it, not to try to organize everything the way you want."

CHALLENGE 2: POVERTY

While rich nations become richer, billions of people live in extreme poverty with short brutal lives. In my interview with Jack DeGioia, the influential head of Georgetown University in Washington DC, he stated emphatically, "The moral challenge of our times is to eliminate extreme poverty." We need to steadily transform our world so that, by the end of the century, it is a clean and decent place for all people. All nations should reach a decent literacy rate and adequate levels of employment. The horrifying situations of today's destitute countries should be gone for ever. Jeffrey Sachs's proposals address ways this can be done.[3]

CHALLENGE 3: POPULATION

Much of the extreme poverty on the planet relates to the population being too high. It is estimated that the Earth's population will soon be increased by 2.5 billion people, most of them in the countries least able to grow enough food. There are now nonoppressive ways to lower the birthrate. The population declines strongly in countries where almost all women can read and full women's liberation is in effect. Population tends to decline when GDP is high. The goal of improving lifestyles equates to the goal of lowering population.

CHALLENGE 4: LIFESTYLES

Most people (almost 9 billion) will eventually want to participate in the affluence of the planet. That cannot happen with 20th-century lifestyles. We need higher-quality lifestyles that are environmentally harmless. Rich, affluent, globally sustainable lifestyles, more satisfying than today's, can be achieved at the same time as healing the environment.

CHALLENGE 5: WAR

All-out war in the 21st century could end everything. No economic or political benefit can justify the risk of an all-out war with nuclear and biological weapons. We have to absolutely prevent war between nations with arsenals of mass destruction. There will be either no war among high-tech nations or no civilization. The existence of weapons capable of ending civilization makes this a very different century from any before.

CHALLENGE 6: GLOBALISM

Globalism is here to stay. The planet is shrinking and bandwidth is increasing, but globalism should be designed to allow local unique cultures to thrive and be protected. The right balance between what is global and what is local needs to be achieved.

Global business will continue to expand and needs to benefit everyone, rather than bypassing some countries and leaving them destitute. Failed nations need to be helped until they become developing nations. Slowly, appropriate laws, trade agreements, tariffs and codes of behaviour will be agreed upon.

CHALLENGE 7: THE BIOSPHERE

We are losing species of plants and creatures at a shocking rate. This represents a major loss of knowledge in the DNA of the species. Many endangered species can be protected by identifying and preserving hot spots, those places with a high density of endangered species (discussed later).

Today 90% of the edible fish in the oceans have been caught. It is possible to create conditions in which ocean life will slowly recover. This requires well-designed marine protection areas combined with global design of well-managed fishing. Laws are needed for transforming the oceans from their appallingly depleted state today to a vigorous healthy state.

Different challenges to the biosphere will come from the Green Revolution's reduction of competing biodiversity to achieve high-yield farming and from farmers' increasing use of genetically modified crops.

Global management of the biosphere is essential. This needs thorough, computer-inventoried knowledge of all species.

CHALLENGE 8: TERRORISM

The dawn of an age of terrorism coincides with weapons of mass destruction that will become progressively less expensive. The capability for terrorist groups to build nuclear weapons must be entirely removed. This can be done by denying them access to enriched uranium and plutonium. All sources of weapons-grade uranium and plutonium must be identified and locked with extreme security.

Above all, it is vital to remove, as far as possible, the reasons why people want to become terrorists. At the start of the 21st century, cultures that were previously separated and potentially hostile found themselves in each other's faces because of the media that accompanied the new forces of globalism. Potentially hostile cultures are suddenly in the same melting pot. It's a critical task of the 21st century to achieve mutual respect and cooperation among these cultures so that they don't want to blow each other up. It will be essential that religions recognize the goodness in other religions and accept the necessity to live with one another (as they do in the First World). It's critical to prevent the perversion of religions with philosophies that promote suicide killers.

CHALLENGE 9: CREATIVITY

The technology of the near future will lead to an era of extreme creativity. Young people everywhere should participate in the excitement of this creativity. Different cultures are likely to accept one another as exciting jobs spread and rich countries help young people around the planet to be entrepreneurs. The world is becoming finely laced with supply chains of electronically connected business that will eventually interlink all countries and become very valuable.

If young people everywhere understand the meaning of the 21st century, the vital role of the Transition Generation, the highly innovative ways in which solutions can be found to the planet's problems and the challenge of creating better civilizations, they will not want to be suicide bombers.

CHALLENGE 10: DISEASE

It is desirable to prevent pandemics in which an infectious disease spreads quickly and kills many millions of people, as has happened many times in history. Such diseases could be introduced as an act of terrorism. We now have sensors that can detect the existence of a dangerous virus in the air and medical procedures to prevent it spreading. We need to be ready, with all our technological resources, to stop bird flu, and future pandemics that are a surprise. We are not ready today.

It is now possible in a cheap laboratory to modify the genes of a pathogen so as to create one that is new to nature. Nature then lacks the protective mechanisms that have evolved over billions of years. Weapons researchers have modified smallpox, bubonic plague and other terrible diseases. Genetically modified pathogens that spread fast could represent an extreme danger. We must build appropriate defences.

CHALLENGE 11: HUMAN POTENTIAL

A tragedy of humankind today is that most people fall outrageously short of their potential. A goal of the 21st century ought to be to reverse that and focus on how to develop the capability latent in everybody. As the century matures, humankind's learning ability will accelerate, aided first by the digital media that we have today, then by powerful computerized tools and then by changes in humans themselves. We are starting to learn how to take care of our planet and treat its inhabitants with decency. This should be the start of an era in which the limitless creativity of people is unleashed. Expanding human potential needs to be pervasive from the most destitute society to the richest high-tech society.

CHALLENGE 12: THE SINGULARITY

Decades from now, computer intelligence that is quite different from human intelligence will feed on itself, becoming more intelligent at a rapidly accelerating rate, until there is a chain reaction of computer intelligence, which we refer to as the Singularity. Humanity needs to discover how to avoid being sucked into a situation that is totally out of control and harmful. Technical controls will be needed for computing, perhaps in the form of hardware design, to ensure that when computers become incomparably more intelligent than we are, they act in our best interests.

The main impact of the Singularity will be that the cleverest professionals will use it to achieve extraordinary results. By the time it happens, the capability for handling the Singularity will be distributed globally, particularly among appropriately educated young people. It will enable many different self-evolving technologies to become "infinite in all directions."

CHALLENGE 13: EXISTENTIAL RISK

The 21st century is the first in which events could happen that terminate *Homo sapiens*. These are referred to as existential risks (risks to our very existence). Lord Martin Rees describes such risks in detail and gives us only a 50% chance of surviving this century.

If we do survive, our accomplishments by the end of this century will be awesome. The magnificence of what human civilizations will achieve if they continue for many centuries is beyond all imagining—so magnificent that it would be too tragic for words if humanity were terminated. To run the risk of terminating *Homo sapiens* would be the most unspeakable evil.

We should regard any risk to our existence as totally unacceptable. We need to take whatever actions are necessary to bring the probability of extinction to zero.

It is desirable to understand all types of existential risks and to either ban the technology in question or devise controls safe enough to prevent the risk. There are many types of safety measures that can be put in place. Long before the end of this century, we will need to have safety rules and procedures in place. They will be designed so that as far as possible they are appropriate for subsequent centuries. The most dangerous time will probably be just ahead of us when we first argue about *whether* we should control science. Lord Martin Rees emphasized in our interview with him, "As science advances, there's an ever bigger gap between what *could* be done and what *should* be done."

CHALLENGE 14: TRANSHUMANISM

This is the first century in which we will be able to radically change human beings, and this fact alone gives very special meaning to the 21st century. Technology will enable us to live longer, learn more and have interesting prostheses. Neuroscience will blossom spectacularly when we can map the brain, recording the transmission of signals among individual neurons and then emulating parts of the brain with technology millions of times faster than the brain. A new world will open up when we can connect diverse neurons in our brain to external devices. We'll connect the brain directly to nanotechnology objects on our skull and to supercomputers far away. This will change human capability in extraordinary ways.

Transhumanism will be highly controversial. It will raise major ethical arguments. We might harm some of the qualities that make humanity wonderful. It will lead to fundamental advances in what humans are capable of but will create extreme differences between the haves and have-nots.

We need to understand where changes to *Homo sapiens* can be made without net-negative consequences. Transhumanism will be a prime enabler of civilizations far beyond those of today.

CHALLENGE 15: ADVANCED CIVILIZATION

The 21st century will experience a major increase in real wealth (adjusted for inflation). Sooner or later, machines will do most of the work. What we do with our leisure will be a huge issue.

The future could be an age with a flowering of great literature, theatre, fine arts and entertainment. Different forms of high-culture civilization will have global networks of enthusiasts. Any civilization of the future will permeate cyberspace. Music, dance, filmmaking, game-creation and diverse forms of new culture may help build high levels of mutual respect in societies previously uncomfortable with one another. Governments may take a deliberate role in emphasizing eco-affluence. The 21st century will make available staggering quantities of new knowledge and computerized intelligence, leading to levels of creativity inconceivable in the 20th century.

A big question that we ought to be asking now is: "What could truly magnificent civilizations be like at the end of the century?" Because of transhumanism and the Singularity, the changes will be more extreme than are generally realized.

The transition from a planet on a self-destructive course to a planet that is intelligently managed is the meaning of the 21st century. If there were no change in technology, this would be vitally important, but technology will transform everything we do. If these challenges are handled with appropriate wisdom, they should put us in good shape to manage whatever challenges technology and human behaviour present in the centuries ahead. In the long landscape of human history, this century has a vital place. It will unleash enormous quantities of human knowledge, represented in ways that computers can use.

It is up to us whether this is humanity's last century or the century in which the nature of great civilizations—with globalism, transhumanism, nanotechnology, the Singularity and so on—becomes understood. It could be more important than the Renaissance.

See the 21st century as a tollbooth after which humanity is on a highway past the time when extreme poverty existed, past the time when there were destitute nations, past the debates about genetic engineering and trans-humanism, past the times when large-scale war was a viable option, into an era when we've learned how to avoid risks to our existence. It will be a time when conventional work is done by machines, and the humans spend their time on things that are uniquely human. Higher levels of happiness will come from higher levels of creativity. Perhaps it will be a time when much of our mental effort goes into glorious music, architecture, culture and magnificent green cities.

If any of the above sounds easy, it shouldn't. We'll make grand-scale mistakes as we progress, and we'll sometimes learn from our mistakes. By the end of this century, there'll be some dramatic monuments to our errors.

By a twisting old apple tree in Isaac Newton's old college, we filmed Lord Rees, surrounded by ancient beauty, his smiling eyes suggesting that he had wisdom he was not revealing. He gave his last word on the subject: "The prime concern of the 21st century is that we should survive it."

CHALLENGE 16: GAIA

Because we need to make sure that we are not close to the limits beyond which runaway global warming occurs, earth system science needs to be a thorough academic discipline that comprehensively measures and models the Earth's control mechanisms. There will be uncertainties in the models, but we should not take the slightest risk of upsetting the immense forces that make our home planet livable.

Perhaps the greatest catastrophe that could befall us would be that we inadvertently push Gaia so that positive feedback causes it to become unstable or to change to a different state. Our Earth may become a roasted planet with tundra, inhabitable by only a small number of humans—probably near the poles.

The forces of the Earth's system controls are immense, and we are interfering with them. Lovelock comments that the idea that humans are intelligent enough to act as stewards of the Earth is among the most hubristic ever. Gaia does its own thing, and we must live within its constraints. If all of the Earth's population wanted to eat meat and have air conditioners, Gaia's constraints would be exceeded. We must learn to live within our planet's means, and do so quickly. All humans should be taught about this and understand the terrible consequences of going beyond what Gaia—the Earth's control system—can handle.

The 21st century must put the science in place to regulate human behaviour to live at peace with Gaia. This will be essential for future centuries.

CHALLENGE 17: THE SKILL/WISDOM GAP

Deep wisdom about the meaning of the 21st century will be essential. A serious problem of our time is the gap between skill and wisdom. Science and technology are accelerating furiously, but wisdom is not. We are brilliant at creating new technology, but are not wise in learning how to cope with it. To succeed in today's world, people will need intricate skills in narrowly specialized areas. Skills need detailed, narrowly focused study of subjects that are rapidly increasing in complexity, whereas wisdom needs the synthesis of diverse ideas. Wisdom requires judgement, reflection about beliefs and thinking about events in terms of how they might be different.

Today, deep reflection about our future circumstances is eclipsed by a frenzy of ever more complex techniques and gadgets and preoccupation with how to increase shareholder value. The skill/wisdom gap is made greater because *skills* offer the ways to get wealthy. Society's best brains are saturated with immediate issues that become ever more complex, rather than reflecting on why we are doing this and what the long-term consequences will be.

University education today is much more pressured than when I was at university. The curricula have become overstuffed, the subject matter intensely complex and the examinations frequent and demanding. The student sticks to the curriculum and can deal with little else. The professors stick to their discipline; they are judged by the papers they publish in the professional journal of that discipline. Most areas of education have almost no interdisciplinary scholarship. As disciplines become deeper and more complex, the brilliance expended on them is formidable, but we don't think much about its consequences or what impact it has on other areas. In specialized areas, computers will become vastly more intelligent than people, but such intelligence is not human wisdom. As computers become more intelligent, with intense self-improvement of non-human-like intelligence, the skill/wisdom gap will widen at a furious rate.

We have vast numbers of experts on how to make the train work better and faster, but almost nobody is concerned with where the train is headed or whether we'll like its destination.

Wisdom is essential and comes from the synthesis of a large amount of knowledge and experience that may take much of a lifetime to acquire. Not everyone can handle such synthesis. We must ask where the broad wisdom about the future will come from. The answer is, *we must set out consciously to develop it.* Wisdom, like advanced civilization, will come when we learn to relax. Our best brains need to stop chasing the most highly paid careers, the fastest boats and the smartest country clubs. A mature society should exhibit deep respect for deep wisdom.

We need to set out very consciously to foster and nurture the wisdom that the 21st century will require. This should be a task for our greatest universities.

These 17 challenges are all interlinked and mutually reinforcing. Together they constitute the 21st Century Transition. As our knowledge improves, our challenges, too, will be refined and added to.

14

A PERFECT STORM

OR A TIME in the 1990s, there seemed to be no limit to the inventiveness of American entrepreneurs or to the funding being made available to them. There was a chain reaction of new ideas feeding on themselves. The Berlin Wall had come down, and it became fashionable to say that we would build a world without walls. The Internet symbolized the ability to pass through walls everywhere. The inflexible power structure of the communist world had imploded, and inflexible enterprises in the capitalist world came under attack by new corporations driven by new people with new ideas. The culture designed to produce new ideas has been wonderfully effective in places like Silicon Valley and within Sony and, perhaps, among 50 million people, but such a culture of innovation is absent among 99% of the people on Earth. Can young people in developing countries become entrepreneurs? Can the enthusiasms and energy of America in the late 1990s be transferred to the Transition Generation everywhere?

FLAT

The high-tech boom of the late 1990s was an overture to the 21st century. It demonstrated that the whole world could become Internet interconnected. Business will increasingly depend on real-time computer-to-computer links among corporations. A person placing an order with a local phone call in Sydney may think he is talking to an order taker in Sydney, but the call has actually been automatically routed—to a person in Sri Lanka, who enters it into an order-entry computer in France, which triggers manufacturing-planning software in New York to place items into a manufacturing schedule in Singapore. The robotic factory in Singapore uses circuit boards from Shanghai that have chips from Japan built into them just in time for final assembly in Singapore, after which a fast software modification is made on the board in Bangalore. The Taiwanese American Express office arranges shipment to Sydney on a Malaysian airline.

Thomas Friedman described this global use of people, factories, laboratories and other facilities in his book *The World is Flat*.[1] A company can decide to employ people anywhere on the planet where labour is the cheapest or the talent is the best. Wal-Mart is one of the most efficient companies at aggregating orders taken at its numerous stores and organizing factories to manufacture the most economic batch sizes and deliver goods just in time. This minimizes the inventory-holding costs in the stores and enables stores to respond quickly to what customers want to buy. Wal-Mart used to be a company making goods mainly in the United States and emphasizing to the public that it created employment for America. Now, 80% of all goods that Wal-Mart sells are made in China, and most of the rest are made in other cheap-labour countries. In this way, Wal-Mart can sell its shoppers goods at the lowest prices possible. Computer networks ensure that goods made far away are delivered "just in time."

Many companies have their accounting done in India. Companies creating complex software for corporations employ a mix of development skills, where the strategic planning is done locally and the coding is done in India. In the 1990s, it was thought that Indians were good programmers; it has since become clear that they are also good at management, film direction, advertising, creative design and research—in fact any form of knowledge work. Even legal work is being outsourced to India.

Fortunes will be made faster in an economy based on knowledge work than in an economy based on physical goods, because knowledge capital can grow much faster than physical capital. Unlike physical goods, knowledge can be replicated endlessly and transmitted to other places quickly.

Great wealth doesn't come any longer from steel, diamonds or conquest of territory; it comes from the mind. The megarich of an idea economy are people who are masters at putting new ideas to work, like Bill Gates and Michael Dell. Value resides more in the intellect than in traditional resources such as capital, land, materials and labour. This changes the map of the world. Ideas, unlike the resources of the classical economy, can come from anywhere. They are more likely to come from an area with good universities. India created its Indian Institutes of Technology (IIT), and these had a large effect on creating an idea economy earning substantial foreign income. The massive poverty of India is now beginning to be transformed.

Unfortunately, the world is not flat; only part of it is.

Developing nations are on a ladder that they can steadily climb. An increasing part of their economy is doing outsourced work from the rest of the world because their wages and salaries are low. Many countries are not on such a ladder. They are so poor that they can't reach the bottom rung of the ladder. To get the tragic destitute nations to the bottom rung of a ladder of development, an infrastructure is needed. This is becoming much easier to build. The poorest areas often have no electricity service or telephone lines. Now they can have wireless phones, solar panels, wind generators and, soon, fuel cells. Large power stations with an expensive electricity grid will give way to localized, small-scale, self-sufficient generators. Radios and mobile phones can be operated without batteries, by cranking a handle. MIT has designed a $100 personal computer with a crank handle for charging the battery. A hand-squeezed flashlight is a pleasant device to use (I use one). There's a drive to bring wireless Internet devices to the poorest parts of the world.

To combat malnutrition, changes in farming practice are needed. But at the simplest level villagers can be taught to grow plants high in vitamins. Uganda has vitamin A deficiencies; many children die from it and some become blind. We explained to a bright young Ugandan that the plant coriander grows like a weed and is rich in vitamin A. Together we found where we could buy coriander seeds on the Web, and wild with enthusiasm, he started a campaign called Coriander for Uganda. A friend who travels in such countries carries as many working BlackBerrys as he can, with wireless Internet links, and gives them to children when they come begging.

Even though there can be an infrastructure for the poorest nations, no nation is likely to develop if its per capita income is less than $1 per day. People can barely stay alive on such an income, and in the future, the price of food will increase. Such countries need enough official development assistance from richer countries to reach out for the lowest rung of the development ladder.

SMALL AND BRILLIANT

As organizations grow old, they become increasingly entangled in their own procedures. Layer upon layer of procedures are added to solve problems that better design would have prevented in the first place. Ageing organizations become inflexible and bureaucratic, with everybody protecting his own job. An economy in which new ideas are rampant is filled with David-and-Goliath stories. Small and nimble corporations attack old arthritic corporations. The old corporation, like Goliath, often reacts with scorn of the newcomer, rather than with appropriate caution. David uses new technology, new processes and excited young people. Goliath is loaded down with the baggage of its past. It often has heavy investments in what is by now the wrong technology. It has old cultures with cumbersome hierarchical structures. The small new companies don't have expensive offices; some employees can work from home. Key players may live in different cities but be linked electronically. A small company can be a virtual company.

In the closing days of the 20th century, a bored undergraduate in his college dormitory wrote software that allowed anyone to download digital music files from other people's personal computers. He took the common practice of swapping and duplicating favourite tunes on cassette tapes and made this doable on the Internet. The service was called Napster, and in its first year of operation, more than 60 million people (20 times the population of Los Angeles) were swapping popular music without paying for it. Napster caused a furor because it threatened the foundations of the deeply entrenched music industry. The music industry eventually stopped Napster with massive lawsuits. Many Internet users were furious about this. Big business was attacking their freedom! Hackers set out to create stealth Napsters—which, they thought, couldn't be sued, and the entertainment industry fought back. But then the world of the iPod emerged.

The motivation of kids who created the music revolution was not to make money; it was to provide digital freedom. They had the conviction that freedom of information must be built directly into the networks. You should be able to find and download whatever you want, the young people felt—low-budget films, education, medical know-how—free of high charges, free of censorship and free from attacks by lawyers. In a similar way, many people in the Transition Generation will innovate *because it's the right thing to do*. There are so many ways they can help solve the problems of the planet.

The music industry story is an American story, but now the world is catching up. Increasingly in the future, the ideas that break the rules will come from anywhere. The motto of entrepreneurs worldwide should be *Small, Brilliant, Changeable and Flat*—"flat" meaning that the components or the competitive target could be anywhere in the world. Instead of being in Palo Alto, new corporations that change the world could be in Beijing, Bangalore or tiny Bermuda. The insurance industry was changed in Bermuda. The computer consulting industry was changed in Bangalore. Wal-Mart was changed in Beijing.

Often the most interesting David will be in a foreign location where a Small, Brilliant, Changeable and Flat group is inexpensive, and young smart people tend to be both inventive and altruistic. They love to attack the Goliaths.

CHANGEABLE

Friedrich Hayek, the Nobel laureate economist, argued for much of his life that centralized economies, such as that of the USSR, must lead to cata-strophically bad management, because central planners cannot have all the detailed knowledge they need to run the economy—it is far too much. Two decades before the USSR disintegrated, Hayek argued that the disintegration was bound to happen. He said that it is a fatal conceit for the central planner to believe that the knowledge necessary for central planning can actually be accessible. Planners can acquire only a summary knowledge about their human subject matter. By the time that that knowledge is com-piled and analysed, it's no longer current. People's motives are known only to themselves and, even then, not properly understood. The basic source of human behaviour—the human mind—eludes central inspection.

Central planners do not, and cannot, use all the dispersed information embodied in millions of individuals. Instead, they try to compress individ-uals into formulas and use far less information than that which animates the actual economy. Hayek demonstrated that central planners inevitably have to disregard knowledge that is vitally significant in the *real* world.[2] The solution to this is to design a system that lets the public decide what it wants and support this with appropriate mechanisms. It has been difficult to explain to power-hungry bureaucrats that the best government may be a shift from centralized decision-making to marketplace decision-making.

In recent years, complexity theory has become a much-studied disci-pline. With computers there have been detailed examinations of complex systems and how they can be controlled. These studies seem to support Hayek's intuition. A basic lesson from the studies is that it is difficult to manage most complex systems centrally, and the faster events change, the harder it is to make central control work.

A major field of study is that of CAS (complex adaptive systems), which continuously adapt their behaviour depending on circumstances. A CAS consists of many individual units, each doing its own thing, obeying certain rules that constrain its behaviour. City traffic is a CAS. Drivers are not following orders from the centre; they drive as they wish. Although nobody is managing the drivers, the traffic exhibits predictable patterns.

The units of a complex adaptive system may be traders on Wall Street, users of the Internet, creatures in the ocean, plants in a jungle, antibodies in an immune system, corporations in an economy and so on. The behaviour of a CAS is the collective result of such individual units each using its own initiative in the context of established rules. If we change the rules, the system behaviour changes. For example, the rules of the International Monetary Fund have a major effect on the economies of poor countries, especially in times of economic crisis.

A marketplace system uses *knowledge* from members of the marketplace to help control operations. The best marketplace systems adapt very sensitively to buyers. Wal-Mart, for example, monitors everything that is bought in each of its stores and uses this information almost immediately to decide what to manufacture and distribute to its numerous stores. This enables it to get the goods customers want on the shelves for the lowest possible price. It is much more effective than members of a central committee deciding what to manufacture and distribute.

A powerful form of decentralized control is to allow participants to use not only their knowledge but their creative initiative. We need "distributed initiative systems" that develop because of the initiative of many participants. In many corporate organizations there has been a shift away from top-down command-and-control towards decentralized-initiative structures.

When governments decided that it was their job to lessen pollution, they did so by setting limits and often decreeing what technology must be used to control pollution. This tended to lock companies into existing technologies when new technologies were needed. More recently, distributed-initiative schemes have been formed to attack pollution by creating tradable permits. A government decides on acceptable levels of pollution and allocates credits to companies for achieving that level. Some companies are able to cut pollution to below the legislated level, and they have credits to sell. Other companies have difficulty achieving the level, and they buy credits from those that have extra. A vibrant market can come into existence, and once it exists, much ingenuity is expended. There are numerous ways to replace the inflexibility of the command-and-control approach with approaches that encourage local innovation.

If you were in a rubber raft rushing into intense white-water turbulence, you wouldn't want a slow centralized bureaucracy controlling the raft. You'd want excellent training so that you could control it yourself; that's the only sure way to survive. The change from central control to local control means that large numbers of participants become trained in how to control the raft. Humankind is heading into a time with two momentum trends: *growing complexity* and *increasing rates of change*. In a world with these two characteristics, central control becomes increasingly unworkable. The environment will have to be designed so that localized units with skill and initiative call the shots.

Having said that, it's still necessary to see the big picture and understand it. There are certain large-scale issues that need to be understood centrally. The big picture may indicate that the world is moving towards disaster or that one of our long-term directions is untenable. There needs to be strategic planning based on the big picture, followed by distributed initiative to implement the plan. Governments can set objectives and rewards for accomplishing objectives, but the objectives are usually best achieved by individuals or competing organizations using their own initiative.

The study of complex adaptive systems shows us that many types of change have remarkably little effect on a complex system—the system is persistent in its behaviour. Some changes, however, can completely transform its behaviour. Computer models show that *sometimes a very small change in the rules can cause a massive change in overall behaviour.* This relates to our comments on leverage factors. Finding the right leverage can work wonders, but the most powerful leverage factors are often far from obvious.

A DIFFERENT TYPE OF PRODUCTIVITY

Economic progress has been concerned with productivity. Almost every year since the invention of the steam engine, productivity has increased in the leading economies. At certain times, it has risen steadily by 3 or 4%

annually; at other times, it has risen more slowly. Developed countries have seen their real wealth increase by more than a factor of 10 each century. This will also be true in the 21st century.

Increasing wealth in the past, however, has been related to increasing consumption of natural resources. The 21st century is different because many of the resources needed will be in increasingly short supply—water, for example. So, the economy will have to use these resources much more efficiently or find substitutes. Increasing efficiency in the use of the Earth's resources is vital to the future—critical to achieving a viable world as the population surges toward its maximum.

The economist's measure of productivity has been *labour productivity*— the amount of goods produced per person hour (in effect, the wealth produced per person per hour). This is what has been going up each year. A fundamentally different measure is needed: *resource productivity*. This is concerned with the amount of wealth produced from one unit of natural resources. As we approach limits in planetary resources, it is important to find appropriate measures of resource productivity and to set targets and incentives for improvement.

The evolution of industrial society was generally designed to improve labour productivity but not resource productivity. As a result, we have low resource productivity. Most rainwater runs down the drain, and other resources are wasted in numerous ways. The improvement in labour productivity has been a slow but steady journey. It typically takes about three decades to double the wealth generated per person. Improvements in resource productivity will come much faster because they start from a low

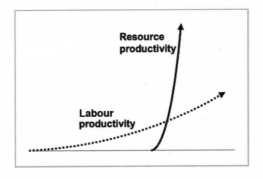

base. Until recently, we behaved as though the Earth's resources were free for the taking. There was no incentive to use them frugally. Because of this, there are now major opportunities to make improvements; many actions can be taken quickly.

FACTOR FOUR

One of the great champions of resource productivity is Amory Lovins. He and his colleagues at the Rocky Mountain Institute contend that, to stop the destruction of the planet, we need to exercise what he calls the *factor of four*: We need to cut by half our use of critical resources and, at the same time, roughly double the total value obtained.[3] He cowrote a book called *Factor Four: Doubling Wealth, Halving Resource Use,* which is packed with detailed, real-world examples of what the title advocates. It showed how this factor-of-four improvement can be replicated on a large scale. His work has produced hundreds of examples of such improvement.

The road to sustainability can be achieved in mundane areas, like the transport of food to your home or the efficient use of water, and in less mundane areas, like hydroponic farming or using computer networks for tightening up logistics. Of course, resource productivity can be improved by using new technologies—fibre optics, fuel cells, fuel-efficient cars, fourth-generation nuclear power. However, most of the examples in the *Factor Four* report don't even need new technology. They relate to things like insulation in buildings, more efficient use of transportation and elimination of waste. Common sense plays a larger role than high technology. Lovins's work achieves repeatable results with today's tools in today's industry. There's massive scope for a four-times improvement in resource use without exotic technology.

There are three main types of resource productivity: energy productivity, materials productivity and transport productivity. Energy productivity is concerned with achieving desirable results with substantially less energy use. Materials productivity refers to achieving results with fewer materials—avoid materials, find substitute materials, build much smaller products, or build products with materials that are recyclable. When fewer materials are used, we save the large amount of energy that would be required to mine and produce materials from ore. Transport productivity is concerned with lessening or avoiding the transportation of people or goods as we carry out desirable activities. Often processes can be reorganized so that goods or materials travel far shorter distances and are routed with more efficiency. Trucks don't make a return journey empty, for instance. The Rocky Mountain Institute describes how a fourfold increase can be achieved in each of the three types of resource productivity.

The authors of *Factor Four* attempt to calculate the total waste of global resources occurring today. They estimate that, in the United States alone, at least a trillion dollars a year is wasted unnecessarily. Worldwide, it is several trillion. The waste occurs because there has been no consistent determination to use Earth's resources efficiently. Society sought labour productivity, not resource productivity. Most of the resources wasted are global common goods; thus, the waste makes everyone poorer. It imperils our environment and creates very expensive problems for the future.

Paul Hawken, coauthor of *Natural Capitalism: Creating the Next Industrial Revolution,* estimates that 99% of the original materials used in the production of goods made in the United States becomes waste within six weeks of sale. When a typical incandescent lamp is lit by a big power station today, only about 3% of that energy actually powers the lamp; 97% is overhead and losses of various types. Eliminating such waste is part of what the *Factor Four* authors refer to as "a new industrial revolution."

At one time, up to two-fifths of the electricity being consumed in the United States was being wasted due to the inefficient design and installation of cooling equipment. Much can be done to lessen this massive waste. A major aspect of "green" buildings is to design them so that they need less cooling in summer and less heating in winter. This can be done by careful positioning of the building. It can use windows with automatic shutters, rooms that are underground and, hence, naturally cool in summer, modern thermal glass, automatic airflow, intelligent insulation and so on. My own house is by the sea, and seawater is used to help cool it in the summer and heat it in the winter because the sea has smaller temperature changes than the atmosphere. A beautiful seawater pond helps in this process.

Unfortunately, it seems likely that unnecessarily wasteful air-condition-ing will continue to be one of the cumulative factors worsening the stress of the midcentury canyon. In China and India, there are major efforts to keep cool with little power use. Much of the ingenuity in resource saving will come from countries that can't afford American-style air-conditioning. In the past, the power companies built massive coal-burning power stations and expensive electricity grids. Today's economies increasingly favour smaller localized units and small towns generating their own power. This will become more appealing when new energy technologies become more widespread.

FACTOR 10

Amory Lovins and his disciples demonstrated how a four-times improve-ment in resource productivity can be achieved *today,* in many different fields. When authorities studied this, however, they concluded that halving the use of the Earth's resources is not enough. In many cases, we'll need to cut their use to a quarter, while at the same time substantially increasing wealth creation. The Factor 10 Club is a group advocating that, with better technology, humankind can now strive for a ten-times improvement. Such an increase, they calculate, is needed in the future to achieve long-term sustainability.

Along with new technology, a factor-10 increase needs major changes in government policy, subsidies and tax structures and in the operation of institutions such as the World Bank and the World Trade Organization. It is essential that corporations are financially motivated to use resources with long-term global sustainability. The processes for improving resource productivity will create many new opportunities for business.

Some countries, mostly in Europe, have adopted Factor 10 as a strate-gic goal. Some are preparing an information campaign to help small and medium-sized businesses design eco-intelligent products. Some leading corporations, such as Dow Europe and Mitsubishi Electric, regard the adoption of factor-10 ideas as a powerful strategy to gain competitive advantage.

Most people don't save resources because it is a good thing to do; they save them if it affects their pockets. Many government regulations and subsidies, though, encourage the opposite of resource productivity. Tax concessions for logging accelerate deforestation. Pesticide subsidies encourage excessive use of pesticides. Electric, gas and water utilities are often strongly rewarded for increasing the amount of electricity, gas and water that they sell. The more electricity is wasted, the more money electricity companies make. Companies are often effectively penalized for promoting efficient use because it would lessen their total earnings. Regulations often prevent a manufacturer's lorries from carrying goods for other organizations when they are empty even though today's Web-based distribution systems can help ensure that trucks almost never have to travel empty. A factor-10 world would require that many of today's government regulations be swept away.

An important part of the future will be that factor-10 capabilities are integrated with the drive for eco-affluence. This integration may use high-bandwidth digital products. The latest fashion goods may be resource-frugal. Families and friends will stay in contact by means of video screens. The largest industries will be knowledge-based. Entertainment and education will be linked. Fuel-cell cars will be simple and elegant while piston-engine cars will be part of a clunky past.

Today much food in the United Statese comes on a thousand-mile journey. It's part of an era of mass systems—mass production, giant factories, big warehouses and big lorries on interstate highways. In the age of advanced automation, the economics change. A local bakery can make fresh products inexpensively and deliver them to your home. Local organic farmers will be concerned with achieving gourmet results. The tastiest strawberries will come from computerized hydroponic units. The world is changing from mass systems with brute-force use of resources, to computer-intensive systems that meet local needs and quality lifestyles.

Just as the Victorian era evolved from steam power and belching chimneys to electricity, the 21st century will evolve from massive centralized facilities to an era of knowledge-intensive networks that allow customers to be served efficiently from physical resources nearby and information resources far away—an era of small machines, nanotechnology, hydrogen fuel cells and eco-affluence.

REINFORCING TRENDS

The film *The Perfect Storm* describes how independent sets of extreme conditions coincided to bring on a uniquely bad storm. The opposite situation can happen: independent sets of extremely *beneficial* conditions can coincide to bring a storm of improvement.

An alternative to anger and unrest among youth in developing countries is entrepreneurship and excitement about the 21C Transition. Rich countries should be helping and enabling young people to exercise the ideas they have. Distributed initiative will spread: Vast numbers of people, especially young entrepreneurs, contribute to ideas. The planet is becoming wired and computing is becoming inexpensive so that work can be outsourced, as it is to India. Distributed-initiative systems can be global. The best management styles of today set the stage for unleashing the capability for innovation latent in most people. Goliath corporations will constantly be under attack from entrepreneurs with better ideas, and the fresh thinking can come from anywhere on the planet. The decades ahead will be a time of massive increase in what most humans can achieve.

As the West encourages democracy and open markets, as it has done successfully since the USSR collapsed, it should be vigorously encouraging entrepreneurship and taking action to help young people obtain the capital they need to start ventures ranging from opening a coffee shop to growing a software factory. When Singapore, South Korea and Hong Kong were transforming their economies, there were posters in public places telling young people that, if they want to get ahead, they should get an education at an American university and then come home. America wasn't the Great Satan; it was the great hope for the future. America needs to return to those days.

This is a time when major forces are coming together, like the perfect storm. These forces, working together, will bring huge changes and need to be encouraged:

- There's a furious rate of new invention.
- Entrepreneurship is becoming global and needs to be vigorously encouraged. It can offer great hope and excitement to youth in the Third World.
- Corporations are being re-engineered for high-bandwidth globalism.
- Scarce resources will be used far more efficiently, with factor-10 improvement.
- The Transition Generation needs to be trained to understand the meaning of the 21st century so that they can apply their abundant energy and enthusiasm to helping bring about the 21C Transition.

15

THE VITAL ROLE OF CORPORATIONS

AS WE CREATE VISIONS of a better world, we need to ask how to get there from here. Vast amounts of rebuilding and redesign are needed both to achieve simple sustainability and to evolve to the high-quality-of-life civilizations that are now possible. Corporations, large and small, are by far the most effective type of organization for making things work. Government may set goals, but corporations get the work done. Huge amounts of money will be needed to correct what is wrong today. Most of that money is likely to come from profitable corporate opportunities. For example, to lessen climate change, we need new types of cars, new methods of generating power and new building architecture. Antiterrorist measures need new types of high-tech security. Numerous new ideas invariably come from entrepreneurs. The role of corporations is critical for the future.

Some people are vitriolic about corporations. They say that they have become massive, antidemocratic, global and out of control. They have their own agenda, the protesters say, far removed from the planet's needs and geared only to profit. Corporations have become skilled at manipulating global bodies such as the World Bank, the International Monetary

Fund and the World Trade Organization so that these organizations tend to support corporate needs more than the larger, more important global needs. Corporations funnel profits into their home countries, and most powerful corporations are based in rich countries.

While there may be much truth to these complaints, corporations are just about the only human organization capable of achieving the complex and difficult tasks that are ahead. It is vital to find ways to make corporations take a greater, more responsible role in world affairs and in solving the problems we describe, and there are many ways to do it.

THE TOP TRANSNATIONAL CORPORATIONS

Wealthy countries give aid to poor countries, but this aid is remarkably small compared to the investment that poor countries obtain from multinational corporations. The largest 500 corporations in the world account for about 90% of the world's investment in other countries. The aid given by governments to poor countries is much less than what those countries could earn if they sold more goods abroad. One of the best things that rich countries can do for poor countries is to help them develop trade.

As the world "flattens," corporations are increasingly wired across the globe so that they can obtain on-time delivery of parts or goods from other countries and, in turn, sell to other countries. The parts needed to build cars arrive *just in time* at the car factory. Worldwide computer networks—which link suppliers, subsidiaries, partners and customers—are becoming ever more intricately engineered to maximize value. This is also true for the service sector.[1] Ten years ago such computerized multinational interaction occurred mainly in North America, Europe and Japan. Now it is spreading to all industrial countries and is vigorous in China.

A corporation is constantly judged by its shareholders; so, it must focus on maximizing its profits and increasing share value. A company selects the countries in which it operates based on how well it can help them reach this overriding goal. Often, it can make no money by operating in the poorest countries; so, it avoids them. Sophisticated countries that can afford and operate reliably in the computerized value chains do well, but the poorest or most troubled countries don't have a chance. Vast, impoverished sections of humanity are ignored because they cannot contribute to shareholder value—a severe problem with globalization.

The largest 200 corporations create 30% of the total gross domestic product of the world, yet they employ fewer than 19 million people. The high-tech sector of India is spectacularly successful, but it employs only about 2 million people out of 1.1 billion.

Rising corporations join the top 200, and other corporations drop out of it, but the top-200 bloc continues to have revenues greater than any country, thus wielding immense financial power. Some of society's major planning functions are shifting from governments, which are accountable to citizens, to corporations, which are accountable to shareholders. The top transnational corporations are becoming far-reaching global organizations driven not by the wishes of global citizens but by the tough mechanisms with which global profits are maximized. Their strategies can be secret. Collectively, they are an undemocratic organism of great financial power. The oil industry, for example, is larger than all but a few countries.

There is a concern that big corporations will become too powerful because they often have growth rates (including takeovers) much larger than those of countries.

MORE POWERFUL THAN GOVERNMENTS?

It is a mistake to compare corporate revenues with countries' GDP. Various authorities have pointed out that only 30 or so countries, today, have a GDP larger than Wal-Mart's global revenue. If we put national GDP and corporate revenue in the same list, more than half of the hundred largest entities on the list are corporations. If today's corporate growth rates and mergers continue for 20 years, it is likely that only four countries will have a GDP larger than the revenue of the largest corporation.

Corporate revenue, however, is a fundamentally different type of measure from national GDP. GDP is a measure of the total goods and services that are sold to the final customers in a national economy. Corporate revenues include many goods not sold to the final customer. If a car company buys an engine, the cost of the engine is included in the price charged for the car when the car is sold and, hence, in the car company's revenue. The engine company also includes the price of the engine in its calculation of revenue. Companies making components of the engine include the price of those in their revenue and so on. Thus, the total revenue of all companies in a country is much larger than the country's GDP.

General Motors's revenue may be bigger than the GDP of Denmark, but the number of employees in GM is much smaller than the number of workers in Denmark. The total of salaries and wages in GM is much smaller than the total of salaries and wages in Denmark.

It is also a mistake to compare the power of corporations with the power of governments. They are very different. Corporations are preoccupied with profits, growth, survival, competitive battles and the very complex issues of global management. They are far too busy to deal in a meaningful way with the issues a government wrestles with—social and cultural matters, schools, welfare, police activities, defence and noneconomic areas in general. Nevertheless, in advanced economies, corporations, collectively, have more impact on people's lives than government. They are the provider of goods and the main provider of employment. Corporations choose where to open factories, shops, research facilities and training centres. They can move critical people from one country to another or conduct a brain hunt and entice away the brightest people. A corporation can shut down its operations in a country with a single decision by its board of directors.

COMPETITION

A vitally important question about human organizations is, "How can they be kept on their toes?" The best answer is usually "competition"—multiple corporations trying to do the best job of pleasing customers. The good ones grow; the worst disappear. Corporations strive to cut costs, find innovative technology and invent new ways to please customers. Competition is essential, and to ensure that it is effective, anti-trust laws are needed, along with rigorous prevention of insider trading and enforcement of independent auditing. The rules of the game need to ensure fair play. In addition, specific rules are needed for specific industries, such as the rules for the US Food and Drug Administration.

Many left-wing spokesmen preach against competition or free trade. Economic Darwinism, they say, is too brutal and leaves too much social wreckage—we need cooperation, not competition, they say. But where no competition exists, the best of human intentions seem to go wrong. Organizations without external competition steadily develop their own agenda, internal politics and inefficient procedures, until the usefulness of the organization shrinks. Some government organizations in the nondemocratic world seem destructive and pointless and nobody seems able to correct them. Great dictators have usually abused their power, and small bureaucrats can carve out a power niche for themselves that is outrageously divorced from any real value.

The collapse of the USSR was sudden and utterly devastating. Russia, one of the best-educated countries in the world, had great scientists and a formidable military. In the 1990s, the brilliance of Russian scientists was wasted because they had little or no link to an efficient marketplace. It would have been unbelievable in 1990 that 10 years later Russia was to have a GDP per head much lower than Malaysia's.

Twentieth-century economies have taken many forms, but those based on ideologies without free competition have usually bitten the dust. The strongest economies are based on open competition, and it will be even more important in the future—because the faster the rate of change, the more it is needed. Joseph Stiglitz, winner of the 2001 Nobel Prize for economics, was involved in the attempts to rebuild shattered economies in the 1990s. He concluded that competition was critical whether corporations were private or not. *Competitive enterprise was essential; private enterprise was less so.*[2]

The concern about corporations being too powerful ought to focus on preventing monopoly. If power corrupts, so does monopoly. A monopoly can charge unreasonable prices, abuse customers and behave with indulgent management whims. A corporation is a viable institution if it operates in an environment of efficient competition.

Globalism will often lessen the danger of monopolization because new challenges can come like a bolt out of the blue from faraway countries. In the early 1970s, Detroit believed that no foreign firm would achieve a substantial share of the US car market, but Japanese and Korean cars firms did exactly that. Chrysler and Ford skated on the edge of bankruptcy in the 1980 time frame, as did General Motors in 1992. The future needs radically different cars, powered by fuel cells. Such innovations may come first from an upstart car industry rather than today's giants. If there is an open global marketplace, globalism will increase the number of potential competitors.

On rare occasions a noncompetitive organization does well if it has a precise and tough goal to achieve. One sees this in war. In the 1960s, NASA (the US National Aeronautics and Space Agency) was the most excellent of organizations because it had the clear and immensely exciting goal of landing a man on the moon by 1970. After the last moon mission, however, NASA became a huge ship without a rudder, vaguely trying to preserve its massive funding by providing circuses for the TV-watching public. Like other government bureaucracies, it became a cash-burning monster that achieved little.

Government departments are generally noncompetitive and often drift into forms of nonproductive behaviour. Recently there have been efforts to make government departments compare their own costs with the costs of doing the work on the outside and select the lowest-price bids. In the 1980s the US Department of Defense spent the staggering sum of $30 billion per year on software. The costs per line of code of internally developed software were outrageous compared to those developed by competitive organizations. I was on the Department of Defense Scientific Advisory Board concerned with software, and it seemed obvious to us how to reform the situation, but no reform happened.

As we look at the solutions necessary for dealing with problems described in this book, many are not happening. There are numerous institutional roadblocks that keep them from happening. It is essential to examine these roadblocks and determine how they can be removed.

NONPROFIT COMPETITION

Competition between corporations has usually been based on the profit motive, but it doesn't have to be. There can be multiple criteria for judging success and lessening the influence of organizations that don't do well.

Some measure of excellence is needed so that organizations compete on the basis of this measure. Competition based on excellence produces interesting results.

Nature has achieved the most beautiful results. It doesn't have financial markets, but it has the most diverse and subtle forms of competition. Nature never normally has noncompeting organisms.

Enterprises without the profit motive need intelligent goals along with mechanisms for making competing organizations progress towards those goals. A manager needs to specify the desired behaviour he wants and establish measures to indicate how well the evolving organization performs. There must be some means of assessing this measurement so that the organization can adjust its behaviour.

Schools compete to attract students on a nonprofit basis. Universities strive for excellence, trying to attract the best professors and students. People select their doctor on the basis of how good he seems to be. Specialized clinics, such as cancer clinics, strive for the highest success rate and reputation. A much wider range of competitive behaviour is possible. Any measure of goodness could potentially be the basis for competitive comparisons. We are now beginning to see new measures of corporate behaviour, including ratings based on sustainable development, or avoidance of pollution. As concerns about correct behaviour grow, such ratings will have a growing influence on where investors and customers choose to place their money.

CORPORATE IMPACT ON ENVIRONMENTAL CORRECTNESS

In spite of many recent television images of CEOs in handcuffs, most CEOs want to be good citizens. The executives who attend the annual World Economic Forum in Switzerland (the Davos set) know that sustainable development is a vital issue, critical to the future of the planet. It would become an urgent political issue if the public understood it better.

In the early 1990s, a Swiss billionaire, Stephan Schmidheiny, organized the Business Council for Sustainable Development, consisting of the CEOs of 48 of the world's largest corporations. In 1995, this council merged with the environmental arm of the International Chamber of Commerce to become the World Business Council for Sustainable Development (WBCSD). By the start of the 21st century, the WBCSD was a highly influential coalition of over 150 leading international companies, and it advocated a shared commitment to the principles of sustainable development. It links to 30 national and regional business councils and partner organizations involving some 700 business leaders globally.

WBCSD's mission is "to promote business leadership as a catalyst for change towards sustainable development and to promote eco-efficiency, innovation and responsible entrepreneurship." It publicizes leading-edge practices that corporations use to achieve these ends, develops policies for sustainable development and establishes ways in which corporations can be measured and rated.

The concept of eco-efficiency is at the heart of the WBCSD philosophy. Eco-efficiency is concerned with ways to create more value with less environmental impact. The WBCSD provides a forum where business leaders can exchange experiences on eco-efficiency. When ecological design rules are followed, it is possible to create products that are frequently cheaper to produce and use, smaller and simpler in design and easier to recycle. A growing number of companies now recognize that eco-efficient practices can improve the bottom line.

Interface, Inc., a carpet company with 26 factories in four continents, is one. Its CEO, Ray Anderson, set out to make his company eco-efficient in every way he could. For example, he leases carpeting to customers rather than selling it so that carpeting is assembled from standard-size squares that can be recycled. As Interface became increasingly eco-efficient, it grew vigorously in revenue and earnings.[3] Anderson believes that most corporations could profit from their own versions of becoming eco-efficient.

At the start of 2006, General Electric announced an extraordinary change in how it will be run. The managers of every GE business unit worldwide will be held accountable for measures relating to GE's impact on the planet—in addition to the usual business measures such as return on capital. This may be the beginning of a fundamental change in which corporations take responsibility for environmental correctness.

GE has set targets for reducing its emissions of greenhouse gases that are much tougher than the Kyoto targets. It stated that it would more than double its research expenditure on clean technologies—from $700 million per year to $1.5 billion—by 2010. This would include research on fuel cells, hydrogen storage, nanotechnology and clean aircraft engines. Before planning this radical change, GE had held many sessions with customers to determine what they might want to buy 10 years from now. It concluded that climate change would cause increasing alarm, and pressures would grow to stop harming the planet. GE will lobby governments to ration carbon emissions and will persuade its customers that they need the new "green" products.

GE might be the first swallow of summer.

THE POTENTIAL IMPACT OF FINANCIAL MARKETS

The stock market has a massive influence on the behaviour of corporations, an influence that is sometimes harmful, because it may make a corporation so intensely concerned with the next quarter's results that it is managed for the short term, not the long term. Now an interesting change is starting to occur. A growing number of people who want to own shares in corporations are becoming concerned with whether those corporations behave in an environmentally correct manner.

Many of the flower children of the 1960s now have big suburban homes and BMWs and care about what is happening to the planet. They won't chain themselves to a redwood tree or join a riot when the World Trade Organization (WTO) meets, but they are heavily invested in the stock market and want to feel that they own stocks of companies that are not doing harm. They tell their broker, "No tobacco companies," and they ask if they can invest in fuel cells.

Responding to this trend, a new type of mutual fund came into existence in the 1990s, known as SRIs (socially responsible investments).

These funds didn't do well at first. There was a view on Wall Street that if you pursue green companies, you are going to pay a financial penalty. That view changed dramatically in the 1990s, however, because some of the socially correct funds performed much better than average funds. By 1999, nearly 3 trillion dollars was invested in stock that was "socially screened." By 2003, there were about 200 SRIs in the United States.

The corporate world has a huge effect on the environment, and it is critical to search for ways to influence its behaviour. Most corporations respond to their investors; so, if appropriate measurements and reporting were in place, investors could influence the behaviour of corporations. This should become normal practice. Corporations will then want to be seen behaving correctly because it will affect their stock price and market capitalization. Managing a company's reputation will become a critical element of corporate management. Improving the capability of investors to invest in companies that help with 21st century problems is a potentially massive leverage factor.

If the World Business Council for Sustainable Development had its way, all corporations would follow a detailed reporting procedure that would enable them to be rated on a basis of the actions they take towards sustainable development. It has developed guidelines for eco-efficiency metrics and reporting. If all companies had such metrics, investors could decide for themselves whether they wanted to own the company's stock. The growing influence of the financial markets can be harnessed in a major way to help steer corporations towards sustainable development. Managing a company's reputation will become a critical element of management because a damaged reputation equates to damaged finances.

The private sector is likely to be the primary engine of the 21C Transition, which is so critical to our future. Trillions of dollars' worth of new products and services will be needed. Putting mechanisms into place to achieve corporate correctness is achievable and essential.

16

CULTURE'S CRUCIBLE

A WONDERFUL HOTEL in Morocco, La Mamounia, in Marra-kesh, has a colourful garden surrounded by a high wall covered in purple bougainvillea, where French and Italian women lie around in the sun topless and almost bottomless. They can see a high minaret outside the wall, where a Muslim muezzin, overlooking the garden, chants the call to prayer five times a day. He seems to do it with an outraged fervency translatable into a plea for these women to cover their flesh with black cloth from top to toe.

Not long ago, this blatant confrontation was a rare intersection between civilizations. Now it happens almost everywhere. Commerce has become global, advertising blankets the world, film and television compete for new extremes in sex and violence, and anyone can seek out the worst of the Web.

The Earth has different civilizations with very different values—as different as Myanmar and Manhattan; as different as Buddhists renouncing all worldly goods and the get-rich intensity of Bangalore; as different as the

Taliban and the *Sex and the City* culture. Different civilizations used to be separate and isolated, but now the wired world makes them starkly visible to one another. This confrontation puts our planet in new danger. Civilizations that were once isolated from one another are now in one another's face. Cultures with intense pressures for change confront religious fundamentalist cultures that resist change. Global corporations, political evangelism, powerful marketing campaigns and new media are laced around the planet with forceful big-money interests. In many countries, one can observe women in full Muslim attire looking at smart magazines with cover headlines such as "50 Ways You Can Have Better Sex." In the souks of puritanical Muslim countries, one is offered DVDs full of the strongest sexual content. Children thought by their community to be isolated from the world know how to get into Internet chat rooms.

The forces of digital media are bringing people increasingly close. In the next 10 years, global networks will evolve from being narrow bandwidth (like Internet chat rooms) to high bandwidth. Wiring the planet with fibre optics, cellular wireless systems, satellites and digital television is one of the great transforming events of our time. Links of extreme bandwidth connecting previously separate civilizations is new in history. People around the world will steadily become used to the idea that there is complete transparency across what have been cultural barriers.

A particularly misleading image is Marshall McLuhan's description of the world as a "global village."[1] In a village, everyone shares a common culture. The wired world is not unified: Numerous global communities with incompatible belief systems exist side by side. It is like the topless women in Marrakesh lying within sight of the Muslim muezzin chanting the call to prayer. No use of the Internet would give them the same belief system.

Today, global media push everyone close together. The closer we are, the more we notice our differences. Media often distort the interaction. Most of the interaction doesn't come from intelligent two-way dialogue; it comes from the Hollywood commercials or PR campaigns designed to sell products. What the veiled women of Islam know about Western women, and their men, comes from glossy magazines that use sex to increase their circulation.

END-OF-CIVILIZATION WEAPONS

At the same time that we heat up this crucible of conflicting cultures, weapons of mass destruction become cheaper, more dangerous and easier to build. A vital part of the 21st-century journey must be to remove the potential reasons for using such weapons. To avoid being enmeshed in increasingly horrifying conflicts, we must instill, as far as we can, a spirit of mutual understanding and coexistence into the civilizations of the planet.

There are many ways terrorists could turn modern technology against itself. They could attack the banking system with computer viruses, or shut down an electricity grid and attack cities. They could blow up chemical storage tanks or the tunnel under the English Channel.

It used to be the view that weapons for massive destruction could only be made by governments with large budgets, but now one individual, or a small team of people, could make biological or chemical weapons. Timothy McVeigh could make a devastating truck bomb used to blow up a federal office building in Oklahoma City out of materials that are found on a farm. A lethal, genetically modified virus aerosol is easy to hide. A renegade creator of biological weapons may not be detected. The potential death toll from such weapons will increase at the same time that the cost of producing them drops.

Enriched uranium, the fuel for an atomic bomb, can be hidden in cities, and if one had enough enriched uranium, it is easier than people think to make a simple atomic bomb—like the one that destroyed Hiroshima. Fortunately, a sufficient quantity of enriched uranium (or plutonium) is difficult to acquire, but some criminal organizations have been obtaining it and secretly offering it for sale.

CLASH OF CIVILIZATIONS

In *The Clash of Civilizations and the Remaking of World Order,* Samuel P. Huntington describes today's confrontations between "civilizations" and how these have become dangerous forces on the world scene.[2] Huntington concludes, "In the emerging era, clashes of civilizations are the greatest threat to world peace." To prevent war between civilizations, he says, there needs to be a new spirit of understanding about what is common and different among civilizations. The only chance for survival is peaceful coexistence, but many cultures resist coexistence.

In past history, "civilizations" have been confined geographically, have had high degrees of autonomy and have tended to have very limited contact with other civilizations. Before the age of colonialism, many civilizations had no idea that others existed. If they came into contact with a different civilization worshipping a different God, they were suspicious and hostile, like people of one planet confronting aliens from a different planet.

Some civilizations held their religious or cultural views very strongly, and any attempt by a different civilization to interfere with these views would be opposed by war. When Queen Victoria's government tried to impose good Christian behaviour on the Zulus, massed Zulus armed only with spears defeated a British army armed with the latest guns.

This is an era when an individual can decide to join a terrorist network. The Internet enables troublemakers to unite and can stir up the passions of many people. In some villages, extremist seminaries persuade young people that the *next world* is what matters and that they can have everlasting glory in it if they perform a suicide killing. After 9/11, mass murder became an act of heroism.

HIGHWAYS OF LIGHT

Extreme-bandwidth telecommunications are an essential component of the 21st century. We'll have fibre-optic cables of astonishing transmission capacity and high-bandwidth wireless links that connect to such networks. Fibre highways and wireless systems need each other. Wireless systems give users anywhere access to the networks, usually over short distances, and fibre trunks, like motorways, carry enormous quantities of bits around the planet.

Until recently all telecommunications channels used electronics—they transmitted electrons. Now fibre-optic channels transmit photons (light), and photons travel at the speed of light. Electrons travel much more slowly because they constantly bump into molecules, as if they were bumper cars at a fair.

Fast electronic circuits can operate at around a billion hertz (cycles per second), but light has a frequency around a thousand trillion hertz. Lasers can transmit at even higher frequencies than that—frequencies beyond the spectrum of visible light—around a hundred thousand trillion hertz (10^{17} hertz). The theoretical maximum bit rate of a fibre is thus in the region of a hundred thousand trillion bits per second. Practical engineering cannot get close to this yet, but fibres are already in use that can transmit more than a trillion bits per second. Technology will soon push that up to 10 trillion— a number too far beyond our imagination to know how to put to use.

Optical fibres make a difference in three ways. First, speed. A signal can go to the other side of the planet so fast that it seems instantaneous. Second, they can transmit awesome quantities of bits. A hair of glass can transmit the *Encyclopaedia Britannica* in a fraction of a second. Third, they are error-free. When data are transmitted with electrons, errors occur because of different types of electromagnetic noise. With optical fibres there is no such noise.

Everything Shakespeare wrote in his lifetime can be recorded with about 70 million bits. We might use the term *shakespeare* as a measure (like a gallon) referring to 70 million bits of data. The machine on which I am writing this has a main memory of about 14 shakespeares. Large data warehouses contain about a million shakespeares. A laser beam can transmit 500 shakespeares per second over today's optical fibres.

This capacity is staggering, but another invention makes it even more so. Laser beams of different colours can be transmitted at the same time over the same fibre. This is referred to as WDM (wavelength division multiplexing), meaning that the bandwidth of the fibre can be divided into many separate wavelengths (colours), and there is a separate laser beam for each. Today, some fibres carry 96 laser beams simultaneously, each carrying tens of billions of bits per second—13,000 shakespeares per second can be sent over one very thin fibre. As the technology advances, the number of simultaneous laser beams will increase to many hundred.

As if this isn't dramatic enough, many of these fibres can be packed into one cable. Some organizations have designed cables to carry more than 600 strands of optical fibre. If each of these carried 96 laser beams of 40 billion bits per second, it would total more than 2 thousand trillion bits per second. We need a new word—a *petabit* is a thousand trillion bits. Cable capacity will eventually be quoted in petabits per second. One petabit is 14 million shakespeares.

All information can be represented by bits—business contracts, television, rules for factory scheduling, engineering drawings, knowledge—everything except love at first sight. Bits can have almost no mass or volume and can be transmitted at the speed of light. The planet is becoming wired to store and transmit near-infinite quantities of bits. Global networks will become radically different from today, with services provided by supercomputers with a non-human-like intelligence eventually millions of times more powerful than human intelligence. The wired world will have properties that will constantly surprise us.

A CRUCIBLE OF IDEAS

Before the world gained this extraordinary connectivity, three sources of order in the world had collapsed: the colonial system, the communist system and the Cold War. However undesirable these systems, they set rules—they told people what to do.

At the same time, management techniques with fixed procedures in many corporations were replaced by techniques in which people were challenged to invent new procedures. It became an age of freedom—freedom from colonial masters, freedom from rigid management, freedom from communism and freedom to have new ideas. Some new ideas were great; many were half-baked. Aggressive corporations want as many new ideas as possible, because that increases the chance of finding good ones. When I interviewed the nanotechnology scientist Bill Parker, he commented that "an individual can marshal the resources with knowledge and science to do what a nation-state could do in the past." It became a time when everything was challenged or reinvented. Governments kicked out their ex-colonial managers. The burgeoning middle class in China, so tightly buttoned down in Mao's day, started a frenzy of multipartner sex. Many women in the Islamic world questioned why they couldn't have the same rights as Western women, and Western women asked why they couldn't be top executives at major corporations. All manner of variations of religions have appeared. The era of designer clothes became an era of designer religions. Unfortunately, some fanatics thought that their religion condoned murder or violence (as did many Christians in the days of the Spanish Inquisition). A man could board a plane to the United States with explosives hidden in the soles of his shoes, try with bumbling incompetence to light the fuse in mid-flight and believe that he was doing God's will. It is a deep irony of human nature that profoundly good religions have been used by fanatics to commit some of the worst evils in history.

For many people freedom is difficult. People not used to freedom want to be told what to do. If they are not told, some develop strange ideas. We have entered an era in which endless ideas coexist. Huntington believes that in spite of the worldwide spread of Western consumerism and popular culture, there is little change in the fundamental beliefs of societies. Innovations in one civilization are regularly taken up by other civilizations without altering the *underlying culture* of the recipient. He writes, "American control of the global movie, television and video industries is indeed overwhelming. Little or no evidence exists, however, to support the assumption that the emergence of pervasive global communications is producing significant convergence in attitudes and beliefs. In due course, it is possible that global media could generate some convergence in values and beliefs among people, but that will happen over a very long period of time."[3] The convergence of beliefs will have happened by the end of the century, but we will pass through a dangerous transformation to get there.

In the wired world, radically different cultures share the same virtual streets and interact in all manner of ways, sometimes stimulating one another, sometimes antagonizing. This crucible of ideas can be full of bulls in china shops. They ride roughshod over ancient beliefs, but the ancient beliefs won't go away. It is a world of multiple civilizations with passionately held and diverse views connected by global networks of knowledge.

Any country that cuts itself off from the new channels will probably cripple its economy and enter a vicious spiral of declining trade and growing poverty, which is what happened in Afghanistan, Myanmar, parts of the Arab world and much of Africa.

A MUSLIM RENAISSANCE?

The confrontation that seems most alarming today is the clash between the extremist Muslim world and the West, brought into focus by the events of 9/11, the Iraq war and the rise of suicide bombers.

When Europe was still in the Dark Ages, over a thousand years ago, the Islamic world was a great civilization, characterized by a thirst for science and knowledge. The world's great trading cities were Arab cities. Sheik Hamad Bin Khalifa Al-Thani, the emir of Qatar, a small state in the Emirates, believes that a new Arab golden age can be achieved through education and research, coupled with creativity and economic development. He and his wife feel that the soul of the Arab/Muslim world can be reinvigorated and its rich culture preserved if there is a major commitment to education. Education, they argue, is essential to stopping the hijacking of Islam by extremists.

In 1995, they created a project called Education City. This is a 24,000-acre multi-institutional campus, containing branches of American universities and joint ventures with American think tanks, such as a RAND-Qatar Policy Institute. There's a Qatar branch of the Cornell Medical School, Texas A&M, Carnegie Mellon and Georgetown universities. These bring American education and research to Qatar. This is in contrast to traditionalists in the Muslim world who dwell on the corrupting influence of Western culture and education.

The emir has a personal commitment to transforming Qatar into a regional educational hub and research centre. If all goes as planned, the youth who benefit from Qatar's Education City will enable the Emirates to build a rich and interesting economy after its oil and gas revenues eventually decline. More important, it could be a seed from which a new Arab renaissance spreads, bringing multicultural tolerance, new ideas and education across the Muslim world.

Education for global integration is absent in most of the planet. The world needs cultural diversity as much as it needs biodiversity, but we must strive for commonality where it is in the common interest.

We need to teach what unites us rather than what divides us. Huntington stresses three rules to minimize the probability of conflicts among civilizations:

- **The Abstention Rule:** If a conflict occurs *within* a civilization, states outside that civilization should refrain from intervening in the conflict.
- **The Joint Mediation Rule:** If war occurs between states from different civilizations, core states from these civilizations should negotiate with one another to contain the conflict.
- **The Commonalities Rule:** Peoples in all civilizations should search for and attempt to expand the common values, institutions and practices they have with people of other civilizations.

There are aspects of civilization that are common—common ethics, the international legal system, networks for commerce, the United Nations agencies and the infrastructure brought by modern technology, intercivilizational universities and so on. All nations signed the United Nations Declaration of Human Rights (although some deviate from its principles).

Instead of one global civilization, think of two layers of civilization.

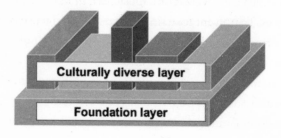

The foundation layer is everything that can be universal. The upper layer is what makes individual civilizations unique.

The content of the foundation layer will steadily grow more comprehensive. It will slowly expand so that a growing number of people will think of themselves as citizens of the planet rather than citizens of the West, or Islam, or Chinese civilization.

In the past, civilizations sometimes held beliefs so passionately that they would go to war for them. By the end of the 20th century, the common foundation layer was growing but was still immature. This common layer doesn't wash away traditional civilizations; on the contrary, the leaders of traditional civilizations become increasingly determined to preserve their uniqueness. All civilizations will be connected to the foundation layer—unless they isolate themselves completely, and the consequences of such isolation would be poverty, probably followed by brutal dictatorship. The challenge is to achieve enough mutual understanding and respect that war or terrorism among civilizations becomes unlikely. If the world is not to fall into increasingly horrifying conflicts, it has only one course of action: to breathe a strong spirit of multicultural tolerance and respect into the upper layer.

Today civilizations are connected by numerous political, financial and corporate links, and we have thousands of mechanisms and modes of behaviour in common. All civilizations are strengthening their worldwide links, through international banking systems with controls on money laundering, increasing air travel, global laws and treaties, common use of telecommunications, common health-care resources and so on.

The foundation layer should include mutually respectful religions with many common spiritual values. Much of the teaching that relates to the foundation layer can be designed to be global, multilingual and cross-cultural—sharing each civilization's music, art, literature, theatre, film and other resources. To prevent fostering hatred among civilizations, every culture needs to teach mutual understanding and respect for everyone.

As the world evolves, certain aspects of Western democracy, or variations of them, may take hold in the foundation layer—the universal right to vote; the rule of law; separation of executive, legal and judicial powers; freedom of expression; the inviolability of private ownership; the rules of business behaviour that provide fair and open competition; and human rights (as defined in the UN Universal Declaration of Human Rights). Great cultural diversity can coexist with this foundation.

Some civilizations regard Western civilization as corrupt—its cult of frenzied consumerism, its unchecked sexual behaviour, its widespread drug and alcohol abuse, its denial of spirituality, its pursuit of immediate gratification and corporate antics that place shareholder value above everything else. We must separate the *principles* of Western democracy from the *behaviour* of Western civilization. The principles and their implementation are a formidable accomplishment, but they have been harmed by behaviour that reveals a general moral crisis and calls for a deep rethinking of what produces quality of life.

Education may be the single most powerful means for minimizing the "clash of civilizations." Education needs to avoid overemphasizing the merits of one specific civilization; it should foster a perspective. Young people should be taught about the harm of chauvinism and propaganda against other religions. They should be taught the inevitable interdependence of the world's peoples.

Much of the education on the Web, TV and digital media could be free of charge and designed for global use. It can be in multiple languages and be acceptable in countries with different forms of government and different religions—for example, acceptable in both Iran and the United States. It must avoid the idea that one civilization is a "chosen" one, benevolently passing its knowledge to the others. Latin America could help teach the world about its music and dance. China could teach about the beauty of its ancient art and literature. Jewish, Buddhist and other religions could make their teachings more easily understandable. What a different view we would have of Iran if we studied the writings of Hafez instead of hearing of its preaching that America is the great Satan.

Life is more interesting if there is great cultural diversity, but it is essential to teach common principles that are in the common interest. We could vigorously identify and teach that which unites us and try to remove causes for terrorism and conflict.

By 2050, or perhaps sooner, most of the world will be familiar with its diverse cultures, made visible by the forces of globalism and high-bandwidth networks. These different cultures will become accustomed to one another. They will slowly settle into stable acknowledgement of one another, realizing that it is not a good idea to fight with 21st-century weapons.

The dangerous time is the next few decades, when the confrontations can still shock and the nightmare of escalating suicide terrorism has not been extinguished. We need to anticipate and defuse potential confrontations wherever possible.

Perhaps the most difficult task will be avoiding antagonisms among religions. The wars of religion in 16th-century Europe cost that continent the lives of 30% of its population. Another round of wars like that with nuclear/biological weapons could become global and would exterminate civilization as we know it. It would be an irony grander than any in great theatre if the religions that evolved from the teachings of the world's saintly prophets somehow prompted wars that wiped out civilization.

HORSEMEN

For many people, the 21st century will be a great time to live because of society's ongoing increase in wealth. Increasing productivity, automation and better technology will enable already prosperous countries to grow even more so—spectacularly. Meanwhile, many new nations will industrialize and have a measured growth rate higher than that of the First World, but they are starting from a much lower base. China and India will create new wealth on a grand scale, and nations that trade with them will benefit. Throughout much of the world, there will be thriving shopping malls filled with new goods and services to buy.

This good news will be accompanied by serious bad news, however. It's tempting to write about the Four Horsemen of the Apocalypse, but there'll be many more than four. The horsemen won't all travel together, but they will be global in consequence, and their grim effects will be mutually reinforcing:

- Catastrophic global changes in climate
- Rivers and aquifers drying up

- Destruction of life in the oceans
- Mass famine in ill-organized countries
- An unstoppable pandemic of a new infectious disease
- Destitute nations slipping into a deepening trap of extreme poverty
- Unstoppable global migrations of people
- Weapons of mass destruction becoming inexpensive
- Growth of shantycities with extreme violence and poverty
- Mass recruitment for suicide terrorism
- Nuclear/biological terrorism
- Religious war between Muslims and Christians
- The possibility of world war with nuclear and biological weapons
- Exposure from extreme science to new dangers—for example, genetically modified infectious pathogens

Problems such as these take us to the midcentury canyon, when the population is at its peak and environmental stresses at their worst. There are ways to deal with all of these problems, and the earlier the solutions are applied, the better. All delays in implementing solutions make the situation worse. There is a reasonable probability that the worst of the problems will be avoided, such as nuclear terrorism and world war with nuclear/biological weapons. As humankind approaches the canyon and the dangers become more visible, there will probably be increased determination to deal with the problems.

China changed in three decades from extreme poverty and the worst famines in history to being an economic powerhouse. The same needs to happen in numerous developing nations. It will not happen in the destitute nations unless they get massive help.

After the Pearl Harbor attack, the United States moved to a war footing with amazing speed. Car production was rapidly switched to armaments production. In the future, there may be a time when the public of the First World at last cares what is happening, and there is a drive to deal with the situation as powerful as America's post-Pearl Harbor drive. A war-footing race to transform the oil, coal and car industries may be triggered by extreme climate change and Category 7 hurricanes raging across Florida and heading north, severe water shortages and crop failures. There will be serious efforts to prevent a pandemic of a new disease, build up food security and stop bioterrorism. A major effort may occur to build mutual respect between Muslims and Christians and to put appropriate education in place for the global melting pot of cultures.

17

A COUNTER-TERRORIST WORLD

BEFORE THE 20TH CENTURY, cities were built that were vulnerable to fire, and some terrible fires swept through cities. At the beginning of the 21st century, cities were vulnerable to terrorist attacks. Just as fireproofing became a mandatory aspect of city design, so terrorist-proofing will become standard practice. Like fireproofing, it won't prevent all disasters, but cities will become difficult to attack. There will be business and tourist corridors around the world that have 21st-century security, but much of the world will be outside the secure zones.

Terrorism of growing ingenuity is likely to be a staple of the future. That sounds grim, but it is much less grim than the world wars of the 20th century. This century's tensions will increase severely as the demand for the planet's resources grows as the resources themselves decline. Terrorist organizations will fan the flames of resentment between the very poor and nations with wealthy modern lifestyles. Very violent technologies, instead of being in the hands of government, will sometimes be in the hands of individuals. Such people could be either organized terrorists or deranged individuals like the Unabomber. An individual genius in his basement could be creating pathogens of mass destruction.

Future civilizations will have to live with counter-terrorist measures, which will become an integral part of advanced societies, as pervasive as air-conditioning. These measures are not in place yet and will take many years to create. It's desirable that the counter-terrorist measures are thorough and powerful, but hidden so that they don't interfere with our lives.

Today's world is one in which it is easy for terrorists to succeed. When our cities were founded and built, there was no thought about suicide bombers, truck bombs, terrorists hijacking civilian jets or hackers breaking into major computer systems. The 9/11 targets were attacked because they were relatively easy to attack. Future counter-terrorist security must ensure that there are no easy ways to succeed.

Most nations could be terrorist targets, and most have an incentive to cooperate in actions to stop terrorism. Particularly important is a determined international effort to make mass-destruction weapons much more difficult to obtain.

DEFENCE IN DEPTH

All good defence is defence in depth. In other words, multiple layers of different types are needed to achieve thorough protection. Counter-terrorism needs the following layers:

Layer 1. Remove the reasons for being a terrorist.

Layer 2. Internationally, establish full cooperation among national intelligence agencies and their computer systems, to help identify potential terrorists and their possible types of attack, prevent money laundering and establish global cooperation to flush out terrorist cells and their support structure.

Layer 3. Internationally, prevent access to highly enriched uranium, smallpox and any means of making severe weapons. Make container shipping secure around the world.

Layer 4. On a national scale, make frontiers and ports secure. Prevent plane hijacking. Make it difficult and dangerous for terrorists to transport explosives or other resources for attack.

Layer 5. Stop the attack before it happens. Use police work and electronic surveillance for detecting suspicious activities or preparations for an attack.

Layer 6. Be prepared for an attack, with blood and medical supplies. When an attack occurs, be ready to minimize the damage and save as many lives as possible.

Layer 7. After the attack, track down the perpetrators and any of their associates. Share relevant information internationally.

GET OFF THE TARGET LIST

Of these levels of defence in depth, the level that has been tackled least well is the first. Serious terrorist organizations advertise for recruits, using modern media. Al Qaeda's recruiting video is on the Web. It shows images of Iraqi children starving under American-led sanctions, Israel bombing women and children in Palestine, and Muslim prisoners of war being brutalized. Osama bin Laden's rhetoric is highly persuasive to many young Muslims, telling them that God wants them to help end Islam's humiliation. Right now, most of the vast target audience for this message doesn't yet have the ability to see Web-delivered TV. As the technology spreads, this means of recruitment will reach far more potential volunteers.

We should make every effort to remove the *reasons* young people want to be terrorists or the reasons extremist Muslims want to attack the West. Poor countries could be made to see America as a country that will help them to grow their economy, carry out rescue operations after a catastrophe, like the 2004 tsunami, or help prevent food shortages. Young people should see America and the West as giving them a chance in life, helping them to be entrepreneurs and providing education, micro-loans and venture capital.

Far more terrorists attack their own country than attack foreign countries. Alberto Abadie of the John F. Kennedy School of Government found that there were 1,536 reports of domestic terrorism worldwide in 2003, compared with just 240 reports of international terrorism.[1] It is often thought that it is extreme poverty that breeds terrorists, but Abadie found that there's often no attempt at terrorism in the poorest countries—terrorism occurs after there is a glimpse *that change is possible*. An interesting conclusion of Abadie's study is that terrorism reaches its highest levels in states that are making the change to democratic governments. Nations with tough autocratic governments and no freedom usually have low levels of terrorism. So do nations with high levels of political freedom. The transformation from autocratic government to democratic freedom can be a dangerous time.

NUCLEAR TERRORISM

Probably the worst thing that a terrorist organization could do is to set off an atomic bomb (not just a "dirty bomb") in a major city. In May 2003, Osama bin Laden obtained a *fatwa* (an Islamic decree) justifying a nuclear attack against the United States on religious grounds.[2]

The likelihood of such an appalling catastrophe happening has been studied in great detail. The conclusions of such studies is that such an event is highly likely unless we greatly strengthen global measures to prevent it. It is much easier for a terrorist organization to acquire and use a nuclear weapon than the public realizes. Graham Allison, founding dean of the Kennedy School of Government, writes that nuclear terrorism is *inevitable* if we continue on the present course, and it may happen sooner than we think.[3] There are steps that can be taken—but are not being acted upon—to prevent nuclear terrorism in the future. If we continue this inaction, we leave ourselves vulnerable.

Several facts about atomic terrorism need to be made clear and taken very seriously. First, if an al Qaeda–like organization had atomic bombs in American cities, it appears they wouldn't hesitate to detonate them. Second, building a crude atomic bomb is not all that difficult if one has highly enriched uranium. It is difficult to make an *efficient* one, but terrorists wouldn't care about efficiency or engineering elegance. Third, the highly enriched uranium for such a bomb could be shipped, with a radiation shield, into most target countries with a low probability of detection, just as drug shipments escape detection. Fourth, if a crude atomic bomb with a radiation shield were hidden in an American city, it would probably not be detected, even if there were an intense search for it.

Now for the good news. A uranium bomb cannot be made without highly enriched uranium. A thriving nuclear power industry can exist without using or creating *any* highly enriched uranium. Converting natural uranium, or the low-enriched uranium used in nuclear power plants, into highly enriched uranium is extremely expensive, slow and difficult. The cost of creating highly enriched uranium is far beyond the means of today's terrorist organizations. A terrorist attack with an atomic bomb is an entirely preventable tragedy. It can be prevented by securely locking up all highly enriched uranium and plutonium with extreme security. Unfortunately, today in many places it is not locked up well.[4]

If a *nation* sets out to create an atomic bomb, it needs one that is safe, reliable, efficient, small enough to fit in a missile and robust enough to withstand the missile launch. This needs special components, major engineering expertise and a vast amount of money. In most countries it would violate their signing of the Nuclear Nonproliferation Treaty. Terrorists, however, want a bomb that need not be reliable, small or robust enough to withstand a missile launch. They would want a bomb cheaper than a *national* bomb, relatively simple and maybe deliverable in a lorry or shipping container. This primitive, inefficient type of bomb is relatively easy to create, but it could have the power of the Hiroshima bomb.

A bomb like the Hiroshima bomb uses a gun to fire a wedge of highly enriched uranium into a target of the same uranium. When brought together, the two pieces of uranium exceed a critical mass so that a fission chain reaction occurs, giving a nuclear explosion of about 10 kilotons. The mechanism is simple but needs to be built with precision. The two pieces of uranium must lock together fast enough. Insane though it may seem, the engineering details of how to build such a bomb have been fully described in the public domain, in documents that were once available on the Internet. Six atomic bombs of this type were made in an ordinary warehouse in South Africa. Neither the South African bombs nor the Hiroshima bomb were ever tested because they were so simple that scientists assumed they would work. (The Nagasaki one, which used plutonium, *was* tested in the Project Manhattan test called Trinity.)

Enriched uranium for making such a bomb could be shielded so that it emits little radiation. Today, if it were shipped to the United States in a container from a seemingly respectable source, it would probably not be detected because only a very low percentage of such containers are inspected in detail.

A simple atomic bomb could be assembled in the United States and the bomb hidden in a city apartment on a high floor in a wardrobe with a tungsten radiation shield. If the police were told that such a bomb was in New York, they probably wouldn't be able to find it. They would use very sensitive radiation detectors, but these would pick up endless false signals because there are many sources of low-level radiation in a city. The bomb could be detonated by remote control, possibly with a mobile phone call, like the 2003 bombs in the Madrid subway. The caller could be in a foreign country. Suicide bombers need not apply. Islam preaches patience over a long period. There might be nuclear bombs hidden in many large cities for some time before they are detonated.

Acquiring such a bomb requires about 100 pounds of highly enriched uranium, which is extremely difficult and expensive to create. The Hiroshima bomb used uranium that was 80% enriched (that is 80% uranium 235 and 20% naturally occurring uranium 238). Fourth-generation nuclear power plants, described in chapter 7, use uranium that is 9% enriched—and this cannot be used in bombs. The only way terrorists are likely to obtain sufficiently enriched uranium is by stealing it, buying it on the black market or acquiring it from a sympathetic government. Unfortunately, there is a thriving nuclear black market. After the USSR collapsed, a vast amount of highly enriched uranium was stolen and sold. Pakistan's Dr A. Q. Khan, referred to as the "Father of the Islamic Bomb," created a massive black market network dealing in parts needed for atomic bomb production. When a Russian Navy captain was convicted of stealing highly enriched uranium, the Russian prosecutor commented that "potatoes are guarded better."[5]

There have been reports of Russian atomic bombs being stolen. In 1997, Boris Yeltsin's national security adviser, General Alexander Lebed, acknowledged that the Russian government could not account for 84 atomic bombs—each made in the form of a suitcase and which had an explosive power of about 1 kiloton. Later, on the US television programme *60 Minutes,* he said that more than 100 were missing. He related how someone could walk down the street carrying a suitcase that is, in fact, an atomic bomb, and it could be triggered by one person. Not surprisingly, the Russian government later denied that any such weapons were missing.

This is perhaps not as alarming as it sounds because such weapons are made with highly secure electronic locks to prevent unauthorized use. Modern versions of these locks are integral to the weapon so that they cannot be bypassed. In other words, the bomb cannot be "hot-wired." If a wrong authorization code is entered several times, or if an attempt is made to bypass the lock, the bomb may be permanently disabled. Stolen nuclear weapons with well-designed security controls could not be used by terrorists. Also, the fissile material is usually stored separately from the bomb. A plutonium bomb without its plutonium is of no use to terrorists.

There is a clear solution to the nuclear terrorist problem. All enriched uranium and plutonium from which bombs could be made must be locked up like the gold in Fort Knox. This requires international cooperation, but all nations have an interest in preventing nuclear terrorism. The work at the Kennedy School of Government describes in detail the actions needed to prevent nuclear terrorism and spells out a budget and timetable. Far too little is being done to bring about these actions. We might worry that, as with some of the other problems we discuss, there will be a catastrophe-first pattern. If a terrorist atom bomb does explode in a city, there will be an extreme reaction. The nations of the world would quickly cooperate to ensure that it wouldn't happen again.

BIOLOGICAL TERRORISM

While atomic-bomb terrorism could be stopped by removing or very securely locking up sources of fissile fuel or existing bombs, it may be almost impossible to prevent terrorists obtaining some form of biological weapon, such as viruses. We did a depth interview with Sergei Popov, one of the top scientists who worked on the Soviet biological weapons programme. In a scientific top-secret town with huge, well-equipped labs, 50,000 to 60,000 very well-trained people worked on this programme. No one could enter or leave the city without having a visa and following a rigorous security procedure. There, the Soviets could create between 20 and 100 tons of weapons-grade smallpox per year. With an expressionless poker face, Popov described, shockingly, how there was never any discussion whatsoever of the morality of doing this.

D. A. Henderson, the hero of the eradication, who fought step by step to eliminate smallpox, breaking bureaucratic rules in order to do so, now has a resigned, deeply philosophical face, like Pierre at the end of the film *War and Peace*. He told us, "A hundred million people died directly or indirectly as a result of armed conflict in the 20th century; 300 million died of smallpox." In an awful death, the body becomes a mass of festering blisters, almost unbearably painful and so dense that the skin separates from its underlayers.

People can be protected from most terrorist viruses if the presence of the virus is detected early enough. A simple and fairly inexpensive device can detect a minute trace of smallpox in the atmosphere and send out a wireless alarm signal. If ever smallpox is detected, that means that a major crime has been committed because natural smallpox was eliminated from the planet. There would need to be an immediate large-scale investigation to find the source and to find infected people. If a person is attacked by smallpox, it may be two weeks before symptoms appear. If the infection is acted upon soon after the attack, the infection can be neutralized in the body before it does serious harm.

An important solution in giving the public protection from biological terrorism is to create machines that will detect, as sensitively as possible, any pathogens in the atmosphere. Machines have been built that can suck up the surrounding air and conduct tests on it to see if it contains anything biologically harmful or suspicious. If something suspicious is detected, the machine immediately transmits details by radio to a control centre. Experimental pathogen detectors were deployed in New York at the time of the

Republican convention in 2004. They were about the size of a small refrigerator. They will become more powerful and smaller—perhaps, eventually, the size of a briefcase. They will become much cheaper when they are refined and taken into volume production. After a biological attack, rapid detection and response could save a very large number of lives.

Today's pathogen sensors are designed to detect many different types of possible biological attacks. Essentially, the same machines can also help protect societies from naturally occurring plagues such as SARS (severe acute respiratory syndrome) or Asian bird flu. A pandemic of such a disease seems inevitable, sooner or later. Widespread instrumentation will give early warning of the possible spread of infectious diseases.

Pathogen detection machines do complex computing to analyse what they find in the atmosphere. They may use gene chips designed for quick recognition of a telltale sequence of letters (nucleotides) in the DNA of molecules. When they detect something and transmit it to a regional or national computer, the computer compares the results of many local sensors or pathogen analysis machines. They look for both terrorist activity and naturally occurring pathogens. The protection from infectious diseases and the protection from bioterrorism will be integrated.

All governments today should be pressured to sign a vigorously enforced treaty with the intent of eliminating biological weapons, while at the same time maintaining a global agency to detect and deal with emergence and limit the spread of disease. Many protective measures are possible, but much work is needed to implement them. If a country is well prepared to cope with a bioterrorist attack, this will deter terrorists from carrying it out.

Shipping containers need to be made highly secure. They will be sealed in such a way that a signal will be transmitted if the seal is broken. They will have sensors that uniquely identify them. The contents of containers can be scanned without opening the container, and the exact journey that cartons go on can be recorded with sensors detected by GPS (the global positioning system). It will probably become law that ships, lorries, shipping containers and shipped cartons have sensors and wireless transmitters. Those that don't will be blocked from high-security ports.

Sources of funding for terrorism will be blocked, and known terrorists will be denied access to the international financial system. There will be elaborate means to prevent the growth of alternative financial networks for moving terrorists' money. Assets of organizations known to support terrorists will be frozen. The uses of banknotes in society may become limited to small sums; serious money will be electronic.

UBIQUITOUS SENSORS

The future is going to be a world of ubiquitous sensors linked to highly complex computer systems, helping business increase its profit, helping the police catch criminals, helping reduce city traffic jams and helping medical researchers lessen disease. This world of cheap sensors everywhere, linked to highly intelligent computer systems, will be an integral part of a counter-terrorist society.

Airports will be crisscrossed with many electromagnetic and other beams, passing through the passengers and luggage. Neutron beams passing through plastic explosives are scattered in telltale patterns, and clever software in computers can recognize such patterns.

The passport is a spectacularly inadequate document. It's trivially easy for criminals to obtain fake passports, although some countries have linked fingerprint detection to the inspection of passports. Car number plates are similarly inadequate; a criminal can easily switch number plates. Instead, all vehicles could be required by law to have a built-in wireless transponder that transmits a self-verifying registration number, the GPS location of the car and the owner's self-verifying identification number. The transmitted number may be used for the automatic payment of inner-city tolls (as implemented in London today, using cameras).

Today's equivalent of the passport could be a smart card, the size of a credit card, which stores a large amount of information, including significant sections of its owner's DNA code so that the card and its owner are inextricably linked. Alternatives could be a bracelet, an ankle ring or finger ring that is rarely removed. The card, ring or bracelet would be interrogated by wireless signals without the owner knowing. The same device may serve as a driving licence, health card and bank card. Today, a US visa is a stamp on a paper passport for which a person is charged $100. For $100, the person could be given a device attached to the passport with 500 million bytes of data.

In Britain, the number of cameras watching the public has steadily increased until now, in the cities, there is one camera for every 15 people. It was once thought that the public would object to such surveillance, but they seem to like it, partly because it helps protect them. Using the camera records, police were quickly able to identify the suicide bombers in the 2005 London train and bus attacks. Because there are so many cameras, criminals think that they will be recorded and may avoid committing crime.

Research labs have demonstrated prototype machines that can identify human emotions correctly 98% of the time. The human face has about 80 muscles, and the computer detects whether these muscles are active. Such computers that recognize emotions have been tested in police interviews. Someone might be interviewing a government official while a computer watches the interviewee's face. The interview proceeds at a calm emotional level, and then the interviewer asks, "Have you on any occasion, or under any circumstance, ever taken a bribe?" If YES, the computer curve of emotional response goes off the chart. The London suicide bombers with backpack bombs were each recorded by cameras as they detonated their bombs. Could advanced technology have detected emotions that would have triggered investigation? As counter-terrorist computing matures, it will become very risky for terrorists to prepare for an attack.

A MASSIVE DETECTIVE PROBLEM

Stopping terrorism in all its possible forms is a detective problem of immense complexity. Today's electronics can generate a vast ocean of data. In earlier times, *humans* collected data. Today, most data are not seen by humans; they go straight to data warehouses where they can be examined by computers looking for trends, patterns, correlations or clues that something of interest needs to be examined carefully. Software will be designed so that it automatically improves its own capability to recognize patterns or find clues that humans may not.

Human intelligence, fieldwork and intuition will be more important than ever. Large numbers of human operatives will tell the antiterrorist computers what to look for and give them feedback when they find something interesting. Human detectives will need powerful computerized techniques to discover patterns and test hypotheses. Computer intelligence and human intelligence are radically different, and each needs the other. The two forms of intelligence will coexist in synergistic partnerships and will be vital for extracting intelligence from the huge and incompatible data warehouses.

Martha Crenshaw, a lifetime authority on terrorists, describes how the most dangerous terrorists don't act alone. Psychologically they need to be part of a supportive organization, rather like a cult. These groups often use religion or politics as a reason for their behaviour, but people become members of such an organization out of psychological need rather than political commitment.[6] The command structure creates a complete psychological framework for the individual, uses powerful mind control and sometimes creates the desire to be a suicide terrorist. The group often uses military-like training camps. There can be many clues that such camps exist, or that individuals have the cultlike behaviour they induce. Computers can be programmed to act automatically when certain clues are detected. When computerized intelligence is used, humans cannot necessarily tell how computers come to conclusions. The computers learn to recognize patterns and relationships that humans could not recognize.

21ST-CENTURY PRIVACY

There is naturally great concern that the surveillance made possible by 21st-century technology will destroy the privacy that we so cherish. It is vital that we design the privacy controls that we want.

The law will, and should, help guard personal privacy. All data collected must be electronically locked. There will be laws saying that the information collected about an individual cannot be revealed or used, except for specifically stated purposes, most of which have to do with preserving security. The majority of a person's life will be of no interest to the security computers, and for those aspects of life, the computers will be required by law to protect privacy. Computers, with their relentless thoroughness, can help to enforce privacy. As computing becomes highly complex, data can be enciphered with codes that are almost impossible to break. Instead of having a five-letter password, machines could use a million-letter password. The battle between the encipherer and the code-breaker will be won by the encipherer. Aspects of privacy that are important to good citizens can be protected much more thoroughly than in the 20th century.

To provide the protection that citizens will need in the future, certain information must be known to the 21st-century police and to medical and other authorities and their computers. This must be achieved without in any sense giving up liberty. You lose the freedom to be a terrorist, but no normal freedoms. Benjamin Franklin said, "They that can give up essential liberty to obtain a little temporary safety deserve neither liberty nor safety."

In villages of the past, everyone knew everything about everyone. In the world of the future, computers will know certain things about everyone. No one will be invisible to the electronic surveillance, but electronically enforced privacy laws will protect everyone. The typical noncriminal individual won't worry about the cameras in the supermarket, or other surveillance devices—he or she may even welcome the feeling of security. He will know that the data they collect cannot be revealed to his employers, bankers, insurance salesmen or ex-wives. Twentieth-century privacy will be replaced with 21st-century privacy.

The public may be divided into security categories. This, today, is a highly contentious subject, but it may not be in the future. It is now beginning to happen with people visiting the United States. There could, for example, be four categories of people.

Category A people are security-cleared and have automatic identification. A wireless beam can interrogate their identity card (which may be in the form of a ring, bracelet or necklace). They can walk through immigration checkpoints or go into the Four Seasons restaurant in New York unaware of the computers that are tracking and validating them. Category B people are essentially good people who choose not to have the automatic identification; they will often be stopped unless they avoid secured places. Category C people will lack full security clearance and will often be subjected to close examination. Category D people will be automatically blocked.

The vast majority of people in a country, perhaps more than 95%, will be Category A. This category may include known criminals because most criminals are not would-be terrorists. Most people will want to be Category A because the constant examination of the other categories will be a nuisance. People will take measures to obtain and retain an "A" categorization. They will *want* to be identified by the surveillance electronics so that they are not stopped by police, guards and immigration officials. Portions of the DNA of Category A people will be on file with the authorities, and their security card or ring will contain details of that DNA. Anyone attempting to use a forged or stolen security device will be caught quickly. The same devices will be of great value in medical situations.

As the 21st century matures, computers and sensors will surround us constantly. They will monitor our activities but also help enforce laws intended to *prevent unnecessary interference with our privacy*. The deluge of nanosensors and data mining will be used for functions other than counter-terrorism. It can help stop drug abuse. It will help bring extraordinary changes in preventive medicine. Machine intelligence, as ubiquitous as the wallpaper, will provide knowledge and communication to aid all human activities.

It is up to society to design computerized-enforced rules that protect our personal privacy, prevent undesirable interference with our freedom and punish officials who cause unnecessary harassment.

18

WORLD SCENARIOS

SOME LARGE ORGANIZATIONS use scenario planning to consider how the future could play out. There can be an infinite number of possible futures, but the number of scenarios is kept to three or four so that the main alternate directions can be explored. Scenario planning was developed at Shell in the Netherlands and proved valuable in helping Shell's management choose the best of the risky and very expensive alternatives in developing new oil fields such as those in the North Sea. It has since been used in many business and military situations.

To describe our scenarios, we'll redefine the terms *First, Second, Third* and *Fourth Worlds.* I find the following categories useful:

- **The First World:** Wealthy industrial nations—the West, Japan, New Zealand and Australia. It has about a billion people and a birthrate below the replacement rate.
- **The Second World:** Vigorous nations striving to climb the ladder from poverty to First World status. The largest examples are China and India. As they become wealthy, they may have

different lifestyles from the West. The Second World may have 3 billion or so people.

- **The Third World:** Developing countries that have long been referred to as Third World. They might become about 3 billion people by midcentury.
- **The Fourth World:** Destitute countries unable to escape from the burdens we described in chapter 6. The Fourth World might become 2 billion people by midcentury.

A model of the world that divides it into four such categories is a severe oversimplification. Most models are—that's why we use them. Models enable us to summarize complex situations. Convention puts both developing nations and destitute nations under the category *Third World*. To discuss the future we need to distinguish between healthily developing nations and destitute nations; so, that's why we use *Third World* and *Fourth World*. There is a great difference between developing nations such as Chile, Brazil, Malaysia or Thailand—where there is vigorous industry and good universities and where young people have hope for the future—and destitute nations such as Angola, Haiti, Estonia or Ivory Coast, where, by and large, young people have no hope. The way we think about the Fourth World should be quite different from the way we think about the Third World. The need for such a distinction will grow stronger as the 21st century unfolds.

Of the many possible scenarios for a world heading towards the canyon, we'll pick four to illustrate the main types of choices. We'll call these scenarios Fortress America, The Strong Nations Club, Triage and Compassionate World.

SCENARIO 1: FORTRESS AMERICA

A Fortress America policy essentially says that, whatever nature has in store for humankind, America will take care of itself. The stresses we have described will increase, and some types of events could make them severe—for example, a major pandemic, war or terrorism with nuclear or biological weapons. We don't know how severe they will become. Bill Joy, a scientist who tries to make us take the potential dangers seriously, stated in my interview with him, "If you put the system under incredible stress, it will break. It's a question of how thick do you think the veneer of civilization is."

Twentieth-century history contains an appalling collection of stories about the veneer of civilization peeling off like roof tiles in a hurricane—societies with elegant sophistication changing from trust to mistrust of neighbours, to secret police, gulags, torture and terror. The French charm of Phnom Penh changed to the killing fields. The intellectuals of China were reduced to being little better off than farm animals. Stalin, from 1935 to 1941, killed more innocent citizens than Hitler did in the Holocaust. We need to ask how we can keep our lifestyle from disintegrating. America's only sensible policy is to ensure that it protects itself from whatever the future has in store.

In September 2002, the United States announced the Bush administration's National Security Strategy. The strategy stated that the United States had become by far the world's most powerful country and it intended to remain so. It declared that it would use force if necessary to eliminate any challenge to its global hegemony. "Our forces will be strong enough to dissuade potential adversaries from pursuing a military buildup in hopes of surpassing, or equaling, the power of the United States. We must build and maintain our defenses beyond challenge."[1]

The US National Security Strategy expresses the belief that the United States is in the vanguard of history, leading the world towards democratic governments, free speech, free trade, human dignity and economic betterment. Because of this global leadership, it must maintain, for the indefinite future, global dominance and unchallenged security.

At various times in history, one nation has been dominant, and because of its unchallengeable supremacy, there has been a time of relative peace. During the Pax Romana and the Pax Brittanica, Rome and Britain, when their empires were in full bloom, tried to spread trade and create better governance in the domains they ruled. A Pax Americana would do the same, now with an emphasis on human rights and human dignity. The American National Security Strategy expresses an intent to improve the lot of the poorest countries and "ignite a new era of global economic growth through free markets and free trade," believing that countries with free trade and prosperity are less likely to go to war.

The strategy document comments that decades of massive development assistance have failed to spur economic growth in poor countries. "Worse, development aid has often served to prop up failed policies, relieving the pressure for reform and perpetuating misery." The document states, "Sustained growth and poverty reduction is impossible without the right national policies. Governments must fight corruption, respect basic human rights, embrace the rule of law, invest in health care and education, follow responsible economic policies and enable entrepreneurship."

A major argument for Fortress America is that the worst scenarios we have described may slowly but inexorably come to pass. The wealthy, complex civilizations of the West live on the same planet as destitute nations with unspeakable conditions. Pakistan already has many nuclear weapons. Global tensions will grow, and some of the most radical nations will acquire weapons of mass destruction. America, then, must have firm but necessary plans to protect itself, probably in conjunction with its close allies.

Until the end of the 20th century, nuclear weapons were considered weapons of last resort. A terrorist organization today might regard weapons of mass destruction as weapons of first resort. The National Security Strategy comments, "We are menaced less by fleets and armies than by catastrophic technologies in the hands of the embittered few. This is a new condition of life. As time passes, individuals may gain access to means of destruction that until now could be wielded only by armies, fleets and squadrons." Furthermore, with the spread of globalism, events beyond America's borders can have a greater impact than events inside these borders.

One aspect of Fortress America thinking that shows that the United States is prepared to go it alone is that the United States has refused to sign many important international treaties, including the Law of the Sea Convention, the Landmine Treaty, the Kyoto Protocol, the Convention on Biological Diversity, the nuclear Comprehensive Test Ban Treaty, the protocol for a UN convention banning biological weapons and various proposed treaties to ban weapons in outer space. The United States has also refused to accept the jurisdiction of the International Criminal Court.

SCENARIO 2: THE STRONG NATIONS CLUB

The Pax Americana, if it existed, would be shorter-lived than the Pax Romana or the Pax Britannica. China is developing and strengthening itself vigorously, expanding its sphere of influence. Twenty years from now, India will probably have a population larger than China's and a rapidly rising GDP. It is inevitable that large nations or blocks of nations will confront Fortress America. We might have Fortress China, Fortress Europe, Fortress India and Fortress Japan. These will be vigorous trading partners. Weapons of the future will be so terrible that the Fortress nations will know that they must avoid going to war with one another (no matter how much sabre rattling occurs). A likely scenario is that there will be a club of strong nations, supporting one another, sharing intelligence and mutually defending themselves from a common threat of terrorism. As in the Fortress America scenario, this Strong Nations Club may be determined to make itself well defended, self-sufficient and wealthy so that it can protect itself from whatever the future has in store.

A Strong Nations Club will have tightly interwoven trade with computerized value chains among corporations making mutual business fast-growing and profitable. Such nations may move to a position in which war among themselves is very unlikely. As politicians assess the risk/reward ratio, they will conclude that, with automated nuclear/biological weapons, war could have *no* political objective worth the risk. Prevention of war will become a major academic discipline.

After 9/11, governments all over the world suddenly started cooperating in new ways. They had a common need to stop terrorism. They moved apart somewhat when the United States attacked Iraq. Sooner or later, terrorists will probably achieve an attack with a weapon of mass destruction. The post-trauma reaction to a terrorist Hiroshima would incomparably exceed that of September 11. The strong nations of the world would join

forces intensively to keep such horror from happening again and to prevent the possibility of any drift towards war with nuclear/biochemical
weapons. They'll need to help one another build global intelligence systems with intense electronic surveillance, massive shared data warehouses
and self-evolving use of supercomputers. The Strong Nations Club could be
the entire First and Second Worlds, wired together with extreme-bandwidth
networks and real-time computer-to-computer business systems. Many
industrial Third World nations—Brazil and Mexico, for example—will
seek to join the Strong Nations Club. In an optimistic scenario, the club's
membership could steadily drift above 4 billion, even though the birthrate
in affluent nations is dropping. One hopes that by then the Strong Nations
Club will have moved vigorously to eco-efficient industry and lifestyles
where a high quality of life is achieved without harm to the planet.

The Strong Nations Club might give little aid to the poorest nations, as
today. Right now, it's difficult to imagine China giving much to Africa or
South America—but times change. Voters in the strong nations may make
only token gestures to the Fourth World. Many today say, "It isn't our
problem." Others regard the destitution of the Fourth World as a virtually
unsolvable problem. Some may express anger that excessive population
growth in poor countries defeats all attempts at pulling them out of
poverty. Television and popular magazines fail to convey the horrors of
life in Fourth World shantycities and generally refer to "developing" rather
than "destitute" countries.

SCENARIO 3: TRIAGE

Triage in a hospital refers to a situation where there are not enough beds
for everybody—so, a decision is made about which patients won't be
admitted. The word *triage* is not used publicly when referring to nations,
but in the corridors of power, there is a growing belief that triage will happen. Aspects of it have been quietly happening for decades. Perhaps it's
fair to assume that triage is not a deliberate policy. If it happens, it is inad-
vertent. It's significant that *Fourth World,* or any equivalent phrase, is not in

our political vocabulary. For most people, the Fourth World is not on their radar screen.

In the first half of the 21st century, the world's economy may grow by more than seven times. The Strong Nations will probably be able to buy the surplus food they need, but scarcity and demand may double or triple the price of food. In countries where people are already spending 70% of their income on food, even low rises in food prices would be catastrophic.

Most of the poorest people are in hopelessly overcrowded shanty-towns, not country villages, and have almost no capability to grow their own food. They are entirely at the mercy of what their dysfunctional government might provide. The capacity to grow food is declining in many of the poorest nations. Climate change and global warming will lower the yield of crops in much of the world. The world used to have food reserves, which could be used if there was a bad harvest, but now the food reserves are much smaller. There'll be a great temptation to sell the food reserves to China or other new areas of massive consumption. Two or more bad harvests in a row could cause devastating famine.

Some government officials concerned with poverty and sustainability say that, in the imagery of the canyon, it is inevitable we will come out the other side of the canyon quite badly damaged. The international organizations we can construct to supply global public goods will not bear the weight we need to put on them, and sustainability will not be achieved.[2]

The new consumer class in China is rapidly growing in number, lifestyle and aggressiveness. Hundreds of millions of people want to eat meat rather than exist on a rice diet, and that requires huge quantities of grain to feed the meat-providing animals. A similar new consumer class is growing in India and elsewhere. Many ships per day will carry grain from the United States and Brazil to China. If left to the marketplace, food prices certainly will rise, even if destitute countries can't afford such prices. Starvation could be a by-product of market forces. To prevent that, the world (or the Strong Nations Club) will have to spend substantial money to create food reserves. Large nations won't do this if they can't buy the food they want for their own people.

Such scenarios may spin too far out of control for financial handouts to stop the damage. China and India, needing to feed their own vast populations, could ignore the poorest parts of the world. Demographics indicate that two decades from now the combined population of India and China will be around 2.6 billion. The Fourth World by that time may be somewhat less than 2 billion people. As global warming becomes worse, the high temperatures during growing seasons will lower the farmers' yield. Water shortages will make the situation worse. The world's grain reserves will probably be too little to take care of sudden starvation. Financial handouts won't solve the problem because there won't be enough grain in the silos of the strong nations.

In the year 2000, the UN created a Millennium Declaration stating goals for helping the poorest people. The intent was to halve the *proportion* of people living on less than a dollar a day by 2015. (If the proportion halves, the absolute number would not because the population of the world will grow.) Sir John Vereker of the British government expressed optimism that, by 2015, this target of halving the proportion would be met. This goal could also be met by raising earnings of the poor in China and India so that everybody has more than $1 per day and is adequately fed. It is possible that there will be little change in the Fourth World, but the Millennium Goal would be met because of what happens in India and China.

The world's wealth and technology today give us the capability to end extreme poverty everywhere. The cost of doing so has been worked out in detail, but not enough official development assistance (ODA) is being made available. If the money available today is spread evenly among all the countries that need it, it will not be enough to put any of them on the ladder to self-improvement. Most would still be too poor to stop their poverty. Should some have a larger share of the ODA? Should selected countries be put firmly on the ladder to self-improvement, while others are not? A triage question.

Jeffrey Sachs explains how official development assistance is a two-way compact. Some poor governments are so corrupt that further assistance would be unlikely to be used for the intended purposes. He says that "the expansion of aid is predicated on a serious plan of action and a demonstrated will to carry it out in a transparent and honest fashion." If such a plan is absent, the money is not likely to be raised to a level that can take these countries out of poverty.

It's morally unquestionable that the rich world should help the poor world avoid starvation. The goal must be that, later in this century, there will be no Fourth World. Fourth World countries should be steadily lifted onto the development ladder so that they become Third World. They should have education, decent farming, food security, entrepreneurs, international trade and no dead capital. Young people everywhere should have hope for the future and opportunities to advance themselves. But unless more money is allocated, the grim situation will get grimmer in some Fourth World countries.

To change this self-destroying situation needs well-coordinated management, social skills and money. Across the poor world, the situation could be turned around with money equivalent to 0.7% of the GDP of the First World. In Jeffrey Sachs's detailed calculations, extreme poverty could be eradicated in 20 years. At present, neither the money nor the management effort is being made available, and the problem is worsening. It could become much worse as the Fourth World population grows, environmental stresses become more severe, water sources shrink and climate changes cause crop yields to drop.

Experts on infectious disease believe sooner or later there will be an influenza or other pandemic, capable of infecting much of the world's population. Humanity could be protected from this if it had a large enough quantity of vaccines, but vaccines cannot be manufactured until the specific mutation of the disease is known. Today, a vaccine would not be available until three or four months after the pandemic started, and it is likely that the quantity of vaccine would be only a small fraction of what is needed. Unless far better preparations are made, the First World would not have enough vaccine for its own people. It would not be likely to give vaccine to the Fourth World.

If a killer flu is spreading, it is likely that it will rage out of control in some of the overpopulated cities that have bad health care and unsanitary conditions—like a forest fire blazing furiously in dry areas of dense pine trees. A nation where this happened would face a stark calculation: If it quarantines the city, it would lessen the rate of spread, like firemen isolating a blaze. Much of the population of the city would be doomed, but the overall national death rate would be lower. The nation would probably confront international pressure to quarantine areas where the disease is most out of control. There may be many such areas. As the flu spreads worldwide, First World countries would impose quarantines. There would be major restrictions on air travel. Much of the global economy would come to a halt. Economic damage would compound the damage from the disease.

In the Triage scenario, not all of humanity would make it through the canyon. If efforts to get the Fourth World into better shape remain as feeble as today, then the massive forces of the approaching canyon will overwhelm the ability to make rescue operations succeed. Humankind will blunder into Triage rather than planning it.

Ironically, there's been no time in history when we are more able to help the poor world. Michael Porter, in my interview with him at the Harvard Business School, lamented, "Today we have massive flows of capital across borders. We have multinationals that can spend billions to bring technology and create jobs. We have knowledge and know-how about how to do things better; about how to deal with health, deal with hunger, deal with bad water. There's been no time in history when the opportunity to deal with the poor and to end poverty has been greater. Yet it isn't happening."

SCENARIO 4: COMPASSIONATE WORLD

The problem with the Strong Nations Club and various fortress scenarios is that the poorest and least educated parts of the world will become violent. If the strong nations are increasingly fortress-like and blatantly wealthier, they will be objects of bitter envy among growing hordes of young people desperate to change their lot.

Jihad advocates will find many recruits in the Fourth World. Islam is spreading in many poor countries—often in the poorest places—and some extremist versions of it are being taught. Poverty and hunger drive young people to the most radical Islamic seminaries, where their parents know that, at least, their children will get free meals. Students in extreme seminaries are taught no lessons in maths, science, geography or history on the grounds that Allah's will, which is recorded in the Koran, is all they need to understand the universe. They are told that there are two forces on Earth: Muslims and infidels. These forces will fight to the death. Glory on Earth comes from killing infidels. Eventually, the forces of God will win, and the world will be put to rights.

If the Strong Nations Club encompasses a population of 4 to 5 billion by 2045, it may leave 4 to 5 billion outside the club. Such a vast population would have many factions whose passion in life is to participate in attacking the West in some way or making life miserable for the fortress countries.

A far better scenario is that young people everywhere are raised to a condition where violence is not their best option. The First World, or Strong Nations Club, could set out to change the Fourth World so that its children are brought up to be able to read and to have adequate nourishment, medicine and clean water, as well as the prospect of a worthwhile job. The birthrate could be brought below the replacement rate, which would benefit the whole world, and measures could be taken to assure food security. A desirable goal is that eventually no nations are destitute and that it is a compassionate world rather than one of violent antagonisms. Such a goal would need a truly massive effort.

The Marshall Plan at the end of the Second World War cost the United States over 2% of its GDP and was remarkably effective. Today the country is providing less than 0.2% of its GDP to other nations. Just 0.7% of GDP from the First World could eliminate extreme poverty and put once destitute countries on a path to a decent life. Rather than being entirely a transfer of cash, it would be partly a provision of services handled with First World management skills. A high level of management skill is needed— both to achieve food security, adequate health care and sanitation and to transform crippling bureaucratic procedures now in place. Rich countries could spread appropriate farming methods, literacy, education and good nourishment and help with population planning. They could help to develop exports and remove trade barriers. A key part of the transformation would be that First and Second World corporations would set up factories and farms in the Third and Fourth Worlds and teach local people the skills needed.

The view that "all people are one people" is gaining adherents, along with the view that we ought to be steadily progressing to worldwide unity. Will this happen? In most rich nations a Fourth World tax would never fly, but today's small foreign aid could easily be doubled and coupled to the management and education techniques spelled out by Jeffrey Sachs. It may be NGOs (nongovernmental organizations) and private foundations that lead the way. There is hope because of the relatively low cost of many leverage factors that could change the lot of the poorest nations. The Bill & Melinda Gates Foundation is doing great work. Goals like the UN millennial goals could steadily be extended.

When corporations invest in a country, the amount is much larger than handouts from governments and is often much more effective. Suppose that Japan made an agreement with the president of Uganda that it would set up factories to make cameras and electronic goods in Uganda instead of China because the labour costs were less, and that it would build roads, power plants and schools; educate women; drastically lower the birthrate; and steadily put Uganda on a staircase to a decent future. There would be a large-scale joint venture between the Japanese electronic and camera industries and Uganda. The goal might be to take Uganda from a GDP per head of $230 today to $1,000 in 15 years. As in the case of Ireland in the 1990s, different corporations would make their own deals with the government of Uganda, but the deals would fit into a coordinated plan. The governments of Uganda and Japan would develop the plan, measure its progress and make adjustments as it proceeded.

Let us suppose that the Japanese Uganda plan is seen to be succeeding and benefiting both sides. Many Ugandans go to Japan for training with an understanding that they can't immigrate to Japan. There are many Japanese communities in Uganda creating an environment that makes their industry work well. It is broadly realized that the Uganda model is the key to putting the poorest countries on a staircase to development. So Germany sets out to build car factories with a similar plan with Namibia. South Africa sets out to rejuvenate a post-Mugabe Zimbabwe. England takes on Tanzania.

Other countries have similar relationships. A feeling of competition emerges about which First World countries are winning the league table with African partners.

As the strong nations evolve towards greater wealth and advanced forms of civilization, it would be a moral scar of horrific proportions to leave the Fourth World to fester and starve. A reason why it may happen is apathy—few people think about it; few people care. People think about the Third World and may go on a tour to Thailand or Argentina, but the Fourth World is not on their radar screen.

When I ask, "What is the right thing to do?" and "What is the most likely thing to happen?" almost all knowledgeable people say Scenario 4, Compassionate World, is the right thing to do, but almost none of them think it will happen. I suspect that it will happen later in the century, after humankind reacts with horror to grand-scale famines and triage. Perhaps it will be part of the less turbulent waters on the far side of the canyon. With any right-thing-to-do scenario, we should explore the stages by which it could be made to happen without a catastrophe-first pattern.

PART FOUR

THE GATEWAY
TO THE FUTURE

19

A GREAT CIVILIZATION?

THE PROSPECT of humankind in the canyon years would seem grim except for one fact: *There are solutions to most of the problems we have discussed.* For most areas, there are many leverage factors. There's a large and diverse set of actions that can be taken to stop the harm being done and put humankind on a different course. When there is a clear and present danger, today's executives and engineers can be highly innovative and pragmatic in finding solutions. The problem is that, in doing so, they may entirely ignore the Fourth World.

The traumas of the canyon years will make it clear that our world has to be made less fragile. As humanity emerges from this period, it will have different rules of behaviour and very different technology. We will realize that because we live on a small planet, we must make its institutions and codes of behaviour robust, geared to the finite resources of the planet. We must make our science and nature's complexities enhance each other. We must no longer fight nature. We must learn how to protect ourselves from terrorists and maniacs, and from scientists with good intent who play with fire. We must be prepared to face the 17 challenges described in Chapter 13.

The types of protection needed to face these challenges entail diverse engineering safeguards. They require rules, protocols, methodologies, codes of behaviour, cultural facilities, means of governance, treaties and institutions of many types that will enable us to cooperate and thrive on Planet Earth. The strong nations need to help the weak nations so that all peoples can be on a staircase to improvement. The likelihood of any group's using weapons of mass destruction needs to be removed, as far as possible. The causes of hunger, despair and extreme poverty need to be removed. As we build a world robust enough to cope with the issues of the 21st century, it will provide a gateway to future centuries.

If we are to survive, we have to learn how to do this. It is well within the capabilities of today's management and engineering. If we do not do it, and our world falls apart, like the rivets on an aeroplane popping under excessive stress, then civilization will be set back many centuries.

We are going to realize that many of the problems of the canyon should have been dealt with before the severe tensions of the canyon built up. It is absolutely clear today that very big problems are looming, but to a large extent they are being ignored—water shortages, destruction of ocean life, global warming and the fate of the poorest 3 billion people. Such problems will become steadily more difficult to ignore. When the public of the rich world realizes that solving the problems will affect their well-being, then action may happen. We are largely ignoring the factors leading to climate change. This will make climate change much worse when it happens, and much of it will be irreversible.

I cannot repeat too often that, the longer the solution is delayed, the more severe the problem becomes. Today there is apathy and lack of awareness of the problems. The public doesn't associate use of air-conditioning with the idea that it might add to global warming, or that such warming will make hurricanes worse. The full consequences of a highly infectious global pandemic are not understood. The public is unaware of how tragic life is in the shantytowns. Extreme poverty on the other side of the planet has been somebody else's problem. There is a deep problem pervading everything we talk about: While there is great inventiveness in technology, procedures and governance, there is also great ignorance about what is needed. Old leaders use old methods; old politicians are committed to old ideologies. Some heads of state in poor countries are destroying their countries. Massive funds that used to go to Swiss banks are now finding ways to avoid the new controls on money laundering. There are massive 20th-century vested interests and powerful men with financial reasons for resisting the changes. There are wrongful subsidies, bogus science and sometimes great corruption in the power structure.

AFTER THE CANYON

The following scenario is one way in which the canyon years might pan out:

The real wealth of the First World nearly tripled by 2045 and that of China went up by a factor of eight or so. Humankind learned how to use resources much more efficiently. In many cases, factor-10 improvements in resource productivity were achieved. Nanotechnology and advanced computing enabled products to be far more interesting and to use far fewer raw materials. The subject of natural capital became understood, and incentives were put into place for corporations to conserve natural capital.

The subject of climate change eventually switched from being treated with almost total apathy to being regarded with near panic. The Gulf Stream began to show alarming signs of irreversible changes. Massive hurricanes and massive insurance premiums inflamed the panic. Wealthy countries in hurricane zones established codes for buildings that would withstand Category 7 hurricanes. Glass manufacturers developed new technology for making large windows that could withstand winds of 220 miles per hour. Dykes or movable river barrages (like that on the Thames) were built to protect cities from ocean surges. Parking garages were fully covered. While major industrial countries could take such actions, the shantytowns in violent-weather zones suffered extreme damage. Television images of dead people in devastated shantytowns triggered large charitable relief efforts and an increase in official development assistance. Often this increase was driven by events and not tied to a plan to put the area in question on a long-term development schedule.

The whole set of canyon problems eventually became clear to the public, and then the enormous resourcefulness of modern society kicked into gear. Fuel-cell cars were perfected and sold everywhere. Large-field solar panels dropped in cost and were mass-produced. Production of large wind generators (between 1 and 10 megawatts) became a major industry. Modular fourth-generation nuclear power became a massive export industry. Clean air became serious politics. Attempts were made to hold corporations responsible for the costs of environmental damage they cause. Plans for recovery of the ocean fisheries were put into effect. Hydroponic food-growing equipment was mass-marketed. Major crops were genetically engineered to do well in water-stressed areas. Hordes of specialists roved the world helping poor countries to install computer-controlled drip irrigation systems and restore soils to a high-nutrient condition.

In some ways, these energetic activities were too late. The measures to reduce greenhouse gases came two decades or so later than they should have. The world realized that it would have to live with irreversible climate change, intense storms and higher temperatures.

Some climate disruptions were slow changes that would go on for centuries. Warm water near the surface of the oceans slowly percolated to lower levels, causing the oceans to expand so that sea levels would slowly rise for 300 years or more. The long-term rise became accurately predictable, and many coastal cities developed detailed long-term plans for building seawalls. The situation was controllable in much of the First World but catastrophic in low-lying countries like Bangladesh. A more alarming effect was a self-amplifying feedback cycle. Tropical forests and their soil, which normally *absorb* carbon dioxide, became transformed by rising temperatures into *sources* of greenhouse gases. This triggered a runaway process that caused temperatures to continue rising. In other words, we had interfered with Gaia, the Earth's grand-scale control mechanism.

In many temperate-zone countries farmers set out to change their farming methods so that the higher temperatures would improve crop production. Canada greatly increased its grain production and shipped huge amounts of grain to China. A vast area of Russia, by far the world's largest country, has a very low population. It stretches from close to St Petersburg to the north Pacific coast. This area, south of Siberia, was green and lush before the global warming. As its temperature rose, it acquired the capability to become the world's largest breadbasket. Terrorism, to some extent, discouraged air travel. Families stayed in closer contact than before with wall screens and video cell phones. Food, which used to come on a journey of hundreds of miles, was often produced by local organic farmers concerned with achieving gourmet results. The tastiest strawberries came from local hydroponic units and local bakeries delivered fresh products to the home. The world changed from mass systems with brute-force use of resources to computer-intensive systems that saved resources.

In the warmer latitudes, on the other hand, farming yield dropped substantially. In many areas, there were droughts. A large section of central Africa had its farm productivity drastically reduced. In some areas, the disruption was so severe that it was almost impossible to have a workable social system. There were mass migrations of people, the largest in human history, some going to unworkable shantycities and many of them trying to break through the immigration controls of the First World. The non-white population of Europe slowly increased, and there were eventually more Spanish-speaking than English-speaking people in the United States. Extremist religious groups, advocating holy war, found no shortage of recruits.

In the 2040s, the first fusion power stations came onstream (possibly from boron-11 nuclei rather than hydrogen nuclei). A fear that the world would be short of energy changed to a realization that there could be abundant clean energy that causes almost no greenhouse gases. A fear that future humans would have to live on rainwater alone changed to a realization that, in a world with abundant energy, fresh water could be produced from seawater. Desalination of seawater enabled water-stressed coastal areas to bloom spectacularly. Some coastal cities planted vast numbers of trees, bushes and perennials. As the planting matured, it changed local climates, sometimes lowering the ambient temperature by 5 or 10 degrees Fahrenheit. Some areas of the Persian Gulf changed from barren, searing desert to an environment of lush greenery with outdoor hydroponics, brilliant architecture and artificial rain at night.

During the canyon period, many of the worst hit countries were those that were in the worst state to begin with. The rich high-tech countries felt comparatively little pain. Many developing countries made it through the canyon bruised but surviving. But the poorest countries, clinging to life by their fingernails, had severe death rates. Shantycities became worsening nightmares of violence, disease, rape, AIDS, unsanitary conditions and food shortages. Global pandemics of new strains of flu and infectious diseases found killing grounds in the shantycities, where they were unstoppable. There were enormous waves of human migration, often with nowhere to go.

The challenges and technology of the canyon era will cause every industry to be radically rebuilt. It will bring new markets of enormous magnitude.

- The car industry will eventually be rebuilt around fuel cells, with totally redesigned cars that are very pleasant to drive.
- The electric power industry will slowly replace today's coal and oil power stations with smaller distributed noncarbon units of different types.
- The nuclear power industry will slowly decommission today's second-generation power plants and build fourth-generation plants, ensuring that the nuclear power is totally separated from nuclear weapons.
- "Green" architecture will avoid today's massive heating and air-conditioning bills.
- Cities will be designed to be environmentally benign, with large traffic-free areas, many of them stimulating to walk in. Intelligent urban transit systems will go directly into the shopping malls and centres of activity.
- The medical industry will be transformed to focus on preventive and regenerative medicine and on enabling people to have a life span decades longer than today's.
- The communications industry will be rebuilt for fibre optics, wideband wireless and switching mechanisms that handle extreme bandwidth.
- The world intelligence community will be reinvented to have global computerized cooperation to combat drug use, money laundering and advanced terrorism.

- The pharmaceutical industry will be aware of the human genome and offer different remedies to people with different genes.
- The entertainment industry will have new technologies that allow maximum cultural diversity and that allow people to download products they are most likely to enjoy.
- Education, worldwide, will be transformed with massive libraries of electronic and computer-aided products.
- The food and agrochemical industries will be designed for a post–Green Revolution world, but with some areas that are severely short of water.
- Businesses of all types will be reinvented for global real-time trade networks, communication with superintelligent computers and individuals with exceptional new skills.
- Aspirations of eco-affluence will bring endless new opportunities for corporate profit.

During the period of the canyon, extraordinary new technologies will mature. To some extent these will help to solve the problems of the era, but some will create a worsening divide between the haves and the have-nots. People worry today about the *digital* divide. How much more worrying will the *transhumanist* divide eventually become?

In the history of technology there are certain inventions that were landmarks in changing the future—for example, in the 18th century, the steam engine; in the 19th century, the telephone; in the 20th century, the computer. In the 21st century an invention that will change the future will be the creation of wireless links that connect our brains directly to external electronics. This may be done with nanotransponders in the brain fluid, eventually with large numbers of such transponders.

The first direct links between our nervous system and tiny computers have already been made. The brain/computer linkage will grow from tentative experiments to robust connections and then to an immense diversity of types of brain/computer connections. This technology will have been maturing for many years before the brain suddenly becomes directly connected to the Singularity.

NEW CIVILIZATIONS

The 21st century could see the emergence of civilizations very different from those of the past and, perhaps, very different from one another. The question "What could a great civilization be like?" needs to be addressed today, not in 20 years' time, because it can help navigate the intense white water we are moving into. If we don't constantly address this top-level question, then the extraordinary advances in technology will probably lead to a stressed-out society with lifestyles far removed from the quality of life for which we now have the potential. Mushrooming complexity may bring a metastable civilization. Now is a time when human wisdom and creativity should be examining the new possibilities in 21st-century civilization.

What characteristics might a great civilization have? This question is unanswerable in detail. Later this century, humankind will achieve what is far beyond our present capacity to imagine. Nevertheless, we can ask what principles should guide us.

As we look at the efforts throughout history to apply sweeping social designs to a nation, too many have damaged rather than improved quality of life. Much of the 20th century was a history of good people being damaged by bad government.

In the 21st century, the rate of change will quicken. The faster change becomes, the less likely it is that centrally planned behaviour will work as was intended. Instead, we need highly adaptive mechanisms that are attuned to the needs of the public. Nonevolutionary, noncompetitive organizations will go wrong. Society's institutions need to evolve faster than they are inclined to. When change is very fast, centralized control can't hope to keep up. The best we can do is to establish principles within which fast and free evolution can occur.

What are appropriate principles for 21st-century civilization?

CIVILIZATION'S MORAL COVENANT

At any time in history, civilization has a moral covenant. Some behaviours are acceptable and some are not. The world's moral covenant has changed in major ways in the last three centuries. Its improvement is an ongoing process. Slavery has been abolished, burning at the stake and other atrocities eliminated. Gentlemen don't have duels any more. The unspeakable ingenious tortures of the Spanish Inquisition have steadily been replaced with total opposition to torture in most countries. There's been a drive for democracy, universal education, the ending of child labour, the recognition of women's rights and the creation in many societies of a social safety net. The term *human rights* came into general use only after the Second World War ended; in 1948 the General Assembly of the United Nations adopted its Universal Declaration of Human Rights, without a single dissenting vote. The rules of the World Bank and the International Monetary Fund are steadily evolving and need to evolve more. New bodies are needed to better support the global community and the environment. All of these changes derive from fundamental shifts in our understanding of civilization's moral covenant.

Many UN agencies and international bodies have their own rules and codes of behaviour—for example, the World Court and the World Health Organization. Such institutions will help build a vast network of treaties among nations. We can automate the implementation of the rules, and this enables the implementation of highly complex global control mechanisms.

This is a time when we are instrumenting our planet. Staggering quantities of instrumentation will funnel data into data warehouses, which supercomputers can put to work. The technology will help scientists model the oceans and the climate and understand the processes that need to be controlled. Elaborate computerized controls will make possible better management of what needs to be managed.

Future civilization must be sustainable in terms of the resources derived from the Earth. It must not take more from nature than nature can replenish; we must not meet our needs by stealing from future generations.

More than that, it must steadily repair the damage that has been done, where possible, helping the recovery of ocean life, the ozone layer, polluted waters, depleted soil, trees and other natural capital and preventing further loss of species. Civilization has to eventually live within its means. The sooner this happens, the less the irreversible damage. Environmentalism is not about tree-hugging; it's about cancer, lung diseases, birth defects, deaths from bad sanitation, food security, violent weather and quality of life. Eco-affluence needs to be explored as widely as possible. Today we are tragically damaging the planet with no good reason.

There should be worldwide understanding that the Earth's moral covenant has improved, is improving and will continue to improve in the 21st century. Chapter 13, "The Awesome Meaning of This Century," discusses aspects of the 21C Transition. The laws, treaties and codes of practice that help bring about these transformations are essential parts of our future. Education about individual aspects of the transformations is essential. Appropriate education of the public often needs to precede the capability to pass the desirable laws.

Perhaps the most glaring example of a need for change in the moral covenant is the extreme poverty, disease and food insecurity of the poorest countries. The gross inequalities that accompany globalization are one of the most difficult challenges of our time, but the abolition of extreme poverty is achievable.

THE HUMAN-POTENTIAL PRINCIPLE

A tragedy of humankind today is that most people fall outrageously short of their potential. A principle of a great civilization ought to be that it focuses intensely on how to develop the capability latent in everybody. The more that is done, the more we all benefit from one another.

Today most human beings are trapped in jobs, lifestyles or social conditions in which they develop only a fraction of their potential. Lives can be wasted in so many ways—drudgery work, watching bad television many hours per day, women being denied the potential that men have, a shop-till-you-drop culture or the conceits of high fashion. Most people could be far more creative. The potentials of human capability will become much greater because of the cornucopia of new technology and fundamental changes in the way enterprises are managed. The most important aspects of new technology are those that make people excited about what they do. Human capability that we now regard as brilliant will become widespread because of the amplifying power of technology. Higher forms of brilliance will emerge, and many of those will become commonplace.

The 19th-century philosopher John Ruskin preached that machines robbed workers of their nobility, freedom and individuality. The machines of the 21st century will be the opposite. Inability to use them well will rob workers of their nobility, freedom and individuality.

In the poorest countries, one can walk among multitudes of malnourished eager-eyed kids who have no hope and know that if any one of them had been adopted as a baby and brought up in a good home in Singapore or Rome, he or she might have been a teacher, musician or scientist. The human-potential principle needs to be pervasive, from the poorest destitute society to the richest high-tech society.

THE MUTUAL-ACCEPTANCE PRINCIPLE

In an era when mass-destruction weapons become cheap, perhaps the most important aspect of any great civilization is that it has a relationship of mutual respect with other civilizations. We need to strengthen the commonality among nations and learn to appreciate the things that are not common. Chinese, Indian, Western, Islamic and other civilizations will be proudly different. What is needed is not cultural conformity but the deliberate creation of a ubiquitous and deep spirit of mutual respect.

Today's era of interlinked civilizations is radically different from preceding eras. The Earth's civilizations are in the same melting pot. Only with a basis of mutual respect can they agree upon a core of common knowledge and common education that makes possible coexistence on a small planet without walls. The building of mutual understanding is a critical part of the 21st-century journey because weapons systems will become so dangerous. Uncontrolled antagonisms can cause us to blunder into a high-tech war.

The great civilizations of the past have often been focused on their own cultures. In the future, they will increasingly study the greatness of other civilizations. The first excursion into a different culture can be revealing and surprising and can give the visitor a great desire to know more. The widespread study and enjoyment of the cultures of other civilizations will help tie the world together and build mutual respect.

Japan, Europe and America have radically different and strongly held cultures. From 1938 to 1945, they engaged in brutal warfare with unspeakable atrocities. After Auschwitz, Hiroshima, Pearl Harbor and Nanking, they had the most powerful reasons to hate one another, but now there is mutual understanding and respect. Their corporations became linked to form the world's great engine of global wealth generation.

COVENANT ABOUT NATURE

Since I was a child, the Earth has lost enormous numbers of species of plants and creatures, and the loss rate is accelerating. Unless we do something to stop it, half the Earth's species could be lost. What leverage factor could prevent the loss of a vast number of irreplaceable species? Norman Myers (the cataloguer of perverse subsidies) realized that certain areas contain an exceptional number of rare and endangered species. He set out to identify these areas and called them biodiversity "hot spots." Thirty-four regions worldwide, covering just 2.3% of the Earth's surface, contain 75% of the planet's most threatened mammals, birds and amphibians. Laws and financing can be established to preserve the hot spots, or most of them. This will keep an astonishing number of plant and animal species off the endangered list.

The tropical Andes range has 20,000 plant species that are found nowhere else. Madagascar is one of the hottest of hot spots. New Caledonia, a small island off the coast of Australia, has a huge number of unique species. The tip of South Africa, an area less than the size of Vermont, has more than 8,000 plant species, of which three-quarters are found nowhere else. Myers asked governments to help preserve such hot spots but received no response. Eventually, the MacArthur and Gordon Moore Foundations provided about $850 million, the largest sum ever assigned to a single conservation policy. Much success is being achieved in preserving the hot spots.

I did a depth interview with Gordon Moore (of Moore's Law). He didn't want to talk about the number of transistors on a chip. He reflected on the special role of the 21st century. The measures available to us for avoiding bad climate change won't be taken quickly enough. So we need a Noah's Ark operation to preserve as many species as we can. How do we make the best of a damaged planet and learn the lessons it teaches so that we manage the Earth well in the future? Ironically, the cost of preserving the hot-spot areas of irreplaceable life is very small compared with the profits of the industries causing their destruction. Again, it wouldn't "cost the Earth" to save the Earth.

RADICAL TRANSFORMATIONS ARE VITAL

Some of the changes that should be part of the 21C Transition represent a break with the past almost as fundamental as the ideas of Nicolaus Copernicus, who concluded that the Earth was not the centre of the Universe but that it revolved around the sun. Copernican transformations encounter great hostility. René Descartes, the father of modern philosophy, insisted that science be based on observation and experiment, and the Catholic Church placed his works on the Index of Forbidden Books. The Church burnt at the stake the scientist Giordano Bruno for believing what Copernicus discovered, and it imprisoned Galileo Galilei for writing about what he could see with his telescope.

Baron von Clausewitz, the early-19th-century philosopher of war, is famous for the widely taught doctrine, "War is the extension of politics by other means."[1] One of the biggest Copernican changes is that there can be no all-out war among states equipped with nuclear/biological weapons. Following the von Clausewitz doctrine, national leaders evaluated whether war, or the threat of war, can achieve a political objective at a cost that makes it worthwhile. Like top business leaders, they did risk/reward calculations. If the calculation says *Attack,* there must be no holds barred; there must be absolute brutality until the war is won. With 21st-century weapons, however, no political objective would be worth the risk of an all-out war. Imagine 10,000 Hiroshimas, each followed up with genetically modified smallpox for which vaccines don't work. If we stumbled into such a war, it would be the end of civilization, perhaps for centuries. The means for avoiding any drift towards such a war must be studied and taught relentlessly and mechanisms put into place for preventing such a drift. There will be numerous lesser wars on Earth, as there have been since the Second World War, but not using weapons of *global* mass destruction. The von Clausewitz logic now says: Keep miles away from any slippery slope that could lead to war.

A Copernican change in religious teaching is essential for the future. It would be healthy if religious people regarded all the major religions as manifestations of one God, but many religious people will not accept this view because they believe that only their own religion has the true God. Nevertheless, it is vital to the future of mankind that religions teach tolerance towards other religions. In a world with end-of-civilization weapons, if some religions teach that other religions must be destroyed, there is no hope for the future of civilization.

In the 20th century, the economy was regarded as central to mankind's affairs and the environment peripheral to it. In the 21st century, the Earth's resources will become regarded as central, in that they have to be sustained. If we go too far in allowing the water tables to drop, the topsoil to be degraded, the rangelands to become desert and so on, we'll create an irreversible situation in which there is devastating famine. Economists have regarded the environment as a subset of the economy; now they need to regard the economy as a subset of the Earth's environment—a Copernican revolution.

Today, enterprises generally don't pay for natural capital. Because we don't include natural capital in corporate or government accounting, we get a false view of what is happening. Somehow or other we must account for the costs of natural capital. The change to such accounting will probably incur intense resistance.

In the 20th century, national sovereignty was generally regarded as something that couldn't be interfered with. In the 21st century, a head of state can't be left free to sell atomic bombs or viruses for mass-scale killing. UN Secretary-General Kofi Annan commented that "just as we have learned that the world cannot stand aside when gross and systematic violations of human rights are taking place, so we have also learned that intervention must be based on legitimate and universal principles if it is to enjoy the sustained support of the world's people." He wanted to establish a formal body of law about when the UN can authorize intervention in a sovereign state, including the capability to remove its leader by force if necessary. He referred to this as *the responsibility to protect* and created an international commission to establish a proposed set of rules relating to that responsibility. The commission established Core Principles of the responsibility to protect, the responsibility to react—including military intervention in extreme cases—and the responsibility to rebuild a state after military intervention.[2] The commission spelled out Principles for Military Intervention.

Again, this is a Copernican inversion of traditional behaviour, which seems necessary for a new era. Dealing with a situation like that of Iraq in 2003 should be something about which the international community has defined rules and a legal basis for action.

In 1997, an Earth Charter Commission was formed by the United Nations to draft a charter that would set forth the principles of sustainable development. The final version of the charter was approved in March 2000. The Earth Charter describes itself as a "declaration of fundamental principles for building a just, sustainable and peaceful global society in the 21st century."[3] The Earth Charter brochure is used in thousands of schools and has been endorsed by thousands of nongovernmental organizations, cities and towns throughout the world. Drafting the Earth Charter involved "the most open and participatory consultation process ever conducted in connection with an international document."[4]

In addition to what is in the Earth Charter, other vital issues need to be understood. Lowering the Earth's population growth rate can be achieved and should have high priority. We need to take urgent action to stop climate disruption, and this requires alternatives to petroleum and coal. The alternatives exist and could be brought onstream much faster than today. It is desirable to develop defence in depth against terrorism. This endeavour ranges from removing the reasons for terrorism to developing the most advanced counter-terrorist technology.

Some of the solutions that are critical to our future are inherently controversial—for example, a new generation of nuclear power stations (Chapter 7) and crops genetically modified so that they do well in water-stressed areas (Chapter 5). The cloning of humans, as was done with Dolly the sheep, should unquestionably be illegal, but the reproduction of pluripotent stem cells will lead to immensely important improvements in medical practice. These cells can't form a complete human being, but they have the potential to develop into numerous different types of cells in the human body. For example, they can replace damaged heart tissue or help rejuvenate the eyes of old people. This is a very important development in medicine—the most important since the discovery of antibiotics.

To address such controversies, top-quality science and committed research are needed, along with ethics based in deep philosophy. We need to be cautious until we have enough evidence that the technology can be trusted. There needs to be a massive and comprehensive collection of data, proof-of-concept projects and thorough long-term field testing. There need to be peer-reviewed papers discussing the concerns. To have a blanket "no" to GM farming in many countries, fourth-generation nuclear power or regenerative medicine is to foreclose options that are vitally important for getting us through the canyon. It would be like saying no to railways in the early days of steam (when some engine boilers blew up).

The 21st century brings the need for Copernican changes. Gentle changes that can be talked about at Town Hall meetings won't save us. We have to escape from some of the deeply embedded viewpoints of the 20th century. Vested interests will try to discredit the Copernican changes, sometimes in brutal ways. The Copernicans need to be sure they understand the journey ahead and need to plan well how to bring about Copernican inversions.

20

VALUES OF THE FUTURE

GREAT CIVILIZATION has two aspects to it: First, its structure provides democracy, religion and an advanced moral compact. Everyone has a clean and decent life with good education and equal opportunity. We heal rather than harm the environment. These are the basic foundations of society about which much has been written. Second, given that such foundations are in place, how do people enjoy themselves? What do they do with life that makes them feel good about it? The future world will become far more affluent than today; how should the affluence be spent? In the future, what will constitute a rich, worthwhile life of the highest quality? When you're at the end of life, what will make you say that you don't want to go—that you've had such a wonderful time?

In a century with computer superintelligence, the ability to modify genes, transhumanism and near-infinite bandwidth, will we enjoy life more? If not, what's the point?

Ironically, while America was trying to persuade most of the world to embrace freedom, open markets and democracy, America itself seemed to have developed a social malaise. Divorce rates and drug addiction had reached new highs. Sociologists carefully measured the proportion of people that were depressed, tense, on antidepressants or generally dissatisfied with life, and they found that the numbers of such people had seriously increased. While technology and skilled management have provided the capability to produce wealth and a vast proliferation of goods, attempts to measure quality of life suggest that it declined, at least for many people. There are increases in stress, boredom, violence, fear, depression and a sense of helplessness. The level of violence among kindergarten children rose alarmingly. According to the National Vital Statistics Report, the most common cause of death in people aged 15 to 34 is car accidents, the second most common is homicide, and the third is suicide.[1]

The incidence of major depression has grown in the Western world in the last few decades. This finding has been questioned because depressed people are more likely to seek help than in earlier times, and the health-care profession is more likely to diagnose depression. Because of these factors, studies of depression have been very carefully controlled to avoid anomalous results. Despite these controls, the studies indicate that severe depression in Americans has increased, even with the effectiveness of new antidepressant drugs.[2] It now affects about 15% of the people in the United States at some time in their life. One of the worst illnesses one can have is severe depression, which can incapacitate its sufferers for years, often rendering them barely capable of working, socializing, loving or sleeping. Major depression often relates to characteristics of a society that produce feelings of helplessness. In a world with 21st-century technologies, will feelings of helplessness become worse?

It would be a grand-scale irony if as a society became momentously wealthy it became trivial, pointless and depressed. This has happened many times before in history. Perhaps the most tragic irony of all would be the Fourth World drowning in worsening squalor and new infectious diseases alongside an immensely rich First World drowning in junk consumerism that makes few people happy. The horror of the poorest people fighting off starvation coexists with the wealthy fighting off boredom. The lifestyle of the West is becoming increasingly expensive but measurably slipping in contentment with life. If just 1% of the junk-consumerism wealth were transferred from the First World to the Fourth World and managed as Jeffrey Sachs wants to manage it, the destitute nations could be put on a staircase to a decent standard of living. Furthermore, volunteers travelling to different lands to help in education or health care would often find their life more interesting, exciting and worthwhile.

A top-level question is "How do we use our scientific and technological advances and increased wealth to make people feel that their lives are worthwhile?"

A difficulty in discussing quality of life is that we have no adequate measure of it. GDP doesn't come close, and if this is used as a measure, it is highly misleading. Some attempts have been made to produce a different index. In 1993, former US Secretary of Education William Bennett produced a study, published as *The Index of Leading Cultural Indicators*, to chart the social decline that has taken place while the economy has grown— divorce, crime, media addiction, cancer, mental illness. Another national gauge of well-being is the *Index of Social Health,* which is published annually by Fordham University's Graduate Center in Tarrytown, New York. Since 1985, the centre has studied the nation's health through the evaluation of 16 indicators affecting children, teenagers, adults and the elderly: infant mortality, child abuse, poverty, suicides, drug use, dropout rates, average salaries and health insurance coverage. The Fordham Index shows steady declines in the United States, from 73.8 in 1970 (out of a possible 100) to 40.6 in 1993.

But neither index comes close to measuring the degree to which life is worthwhile and enjoyable. For society to provide a good life, several foundations are desirable, such as a trustworthy government, a decent economy and a welfare safety net. Asked what constitutes a good life, people say things like "a healthy, safe life for children," "closeness to good friends and family" and "confidence in one's own self-worth." After family issues come concerns about the economy and the environment: "healthy food," "safe environment," "freedom of speech" and "job satisfaction."

If items such as the above are regarded as foundations for a decent life, what characteristics are desirable, over and above these? I have asked that question in lectures and discussion groups, and the following is a typical list of responses:

- Excellence in education; education for everyone
- Great TV, films and theatre
- Good sports events
- Great music
- Attention to spiritual values
- Plenty of leisure time
- Excellent *preventive* medical care
- Get-togethers with great conversation
- Beautiful parks and gardens; flowers and trees in the city
- Good hiking, or time with nature
- A society with more fun

There are four interesting aspects to such a list. First, what is on the list can be achieved without harming the planet. Second, the items on the list are not outrageously expensive. Third, while today we worry about automation putting people out of work, many aspects of quality of life require nonautomated work—for example, talented entertainment and beautiful gardens in the cities. Fourth, there is almost no correlation between the major technology thrusts described in Part 2 "Technologies of Sorcery" and the items on this list.

HIGH CIVILIZATION

The term *civilization* has two meanings. It is commonly used to refer to a large body of people with coherent beliefs, customs and behaviour patterns that have accumulated over centuries. Our history is a history of civilizations. Civilizations used to be based on geography. We speak of Roman, Greek or Chinese civilization. Now people with common beliefs, customs and behaviour can be scattered around the world, united by the forces of globalism.

A different meaning of *civilization* implies a highly advanced state of human culture—high civilization. It is generally agreed that there were certain periods in history when society achieved an extraordinary intensity of cultural interaction. Historians usually cite three eras as being the paragons of civilization: the Renaissance in Italy, Athens in the age of Pericles and Paris at the time of Voltaire. These periods were characterized by an intense attention to the arts, literature, theatre, debate, reason and architecture. Other historians would add the first two centuries of the Roman Empire and certain periods in Chinese and Persian civilizations.

In the eras of high civilization, being able to achieve and appreciate excellence in culture gave a much higher and more refined level of pleasure than a noncultural life. To be able to fully appreciate the best literature, poetry, music, theatre and so on, a person needs focused education. Acquiring it requires time and attention. In a time of high civilization, numerous people have such education, participate in cultural activities and stimulate one another, as they did in Leonardo's Florence and Shakespeare's London.

Presumably, periods of high civilization will be achieved again. Future civilizations, when machines do much of the work, could be an age with a flowering of great literature, art, classical music, philosophy, science, computer games, martial arts and sport—an age with the managerial competence and political skills to know how to make things work smoothly. The skills of great theatre would be represented in film and electronic media. Perhaps there will be two levels of film industry, one aimed at the masses with large production budgets and low ticket prices and one aimed at high civilization with low production budgets and ticket prices akin to those of the theatre.

The radiant periods of high civilization in the past existed for a century or two. In the future, these periods may be longer, even permanent, because the learning upon which the high civilization is based will be captured and enhanced by technology. Civilization may be global and deeper because it is aided by artificial intellect and Internet resources.

It is important to distinguish between an age of greatness and an age of high civilization. Nineteenth-century England and 20th-century America are among the great ages of human history, but they were well below the high-water mark of high civilization. An interesting question is whether the 21st century will be remembered for high civilization, as well as its critical transformations.

There seem to be good reasons why the high-water mark will not only be reached again but greatly surpassed. The high-civilization participants, however, may be only a small fraction of the people. That was true in the past. At the peak of Athenian civilization, Attica had a population of about half a million, but only 22,000 citizens had the right to vote. By today's standards, Athens would be condemned for civil rights violations.

A NEED TO WORK?

In 1970, various authorities tried to predict the impact of computers. It was expected that computers would increase in power at a Moore's Law rate. By 2000, a computer would be a million times more powerful than in 1970. It seemed that much of the work of 1970 would be automated by the year 2000. Automation, it was thought, would buy us more leisure. In 1970, it seemed that a 3 ½-day workweek would make sense (3 days one week and 4 days the next).[3] Expensive factories and office blocks could then be in use full-time, while people worked half-time. An important question seemed to be, "What would people do with their increased leisure?"

In reality, the predictions about computer power were correct, but the predictions about leisure time were completely wrong. Most families had less leisure time in 2000 than in 1970. Many men were working 10-hour days, and their wives were in the workforce, too. Day-care centres were needed to look after their children. Often both husband and wife were stressed out. Perversely, the greater the use of automation in a society, the more overworked people seemed to be.

They *could* have had more leisure. As technology gives society increased wealth, we can spend that wealth on more leisure or more goods. The public has opted for more goods. In 1970, a typical family was content with one car, often costing $3,000 (as mine did); in 2000, such a family wanted two cars costing $30,000 each. A family receiving one letter a day in 1970 might receive 50 e-mails a day now, many of them demanding an immediate response. Television became very efficient at making people spend money on things they didn't need.

In the next 30 years, computers will grow in power by a factor of a million again. Robots will steadily acquire intelligence until, eventually, the robot industry explodes as the Internet did in the 1990s. Again we might propose a 3 ½-day workweek. It makes sense, but will it happen? Will we continue the trend of the last 30 years and have more goods, or will we break the trend and have more leisure? The trend to increased consumption seems to be associated with increased stress. There is a limit to how far it can go. We are so steeped in our own work ethic that we don't realize that it is peculiar to our own time.

Perhaps the greatest gift we will receive from technology will be that the need to work will lessen, and most work will become interesting. Today's factory worker or accountant spends the most active days of his life working. Soon, largely automated factories will churn out most of the goods we want, with fully computerized accounts. In the future we'll have longer lives and greatly expanded leisure.

For past generations, quality of life was strongly affected by work. In the future, quality of life for many people will be mainly affected by what they do with their leisure. Western society may have difficulty adjusting to increased leisure. People will range from those who do no work to those in such demand that they never stop working. There'll be intense political debate about how those who work hard support those who don't.

Future society needs education for leisure. As with the idle rich of the past, some people will be bored or drugged or find ways to get in trouble. A deep sense of ennui may pervade a society with too much leisure. The grand aristocrats of the past were often bored out of their minds. One could imagine a leisured class of besotted hedonists in pursuit of meaningless sensations of ecstasy. Some Chinese emperors had tens of thousands of concubines—enough to sleep with a different one every night of their life. The future may have designer drugs producing nonstop sensations of rapture. In the past, the idle rich were 1% of society; in the future, they may be most of society. A grim scenario for the future is one in which people disintegrate mentally through boredom, lasciviousness and drug use.

THE VALUE OF TIME

Increasingly, as the opportunities of life become richer, one of our most valuable commodities will be time—time for romance, time for adventure, quality time with our children, freedom to develop passionate avocations—perhaps particularly important, time to acquire the education needed to be civilized. *Time becomes more valuable than money.* A critical aspect of the future will be learning to use our time much better, maximizing how worthwhile our life is.

Badly designed technology robs us of time; we can waste hours with software that doesn't work well, or watch endless sludge on television in the hope of being amused or informed. We are seduced into spending time in decivilized ways. In the future, we'll want to communicate instantly, find information instantly and watch television at times of our own choosing, not the broadcasting company's.

People will want media that entertain them when *they* want to be entertained, find the news that *they* are interested in whenever *they* want it and that allows them to drill down into detail. The public will have intensifying impatience with media. The generation of teenagers 10 years from now might be referred to as the "fast-forward generation." They will look back on today's television as being time-wasting and excruciatingly boring—all those tedious talk shows, repetitive irrelevant commercials, news stories you don't want—but worst of all, you couldn't press the fast-forward key. Instead of being tied down by broadcasts, people will watch television with a unit like TiVo boxes that capture the television they want and allow them to watch with their finger hovering over the fast-forward key.

Our television set will be a computer that helps us to find items we might be interested in and stores them. We can play the items that interest us and skip those that don't. Broadcasting will transmit not only television but anything digital—text, photographs, music and large bodies of software. We will interact with some broadcasts the way we interact with DVD-ROM disks today, and we'll want to do so with the fastest response times. Users of Yellow Pages will be able to expand click by click into a gigantic global mall of full-motion video. We'll visit online doctors' surgeries, or use portfolio-management facilities offered by hedge funds.

Today's TiVo boxes will evolve into personal media machines designed to learn as much as possible about their owner's likes and dislikes and record what he or she wants. Such a machine will store thousands of hours of television so that it can always keep its owner entertained or informed, constantly finding the best material for its user. Personal media machines will give the public the ability to surf television as they surf the Web, hot-linking from one item to another. News channels will become very different when users employ a fast-forward key enabling them to skip from one TV or text item to another, searching for explanations or commentary. It will become expected that news will blend television, newspapers, editorials and education with the ability to explore deep archives. Perhaps all mass-produced television sets will become forms of personal media machines. These machines will be the subject of highly targeted marketing campaigns from which the machine will select what it knows its owner might want to see.

Perhaps many people will want to escape from the frenzy of the future and find a higher-quality lifestyle in high-civilization cultures. Electronic education, accessible anywhere, can have a major effect on helping someone to appreciate music better, or find enjoyment in great theatre. With appropriate guidance, a person may find immense entertainment in the wit of Ben Jonson or Molière and in watching how different actors have created outrageous characters from these plays. Some Shakespeare disks are more fascinating than books, films or theatre because they allow the user to explore the nuances of the language used and to see how different directors have interpreted the work. Great plays can be staged with radically different social implications. There is endless scope for creativity in exploring theatre.

Electronic education will become much deeper as it becomes aided by superintelligent computing. Imagine the History Channel evolving into an electronic warehouse of television linked to software that specializes in the different explanations of history, with input from philosophers and historians in civilizations that see things differently. Education for leisure will do more to improve quality of life than education for jobs.

DRUG-CHANGED VALUES

In Thomas Jefferson's view of civilization, the pursuit of happiness is an inalienable right, but this pursuit can be done in destructive ways. As we described in Chapter 12 ("The Transhuman Condition"), we have many drugs today that affect our brain chemistry, and a bevy of better-targeted drugs are on their way. Before long, we'll have tiny capsules in the brain capillaries that can release a precise drug at a precise place in the brain.

Movies may be designed so that you take a capsule when watching them. A digital movie may be designed to transmit wireless signals to the capsule that make it release chemicals in your brain to give intense feelings of alarm, sexiness, tension, excitement or euphoria at appropriate moments in the film. Some pills will make you laugh unreasonably throughout a comedy.

For most aspects of our brain's functioning, feelings and brain chemistry go hand in hand. When we fall in love, for example, a certain chemical, perhaps located in the hypothalamus, soars in concentration.[4] When Romeo says of Juliet, "The all-seeing sun ne'er saw her match since first the world begun," it's not logic talking, it's chemistry—a specific chemical, which we'll be able to deliver directly to the hypothalamus with a microcapsule. Some pills will work wonders on a first date. Some will be part of the armory of a modern Don Juan. Some will be used by women uncertain of their allure.

Intense religious feelings, like those of mystics, have been produced by stimulating the brain. Neuroscientists think they know what abnormal chemicals may have been in the brain of Joan of Arc when she heard voices. Five centuries after her "commune," the Roman Catholic Church made her a saint. One night in August 1951, in the village of Pont-Saint-Esprit in southern France, about 300 people poured into the streets screaming, overwhelmed by visual hallucinations and religious sensations. They had all eaten bread from the embarrassed village baker, who had used flour that was diagnosed to have a mold that caused the problem. In history and folklore, there have been many such stories of collective hallucinations, some giving rise to religious legends.

The argument for high civilization is that when we make ourselves very knowledgeable about music or art, we can achieve intense feelings of pleasure, but we also obtain pleasure from psychotropic drugs. Is there a difference? Surely, you might say, intense pleasure induced by chemicals can't be the same as intense pleasure from listening to Mozart. Suppose that, when you go to a four-hour Wagner opera, you are given pills with your ticket, and these keep you in a state of euphoria throughout the performance.

Hegel believed that the historical process was fundamentally driven by the desire for recognition. Much of the competition for corporate dominance derives from CEOs struggling for self-esteem. Now a person can be given a high sense of self-esteem by using Prozac. Soon there will be far more effective drugs than Prozac. Much of life's achievements are driven by constant struggles for certain mental states, which will be obtainable with time-release microcapsules. The bioethicist Leon Kass asks, "If life's most intense satisfactions, even love, come from the chemist, are we truly human?"

In my depth interview with Freeman Dyson, he said he is scared of future neurology. "The present wave of biotechnology is limited to what you can do with genes, but neurology will enable us to start monkeying around with brains. There'll come a time when we'll understand brains as well as we understand genes. I find that very scary because we could easily produce a complete disconnect of people from reality." He commented that artificial mental experiences may be fatally easy to induce. "A society addicted in this way to dreams and shadows has lost its sanity."

Understanding how to improve civilization when we can interfere with the brain is a deep subject that deserves much philosophy, experimentation and knowledge of the state of the world.

THE QUALITY PROBLEM

Electronic technologies are causing humankind to accumulate an overwhelming deluge of digital material. The deluge will become larger at a rapidly accelerating rate. This global deluge will be searched with Googles of the future. As it grows, an important question becomes, "How do you find the good-quality items in the vast ocean of sludge?" We will need the ability to separate the sheep from the goats, because as the deluge progresses there may be one sheep for a million goats.

There are wonderful research papers, accessible to anyone on the

Internet, worldwide, but most of what is on the Internet is junk. The search engines search the junk. It's as though they could instantly search through every rubbish tip of every town on the planet. How do you decide whether you want to see a movie if the Hollywood PR people are masters of hype and every village idiot writes film reviews?

There are two good answers to this question: First, the quality system needs not only to review products but to review the reviewers. Reviewers who are trusted by a specific clientele need to have quality ratings themselves, and to make sure that they remain worthy of those ratings. The red and green Michelin Guides are wonderful examples of this. They give unbiased quality ratings of restaurants and travel locations and have earned respect over many decades. We need digital Michelin Guides for digital media.

Second, the advice you accept may not come from one reviewer but from the aggregate opinion of a large number of like-minded movie fans. Which films you would enjoy is a matter of personal taste. You would like to access the views of people who share your taste. If you produce a list of your favourite films, computers can compare this with the lists of a large number of people. The computer identifies people with similar tastes and suggests that you watch what they liked. If, every time you watch a film, you fill in a simple chart indicating what you liked or didn't like about it, computers will continuously refine a profile of what you enjoy. They will find people with a similar profile. By comparing vast numbers of such profiles and continuously refining them, they will be able to tell you, with surprising accuracy, how you will react to movies you haven't yet seen. These two types of quality assessment can also be used with books and music and other things we search for on the Web.

How do you separate respectable science from tobacco science? *Nature* and *Science* are publications that check the validity of the research material that they publish. This is essentially a human task. In science, the peer-review process works reasonably well for academic papers. Something like it needs to exist for digital media in general. A huge body of medical information is becoming available to the public. Increasingly, the public will monitor their own heart and blood pressure and do their own blood tests with laboratory-on-a-chip devices wireless-linked to their computer. Your car, toilet and pyjamas may monitor various aspects of your "health" and may transmit the results to your computer, but if your computer gives medical advice, who will scrutinize the respectability of such advice?

A characteristic of film and television all along has been low cultural diversity. In the past, television had very few channels, and films were expensive to make and risky to invest in. To achieve a decent return on investment, they were often aimed at a low-level mass audience—close to the lowest common denominator. New technology, however, will enable maximum cultural diversity—DVDs, video-bandwidth Internet and peer-to-peer networks for exchanging uncopyrighted material. The future will bring low production costs so that low-budget groups can make films and television. Musical groups who have no chance of finding a publisher and filmmakers who have no chance of traditional distribution will publish their own work electronically.

Digital civilization will be global and will have an ever-growing and ultimately overwhelming quantity of data and products. This is tenable only if there are strong means of quality control. Processes of refinement and enhancement need to go on continuously, distilling out the best and improving it and hiding the worst. Without such processes, we'll all be drowning in an ocean of mediocrity.

In the world of high civilization, there will be guides to what are the greatest performances.

BEAUTY

Beauty, high civilization, cultural diversity and the attractiveness of a leisure society can greatly add to the richness of life without causing pollution, global warming and a stressed-out society. Beauty is something that we can't measure directly, but beautiful surroundings, homes and cities increase our feelings of happiness. A great civilization needs a great sense of beauty—an aspect of society that's astonishingly absent in most academic treatises, perhaps because beauty has no objective measure. We love to walk in Paris or Venice—yet we build cities like those to the west of Chicago, in which there are no pavements.

I hope that our civilization, as we rebuild it, will regard beauty as of prime importance—beauty in cities, gardens like those of Persia in the 16th century, sculpture more like that in Italy than that in Switzerland, and extraordinary stonework like the Taj Mahal, even if it's put in place by robots. Future cities may be designed for people, not cars, with traffic-free precincts, open-air cafés, rose gardens, elegant bridges like in Paris, woods and walks by lakes. Future cities may compete for the most beautiful architecture with sculptures and bird life, mosaics and tapestries and long pergolas draped with flowering vines.

Freeman Dyson described to me, with a puckish grin, his vision of the Philadelphia Flower Show. "There are tens of thousands of people who are dedicating their lives to breeding flowers and bushes and creating these beautiful gardens, and I can just imagine how much more beautiful that will be in 10 or 20 years. When biotech is available to these people, they will all have their little take-home, do-it-yourself kits so that they can manipulate the genes and create their own roses and other new species—a huge explosion of biodiversity." This, he said, is the opposite of what people fear—that biotechnology is going to result in monocultures because big companies, looking for an optimum species, bioengineer everything to be the same. He commented, "When the individual rose enthusiast and orchid grower gets ahold of it, it will be the new art form of the 21st century. You'll have a whole lot of these artists, good and bad, producing works of art." The creation of genetically modified orchids could become a hobby to rival the orchid-collecting mania of the 19th century.

Experimenters with GM have discovered that a gaseous hormone called ethylene triggers the ageing process in flowers. They have learned to minimize or switch off the effects of ethylene so that beautiful blossoms last much longer than normal.

I used to think that Iguaçu Falls in South America was the world's most beautiful waterfall, but a man at Iguaçu told me that he knew of a "secret" waterfall, even more beautiful—the largest volume of falling water in the world. He said it was the most beautiful place on Earth, but if I wanted to

see it, I had better do so quickly because it was going to be destroyed—
with money from American banks.

I found his falls on a map. It was called Sete Quedas and was indeed
listed in the *Guinness Book of World Records* as the greatest waterfall on Earth.
There seemed to be no road to it and no airstrip. The Paraná River, crossing South America, has a stretch in deep virgin jungle between Brazil and
Paraguay where it is 5 miles wide, and that is where Sete Quedas fell—
about the height of Niagara, with three times as much water as Niagara.
The man took me there. The rugged islands through which the cascades
thundered had numerous exotic flowers and vines that thrived in the constant warm mist from the falls.

Engineers were busy changing the course of the Paraná River and
building a dam 4.8 miles wide for electricity generation. The concrete
poured for the dam was 15 times that in the "Chunnel" under the English
Channel. The steel used would have built 380 Eiffel Towers. The world's
biggest waterfall is now gone. Nobody protested because potential protesters didn't know it existed.

There have since been detailed studies of the impact of the dam. The
river water was raised over 600 feet and created a vast flooded area
125 miles upstream. Almost 85% of the forest along the Paraguayan portion of the Paraná River was destroyed. Vast numbers of species, including
rare orchids, became extinct. Paraguay has only 3 million people and little
industry; it uses less than 2% of the electricity generated. Ironically, this
giant facility started to generate electricity at the same time that the electricity industry was entering an era of downsizing. Small distributed power
units were becoming more economical, especially in areas of sparse population, like that part of South America. The Itaipu Dam need not have
been built.

An aspect of the story not mentioned in the academic studies was that
it was the most beautiful waterfall on the planet. How much was the beauty
worth? We can't give it a numeric value. It should have been a World Her-
itage Site that could not be destroyed under any circumstance. Tragically,
local folklore says that one day, when society develops better ways of generating electricity, the Sete Quedas falls will be brought back to life.

Nineteenth-century pictures of medieval universities show astonishingly beautiful environments where university members met one another in the streets and debated ideas. Traffic destroyed this beauty. In the future, a civilization without a sense of beauty should be regarded as barely civilized.

21

CATHEDRALS
OF CYBERSPACE

THE AWESOME CATHEDRALS of the early Gothic era sometimes took over a hundred years to build. Notre Dame cathedral was initiated in 1160. The choir, the Western façade and the nave were completed in 1250. Porches, chapels and other items were added over the next hundred years. The large clerestory windows of the apse were built from 1235 to 1270. Reims, the third of the classic Gothic cathedrals, was begun in 1210 and completed in outline in 1311. The men who designed them and who first worked on them knew that they would never see them finished.

The sculpture, the stained glass and the capability to convey the soulfulness of religion have not been equalled since. The detail of the stained glass and sculpture high in the tall cathedral is exquisite, but you can't see the detail without a telescope. The sense of values had nothing to do with profit. This craftsmanship was not just for humans (who didn't have telescopes); it was for God.

A great civilization needs the 21st-century equivalent of cathedral designers—people who can start projects that extend far beyond their lifetimes—and who ask the question, "What goals, today, could be so grand that the work would continue with passion long after those who originated it are dead and gone?" What will be the cathedrals of the 21st century?

Twenty-first-century cathedrals may be cathedrals of the mind, far more complex than cathedrals of stone. When we die, the memories and intelligence stored in our neurons and synapses become dead meat, but computers act as an ever-growing accumulator of memories and intelligence. They provide a new form of immortality for human knowledge and creativity.

We'll contribute to software that will eventually evolve at staggering speed. Most of the information on the Internet is encoded so that humans can use it; in the future it will be encoded so that computers can use it with their relentless logic. The world's textbooks need recoding for computers so that their knowledge can be put to use automatically. There will be a vast, ever-growing body of knowledge designed for machines to use.

GLOBAL COMMON GOODS

There are certain goods that are public, in the way the village commons is public. If they are global, they are referred to as "global common goods." The term could refer to the oceans, atmosphere, ozone layer, food security, international forces to help with earthquakes and so on. We can create numerous types of global common goods that haven't existed before.

Common goods are interesting when their aggregate benefits are much greater than their cost. Medical knowledge can be in such a form that it can be automatically used by computers. The Internet and software give many opportunities for such global common goods. The radio spectrum has been reallocated so that cheap portable devices can be used everywhere. The "grid" refers to applications made possible by the Internet but implemented with computerized intelligence far beyond that of the Internet. As computer systems multiply and drop in cost, shared services will become more and more valuable.

Many types of software and databases will become globally shared goods. "Open" consortia agree to share and maintain such resources. Three-dimensional coloured maps have been standardized to make complex information usable—for example, information showing soil types or crops being grown. The Earth is becoming instrumented in numerous ways that will help manage its complex problems. Numerous evolving databases are becoming public and free.

The Lawrence Livermore National Laboratory, which sequenced (mapped) the human genome, is now sequencing the genomes of everything it can get its hands on—other creatures, plants and microbes—creating "instruction manuals" for life on the planet. This extraordinary quantity of new knowledge is posted on the Internet as soon as it is obtained so that anyone in the world can access it free of charge. As researchers learn what to do with it, it will be a "commons" of very valuable knowledge.

Perhaps the most valuable future "commons" will be education facilities on digital media that can be accessed almost everywhere. Unlike books, digital products can be replicated and distributed globally at almost no cost. There is nothing *physical* either to manufacture or to distribute. The huge body of education accessible on the Web will range from primary education to the highest-level courses for professionals. The quantity of educational material will grow relentlessly and will steadily become better in quality.

Most of the world's great books for which copyright has expired are available free on the Web. There is a massive amount of educational material on the Web that is free of royalties. It may make sense to have treaties that make royalties expire on globally marketed digital education earlier than they would with conventional copyright law. A computer network of an advanced civilization should have the best teaching modules possible, with superintelligent computing.

The grand total of human intelligence becomes a "commons" of overwhelming value if it can be shared. As the planet shrinks, it will have enormous quantities of shared resources. A massive 21st-century opportunity will be the creation of "cathedrals" of global common goods.

ARTIFICIAL INTELLECT

The term *artilect* has been used to mean an "artificial intellect." Artilects will be designed for specific purposes, such as being an expert on horticulture, or making money in highly specialized types of hedge funds. They'll be designed to improve their knowledge of their subject and their capability to reason about it, constantly adding facts and rules to their knowledge bases and automatically learning so that they can achieve results better.

There'll be numerous types of artilects, expert in different subject areas. Each artilect may have a defined purpose, and tight security may be used to make sure that it can't deviate from that purpose. Artilect capability will be used in conjunction with data warehouses. Previously imprecise subject areas will be converted into disciplines based on massive bodies of precise data and computerized ability to recognize patterns in those data. Homeopathic medicine today, for example, is largely a black art, but it can probably be transformed into a precise computerized practice. Machine intelligence will impose a discipline in areas that were undisciplined.

There are endless ways in which artilects will help improve society. Consider personal health care, for example. An artilect, available to anyone via the Internet, will be designed to have thorough knowledge of medicine. It will constantly add to its knowledge. It will also have knowledge of vast numbers of individual patients, knowing their history and DNA (but obeying strict privacy laws). It will understand the capabilities of new drugs, sensors and nanocapsules that we swallow or inject. It will recommend tests, look for symptoms and recognize patterns that human doctors can't recognize. Medical knowledge will be encoded in such a way that computers can make deductions about a patient and come to logical conclusions. The artilect will be designed to learn from experiences with many millions of patients, worldwide, thus acquiring capability far beyond that of any one human doctor.

Machines will enable the intelligent public to take charge of much of their own health care. Electronic devices will monitor your heart, blood pressure and blood tests. If you feel ill, you may have a dialogue with your computer, and it will tell you whether a human doctor is needed.

The creation of highly capable artilects will require an enormous amount of work. These will be some of the so-called cathedrals of cyberspace— works that can be useful to or admired by future generations. The work of creating capable artilects could be parcelled out to vast numbers of personal computer users around the planet. There'll be all manner of projects that harness the collective power of many millions of home computers and the collective intelligence of their owners.

The very primitive precursors of artilects exist today. People have become used to the idea that machines play chess better than they do. Machines excel at optimizing the wiring layout of an ultracomplex chip. Artilects have applications in genome research, homeland security and financial investment and are put to intensely competitive use in electronic commerce. An artilect might be an expert on terrorism, Chinese medicine, 12th-century stained glass, skin cancer or the dynamics that lead to war. It may be familiar with all money managers specializing in foreign exchange. It may make the best prescriptions in homeopathic medicine. Artilects will inevitably develop that are quite different from human intellect.

Artilects will soon be designed to intercommunicate. This will be important for certain applications—for example, money management. Perhaps the most important aspect of money management is knowing when to switch from one category of investment to another—when to switch from high-tech stocks to commodity trading, for example—and how best to spread a portfolio among investment categories. Many brilliant financial analysts are poor at making this decision; they are expert in one investment category. An integrated artilect system for investment would estimate the expected return and risk of many investment categories and be able to assemble baskets of fundamentally different types of investments that have different performance estimates. As artilects become very powerful, they will link together on global networks. Twenty or 30 years from now the world economy will be more dependent on artilects than it is on petroleum.

Artilects are merely a collection of bits. So, in a sense they are immortal. They can be transmitted, updated and replicated so that when physical damage or failures occur, the bits won't be lost. They won't grow old, as we do, but will grow smarter at an exponential rate. Artilects of rapidly growing power will become integrated. Once they mature, they will evolve like a chain reaction. Part of the 21st-century journey is learning to live with artificial intellects incomparably beyond our own intellect.

JEEVES

Early in the 20th century, wealthy Englishmen achieved something close to perfection in the art of having a personal butler, like the legendary Jeeves of the P. G. Wodehouse stories. Jeeves knew about his employer's desires and troubles and was endlessly discreet. In our future, it may make sense for each of us to have a computerized Jeeves.

As many 21st-century technologies become "infinite in all directions," our world will become one of overwhelming complexity. We'll need a guide. The guide will have to be a highly personal assistant that knows its owner intimately and understands his or her requirements because each individual's needs are different.

Today, people worry about privacy—about computers knowing too much about us. Tomorrow, they'll worry about the opposite: We'll want a very personal machine that knows everything about us so that it can provide the best possible assistance. Of course, we'll want it to be as discreet as Jeeves. A golden rule for such a machine or service will be that it keeps information about its owner absolutely private and uses tight security that is almost impossible to break. In some ways, it will be like a trusted family doctor, who knows intimate family details but does not pass them on to other people. Its owner will learn to trust it as he or she would a family doctor.

The automated Jeeves will be in a totally different league from today's personal computers. It will be designed to be a close companion, more perfect in some ways than human companions because it won't have human foibles. It'll be dedicated, with computer relentlessness, to the happiness and well-being of its owner, but it will also be able to contact millions of other such Jeeveses on the Internet.

An electronic Jeeves would not hover around like an English butler; it would be built into your clothes, wallpaper, car and jewellery, and it would be in wireless contact with you anywhere you needed it, constantly aware of whether you are happy or frustrated, sensing when you need something. Sensors in the walls will detect your presence and anticipate your wishes.

Computers may be designed to display subdued human-like emotions, like an English butler.

We tend to think of intelligent machines in terms of whether they beat us at chess or solve complex problems, but they may have a greater impact on our lives through their ability to understand our emotions. Where a computer's reason for existing is serving a human master, psychology and psychotherapy will be among its primary bodies of knowledge.

Machines that speak to us with a synthesized voice are often insufferably boring. An automated Jeeves will be designed to become a sensitive conversationalist. Since a lot of people live alone and get lonely, there will be a market for "talkies" that, over the years, will get smarter and have a richer vocabulary, a greater learning ability, a larger memory and so on. Intelligent conversational computers will be able to adapt to their owners by building up understanding of their owner's interests and knowledge levels and will behave in a comfortable, familiar way, as a spouse does after years of marriage.

Today we say, "How could we ever have lived without the telephone?" Tomorrow, we will say, "How could we ever have lived without our computer servants that know about our likes, skills, phobias, needs, taste, medical issues, garden, kids and the rest?" Jeeves software will understand the fundamentally different ways in which people can achieve a high quality of life and will help them to do so. It would know what music and films its owner would like and what would make him or her laugh. It may assist its owner in creating poetry, or composing music. If you go for a country walk in a new place, it will know what wildflowers are in bloom.

Do you wish you could find a perfect (human) mate? Your Jeeves machine can find candidates at the speed of Internet search engines and get to be on intimate terms with *their* Jeeves machines. The Jeeves machine may help people form and maintain healthy social relationships with other people. It may provide wise counsel and help in resolving conflicts. It may act as a supertherapist to people who need one, detect prejudice, explain injustice and have important insights to share with human shrinks. This capability to bring people together may become one of the most important effects of such technology. It will fundamentally change society.

GRAND ADVENTURE

With the exception of the deep oceans, there are no physical new frontiers left on Earth. There are plenty in space. If space is to provide adventure for a substantial number of people, we'll need major human colonies beyond the Earth. One candidate for a 21st-century cathedral that will take a century to create is a civilized human settlement on our next-door planet, Mars.

Freeman Dyson talked to me about expeditions to Mars: "The expeditions to Mars up to the present time made no sense because they were always thinking in terms of 10 years or something like that. In 10 years, it doesn't make any sense, but on a hundred-year timescale, it does. You've got to probably spend 50 years engineering your plants and then another 50 years developing an ecology and finding out how to do the farming so that they actually survive. So, it's that kind of a project." Mars is slightly smaller than Earth and has a day of 24 hours and 37 minutes. It rotates with an axis tilted about the same as the Earth's—thus, it has a somewhat similar day and night. Because its year is 669 Martian days, the seasons last almost twice as long as on Earth. This length of day, night and seasons will allow animals and plants to live as on Earth. Many plants will need to be species especially created to adapt to the long winter. The prospect exists of large-scale colonies and agriculture on Mars.

The key to putting people on Mars at reasonable cost is to let computers prepare the way. Long before humans land, robotic devices should have explored the place and created as much of the infrastructure as is economic. There are two critical factors in keeping the costs down: First, *avoid the use of people for activities that can be automated,* because the life support systems and safety engineering needed for humans are very expensive. Second, *live off the land.* Manufacture fuel there for the return journey,[1] grow food on Mars and exploit Mars's mineral riches to build the bases needed.

Going to Mars is not about doing something as unproductive as the moon shots. It's more like the British establishing the first base in America at the start of the 17th century. The goal is to develop a civilization on Mars. It will take less time than the development of America took. The time from Jamestown to Jefferson was 160 years. The time from the first manned base on Mars to having another home planet for humankind might be 40 years.

The colonization of Mars will provide extraordinary lessons in sustainability. In the beginning, there will be freighters bringing large gobs of the Earth's biosphere with them. Quality of life for the settlers will be consciously designed to fit in with care of the environment. One couldn't imagine a much more interesting laboratory in which to learn lessons about planetary maturity.

Much of the public thinks Mars is boring because of photographs showing the early machines there on rock-strewn plains. There are mountains on Mars, however, that are three times as tall as Mount Everest, vast ancient volcanoes and canyons three times as deep as the Grand Canyon. One rugged canyon, Valles Marineris, is almost as long as America is wide. Olympus Mons is the largest mountain in the solar system—16 miles high. Now, there's a challenge for mountaineers.

We will eventually have magnificent photographs of Mars. The Ansel Adams of Mars may use many flying cameras propelled by tiny jets, with servomechanisms providing virtual tripods and camera settings adjusted by wireless. Crowds on Earth may flock to the IMAX cinemas to watch the spectacular progress of exploring the solar system.

Long before human explorers go forth, trillions of bits of exploration data will be examined. Machines will search relentlessly for the best potential crop-growing soil and for minerals needed for manufacturing. One of the most important concerns will be to find places where water is accessible.[2] If there are underground aquifers, as seems likely, these will become locations for manned bases. It will be desirable to search for areas where there might be underground geothermal activity.

The economics of automation on Mars will be so different from those on Earth that it will have its own engineering disciplines. Computers will be cheap, telecommunications ubiquitous and people scarce and expensive. Gravity on Mars is 38% of that on Earth, and most of the Earth's sources of corrosion are absent; so, it will be possible to build spectacular structures. Mars has fierce gales, but the wind exerts little force because the atmospheric pressure is 1% of that on Earth—making a 100-mile-per-hour hurricane on Mars feel like a 1-mile-per-hour breeze on Earth. There could be giant suspension bridges built with carbon nanotube hawsers, spanning gorges far larger than any that could be spanned on Earth.

Particularly important will be enclosures for farming. The atmospheric pressure on Mars is far too thin for humans, but it will be suitable for farming; so, large greenhouses are practical, with plants being tended by machines. Reflectors that move with the sun can be positioned outside the enclosure, reflecting additional sunlight into it.

Interestingly, the best Martian soil is thought to be considerably better than most land on Earth.[3] It has plenty of carbon, oxygen, nitrogen and hydrogen and is rich in the minerals needed for crop growing. Mars's soil quality varies so greatly from place to place that much exploration is needed to determine which would be the best crop-growing areas. Natural nitrate beds probably exist on Mars, and Martian soil has more phosphates than soil on Earth. Much experimentation will be needed to determine the content of Martian fertilizers and how they could be best produced.

Mars has long periods of sunshine because there are no clouds. Occasionally, there are lengthy dust storms, but they reduce light only as much as an overcast day on Earth. The mass production of solar panels and fuel cells on Mars will be important. A key to populating Mars will be to have profuse vegetation around the Martian colonies. Much of this will be genetically created for Mars, along with bacteria and insects that will form a whole ecology that can flourish there.

Life on Earth is becoming more artificial, dependent on man's inventions. Life on Mars will be almost totally artificial. On Earth, we need to preserve and respect nature's complexity; on Mars there is almost nothing to preserve. Mars is an empty laboratory in which we would create the biosystem largely from scratch. We'll make mistakes, but Mars is vast enough that we have space for trial and error. Martian science will develop steadily.

In practical terms, we couldn't have a much better gift than having another inhabitable planet, even though it will only be comfortable in well-designed habitats. One day, civilization on Mars will be very different from civilization on Earth. Developing the most useful vegetation for Mars may be a "cathedral" that takes a hundred years, first breeding the plants, then developing the diverse components of an ecology, then finding the best ways to do automated farming. Hydroponics and genetic engineering will evolve together. Martian technology may become one of the most inventive subjects in universities on Earth, most of it done by academics who would never dream of leaving the Earth.

BRAIN/COMPUTER LINKAGE

Much more valuable than going to Mars will be the enhancement of our own capabilities. Human transformation may be the most important of 21st-century cathedrals. As we have commented, one of the landmark inventions of this century will be the capability to link the brain directly to computer power. Like the steam engine, this will change the world.

Today, the links we have between our brain and computing power are normally *offline*—we use a computer screen, mouse and keyboard to communicate with an endless assortment of software and information. A key breakthrough will be to make *online* connections relatively easy so that the brain will be linked directly to electronic devices. This has already been done with cochlear implants, which connect tiny computers directly to our nervous system as a means of allowing stone-deaf people to hear. Many thousands of people now have successful cochlear implants. There have been other research efforts that have connected the human nervous system and chips.

We could make a direct connection to our brain with a very thin wire through the skull. This requires surgery and is likely to be unappealing to most people. The more appealing alternative is to have wireless connections to tiny devices in the blood capillaries of the brain. Already, various types of systems have been used in the bloodstream and digestive system. They are bigger than nanodevices but can be experimented with today. Twenty years from today, nanotechnology transponders that transmit and receive signals will be used in our bloodstream. Nanotransponders in our brain will be a natural extension of the use of nanotransponders in our bloodstream—for example, for precisely targeted chemotherapy. Each transponder may be about the size of a blood cell—much smaller than a neuron. The transponders may transmit to a relay station on the outside of our skull. This may send their signals via networks to the entire world of computing technology. These networks may be designed specifically for brain communication and have a suitably fast response time.

It will probably be possible to have such transponders in our brain fluid, or possibly attached to individual neurons or synapses—with no need for surgery. This hasn't been done with humans yet, but many experiments are under way with chimpanzees. The US DAPRA (Defense Advanced Projects Research Agency) has a chimpanzee that can use *thought* to operate a remote device by means of a tiny brain implant.

When connected to electronic devices, the brain will be able to do things that an unassisted brain can't possibly do. In order to use such devices, a person will have to have repetitive training, like learning to ride a bicycle. This training will establish, among the person's neurons, the connections that enable him to do new tasks. When the learning has occurred, the person will become completely comfortable with the new facilities. It will take much training to become good transhumans.

Once brain-computer connections are standardized and work well without the brain rejecting them—and when the training is made easy—floodgates will be open to enormous numbers of applications. There may be many generations of them in rapid succession. The transponders will become mass-produced, and there will be large numbers of them in one brain. Different people will build different applications, as they did with the Internet. College students everywhere will be exchanging the latest enhancements.

Ray Kurzweil, the inventor of reading machines for the blind, described to me the future as he saw it: "We'll have millions of nanobots, which enter the brain through the bloodstream. They'll be communicating with each other because they'll be on a wireless local area network. They'll be communicating on the Internet, downloading new software; they'll be interacting intermittently with our biological neurons." In 1950, there were very few computers and most people regarded them as irrelevant to their lives. Fifty years later, microprocessors and the Internet had changed their lives. It will be similar with brain enhancement. At first, many people will think of the film *The Manchurian Candidate,* then there'll be medical applications, such as curing depression, then a flood of different uses. As with computers, the diversity of applications will grow endlessly. In the lifetime of young readers of this book, it will probably become unthinkable to have an advanced society without brain enhancement. Enhanced brain function will start with young children. In fact, someday it may be regarded as unethical to bring up a child without it.

The digital divide came about originally because computers were expensive. Wealthy people had them; poor people didn't. Then the computers dropped in cost, and the digital divide was between people who had learned how to make good use of the Internet and those who hadn't. Some intelligent people resisted using computers, or the Internet; so, for them the digital divide was self-inflicted. The same will be true with the transhuman divide. In the beginning a brain enhancement will be expensive and will have problems. Only the serious or eccentric enthusiast will persist with it. Once some benefits of the technology become clear, more people will learn how to use it, and it will start to drop in cost and be made easier to use. The first applications (the easy ones to implement) will be a tiny fraction of the applications that will eventually be available. The first applications may be related to health care.

There was a time in the late 1990s when the number of Internet users grew explosively, doubling every month. The same will probably be true of brain enhancement. Not only will the number of users double each year, but the number of transponders per brain will double, as well, and there will be an endlessly growing number of applications. There will be brain-transponder kits for different purposes—some for artists, some for hedge-fund money managers, some for Special Forces, some for mathematical physicists and some for musicians. Brain transponders will become integrated with other transhumanism technologies. Once we start to make brain transponders work well, we'll have extraordinary new worlds to explore.

HIGH-CULTURE TRANSHUMANISM

Brain enhancements may become an integral part of high civilization. Today, people who are very skilled musically use very small areas of their brain. Their learning processes since childhood have established connections among the neurons in these areas, making the person skilled at understanding or playing music. Music appreciation uses a nodule about a millimetre across, which spans the average 3-millimetre thickness of the cerebral cortex. It uses about as many neurons as the number of transistors on a large Intel chip (though the interconnections are much more complex). Some other specialized activities in the brain use only a tiny portion of the cortex. When the brain can be scanned with enough resolution, neuroscientists will learn to replicate these in electronics.

There'll be international conferences on how the brain responds to music and how we learn high levels of musical skill. We'd like to know what is happening in the brain when we get intense pleasure from Mozart or Miles Davis—or for that matter, from poetry, art and great theatre. A music lover constantly refines his or her brain wiring, becoming more able to enjoy music or to master the difficult task of playing a violin.

Even with today's low-resolution brain scans, scientists observe certain small areas of a person's brain lighting up on the scan when that person has an intense response to classical music (for example, in one set of experiments, Rachmaninoff's Third Piano Concerto). Such responses may be associated with physical symptoms such as shivers down the spine. This seems remarkable, because complex music was not necessary for our ancestors to survive in the jungle; it is a modern artificial form of culture. The reaction can be intense, however, and is related to the depth of the person's cultural training.

The technologies for brain-scanning machines are rapidly improving. In 20 years or so we'll create images showing individual neurons and their synapses sending signals to other neurons. Nanodevices in the brain will be recording what's going on. Step by step, neuroscientists will be able to record the brain while it's performing its tasks and steadily combine many such recordings to create an integrated model of the brain. Neuroscientists will compare the behaviour of such models with the behaviour of the brain itself, steadily refining their understanding. Once we can map in detail the brain interconnections, then we'll start to replicate such interconnections in electronics. As we build electronic versions of that wiring, we'll be able to connect them, probably, to our cerebral cortex. This is an intriguing idea because the electronic replicas of the nodule will operate millions of times faster than our biological nodule. Neuroscience is going to become one of humanity's most exciting research areas.

The human brain is almost the same size as it was 10,000 years ago and has almost the same number of neurons, but the way we use our brain is astonishingly different. It is almost as though we have a computer that we have only recently learned how to program. The programming is far from perfected yet; we are only a small fraction of the way towards using our brain for what it is truly capable of, but soon we'll have electronic expansions of the brain. When humans don't have to spend their lives running factories and doing accounts, they'll be able to use their time on the wonderful new forms of creativity that are beckoning to us.

A human baby rewires its neurons all the time; this is the process of learning and developing human capability. At a slower rate (because he's ageing), so does an orchestra conductor—we continue to rewire our neurons when we are older. Since the capability to appreciate music and art uses a relatively small number of our neurons, it is likely that we'll map those neurons and be able to replicate them with technology. An intriguing question is, "As we make narrowly specialized functions of our brain operate at electronic speed, will that enable us to learn classical music much faster?" It does seem likely that we'll be able to learn far more complex music. Eventually, there may be radically new music for brain-enhanced people. What higher levels of culture, or aesthetic pleasure, will be achieved with brain enhancement? What will a brain-enhanced high civilization be like? At first it may involve a relatively small number of people, but they would be in touch over worldwide networks. Transhuman civilizations may *start* globally, with people around the planet sharing what is created. Brain-enhanced people will live in the same cities as non-brain-enhanced people. Will they be accepted, just as hedge-fund managers or film animators are accepted?

Linking the brain to electronics will give people an ability to do far more complex tasks. The learning process will be difficult for some new capabilities, and only a portion of society will make that effort (as today). The differences between highly enhanced and nonenhanced people will be huge. We don't know how the more extreme potentials will play out. Maybe one in a thousand people will have brainpower like Stephen Hawking's.

Some authorities believe that when we can map *all* the brain's neurons and their interconnections, we'll be able to create an electronic replica of the entire brain. Some authorities, such as Ray Kurzweil, believe that we'll reverse-engineer the entire human brain and upload it to computers. He believes we will create silicon humans. Most neuroscientists don't believe such scenarios, maintaining that they naïvely underestimate the complexity of the mind and the subtle effects of the brain's chemicals. We won't, they say, find immortality by uploading our mind to a computer, or to the Internet. This crosses the boundary between what is practical and what is science fiction.

Here is the conservative scenario: The *essence* of the mind is not uploaded to a computer. The mind remains biological and happy, but it has many transponders in the brain linking it to an external network. Some of the brain's functions are replicated in electronics and run at a million times biological speed. Some of these components are attached to the skull, and some are far away. Just as a computer system can have numerous faraway components, so will the human brain. Some of these will be artilects— artificial intellects—that are designed for direct coupling to the brain. Some will be NHL (non-human-like) intelligence that, in a few decades' time, will be of extreme power. The transponder-enhanced mind will become linked to a vast collection of resources and may also be linked directly to other human minds.

TRANSFORMING OURSELVES

The brain/computer link will probably become one of the most transformative inventions of the 21st century. The brain will be linked to multiple types of electronics. First, it may be connected to an electronic unit that is an electronic replica of a small part of our brain, which, because its operation is familiar, our brain learns relatively easily, although probably not at full speed because it will operate millions of times faster than the biological original. Second, we may eventually map *all* of the thinking brain (the cerebral cortex) and connect to an ultrafast replica of it. Third, there may be a link to conventional computing, with massive storage facilities. Fourth, by the time brain/computer linkages are common, many forms of NHL intelligence will have developed, fundamentally different from human intelligence, many of which will be able to automatically increase their own "intelligence." Some NHL intelligence will be far beyond human intelligence. There will eventually be numerous types of artilect, as well.

Once brain/computer links are established, they are likely to use any or all of these types of computing. This represents a huge break with the present, so large that we can't predict the details or the consequences. A typical human lifetime is about 40 million minutes. Silicon circuits operate at more than a million times human speed. So will something as complex as a human lifetime happen in 40 minutes? Eventually, yes, as the brain becomes connected to many different types of supercomputer.

Human capability has reached a metastable state in which it can flip from biological operation into semi-electronic operation, which will increase in capability at a great rate.

A musician might say that human intelligence was important because it made Mozart possible. Others would say that it made possible Shakespeare, Einstein, Picasso or Socrates. The combinations of humanlike and non-human-like intelligence will make possible humans of far more extraordinary capability. The essence of their greatness may come from the humanness, or it may come from computing rather than human capability. Even if its primary essence is nonhuman, it may need humans to recognize the greatness.

The masters of this technology will make vast amounts of money. The wealthiest individuals of this era will be wealthier than those of today because they will be able to add real value at a greater rate—and for every such person, there will be an army of people who benefit from his activity.

Brain/electronics coupling will probably have matured and become highly complex before the Singularity occurs. The combination of the brain enhancement and the Singularity is awesome to imagine.

By the end of the 21st century, we may know what is practical and what is not in transhumanism. Strong transhuman technology will take us into ethical problems that need rigorous philosophical reasoning. Ethical problems will be discussed endlessly but probably without global agreement because the economic benefits of transhumanism will be huge. One of the patterns now emerging is that professional philosophers applying rigorous logic to ethics are coming to conclusions that contradict religious doctrine in certain important areas. Indeed, the major religions themselves take different viewpoints on some subjects where action is vital. India seems able to adapt Hinduism to what its scientists want to do. China will probably dismiss religious arguments and do whatever China wants to do. The West can hardly block transhumanism if China and India race headlong into an economy based on transhuman capabilities.

The term *paradigm* refers to a broad pattern of behaviour or approaches to dealing with situations. For billions of years, paradigm shifts (fundamental changes in approach) occurred very rarely on Earth, but as evolution became more advanced, paradigm shifts became more common:

PERIOD	TIME NEEDED FOR MAJOR PARADIGM SHIFTS
When life on Earth was single-celled creatures	Hundreds of millions of years
When the Cambrian explosion occurred	Tens of millions of years
When prehuman creatures developed	Hundreds of thousands of years
When *Homo sapiens* had axes and fire	Tens of thousands of years

PERIOD	TIME NEEDED FOR MAJOR PARADIGM SHIFTS
When mankind had writing	Centuries
When mankind had power-driven machines . . .	Decades
with computer-driven technology	Years
When automated evolution becomes mature	Months
The Singularity	Weeks, perhaps days

Can technology continue to speed up indefinitely? If it depended on unassisted human comprehension, the answer would clearly be no, but the intelligence of some humans will be massively augmented. The advent of immensely smarter-than-human intelligence will make the future fundamentally different from the past.

22

RICH AND POOR

I F A NEW and great civilization evolves, will it be only in the First World?

Certainly not. It is possible to evolve a great civilization with a much lower GDP per head than today's First World and with much less environmental damage. Athens in its prime was so different from today's societies that it is difficult to estimate its GDP per head in today's terms, but it would probably be in the lower half of today's Third World.

Different civilizations on Earth will reach their own form of global maturity at different times and in different ways. A Buddhist or a Muslim civilization might be quite different from Western civilization. The world will have a tapestry of civilizations, with radically different rates of change. Some will be global, linked by high-bandwidth networks, some local and proudly insular, some with immense wealth and some where wealth doesn't matter much because culture dominates. Some civilizations of the future may be free to spread their wings because they've avoided pressures for extreme consumerism. It's desirable that future civilizations accept one another's differences and enjoy the diversity.

While it might seem logical to predict that a great civilization will develop in the wealthiest countries, taking advantage of the most advanced technology, there are reasons why it might be otherwise. High-culture civilization, to a large extent, is based on things of the mind and things that can be represented digitally, such as art, film, music, education and reason (which can be represented in rule-based computer languages). Physical goods take time and money to manufacture and distribute, but digital things can be reproduced and distributed almost instantly. A high-culture civilization of the future could spread widely and rapidly at a reasonable cost. High-culture civilization is also concerned with the magic of live performances, but these don't need a high GDP.

In a consumer society dominated by intense marketing of material goods, it is unlikely that most people will develop the education for high-culture civilization that we associate with classical periods of civilization. This is not only because of the distraction of material goods. It is also because people are pressured to work hard to achieve the advertised lifestyles rather than to settle into the learning patterns that high-culture civilization requires. Intense consumerism, which could have brought high quality of life, tends to bring a tense, overworked, media-saturated society dominated by clever advertising and pressure to keep up with the Joneses. This is not the fault of technology; the same technology could be the enabler of a highly advanced civilization.

Some Third World countries have creative energies quite different from those of intense consumerist societies. Brazil has joyous energies for music and dance. In earlier centuries, China was an amazing civilization. It is too early to guess what China will be like two or three decades from now, but it might regard the consumerism of the West as pointless, decadent, outrageously expensive, wasteful and psychologically destructive. An interesting characteristic of China today is its remarkably low level of advertising. In 2004, the United States had 22% of the population of China but 65 times the expenditure on radio advertising, 41 times the expenditure for magazine advertising and 14.4 times the expenditure for television advertising.[1] China's energy may make it a powerhouse of new forms of creativity, with interesting ways of being civilized at a fraction of the costs per person of the West.

India today is succeeding spectacularly in some forms of intellect-intensive industry, such as programming, filmmaking and the development of electronic education. As societies in Bangalore and Mumbai become very affluent, they may develop quite differently from that of the United States. In the West, the writings of the Bloomsbury School and their innovative terms of expression on civilization are largely out of print and unavailable, but they are still in print in India and are being debated by many bright Indians asking what India could become.[2]

Schools and universities have a vital role to play. There can be a deliberate emphasis on education for quality of life and education about other cultures to help spread understanding among civilizations. When at university, many people enjoy more live music, performances and contacts in the world of literature and arts than they do in all the rest of their lives.

VENTURE CAPITAL

I sometimes spend time with the graduating students of Third World universities. Like students everywhere, they are full of bright ideas. They ask about how they can raise funds for their ideas or if it would be best to go public on NASDAQ. It is tragic to listen to their schemes and enthusiasm and know that they'll meet a brick wall.

A vital part of the change in poorer countries should be that bright kids are given a realistic dream of being entrepreneurs. Hernando de Soto, the immaculately suited economist from Peru, wagged his finger and told me, "The future of the Earth is not an oppressed proletariat; it's an oppressed entrepreneurial class." There's enormous entrepreneurship in the Third World, he said—it's a force "thousands of times greater than the United Nations, or the World Bank or the IMF [International Monetary Fund]." Given the opportunity, rebellious youth want to be entrepreneurs, not suicide bombers. When American officials go to developing countries, he said, they concentrate on the elite. "In other words, they're really catering to George the Third rather than to the Daniel Boones." The resentful people who might be jihadists could be the West's best allies if the West enabled them to be entrepreneurs. They have vested interests in destroying the status quo. De Soto stated, "The crucial thing to understand is that the poor people of the 21st century in the Third World are no different from the poor people of the 19th century in the United States. They want a chance to get into the system, and if they're kept out, they're going to do like you did in the far west—they're going to shoot their way into the system."

Access to capital is essential to people smart enough to create new ventures. The spread of micro-loans—small loans that enable an entrepreneur to start a business—has been extremely valuable in much of the Third World. Venture capital, mainly in America, grew spectacularly in the 1990s. Of the corporations funded, many failed (as always), but the growth of the best ones contributed in a major way to the booming economy of that period. One of the extraordinary differences between highly developed countries and the others is the volume of venture capital. Most of the world other than the First World has no venture capital. In the year 2000, the United States (population 273 million) had $103,170 million in new venture capital, whereas South Africa (population 42 million) had $3 million. Hong Kong (population 7 million) had $769 million of new venture capital, whereas the rest of China (population 1,250 million) had only $84 million.

A measure that I believe should be kept about the countries of the world is the *annual new venture capital per person*. In 2000, in Singapore this was $217, whereas in South Africa it was 7 cents.

The following table shows this ratio in various countries in the year 2000:

ANNUAL NEW VENTURE CAPITAL PER PERSON COUNTRY	(IN US DOLLARS)
Singapore	217.00
Hong Kong	110.00
South Korea	1.38
India	0.34
Philippines	0.12
China	0.07
South Africa	0.07

Over 150 countries have zero annual venture capital per person (if rounded to the nearest cent).

Developing nations, especially those with a GDP per head over, say, $3000, have many moneymaking start-up corporations, ranging from wineries to software. Their most successful corporations are often those designed to export goods to wealthy countries. Contrary to what is often assumed, exporters of high-tech goods—for example, software factories and manufacturers of electronic products—generally do better than those based on farming or mining.

Aggressive young people are much less disposed to troublemaking if they are fully challenged to participate in an *idea economy*. If their conversation is, "What ideas can we dream up?" or "How can we get a piece of the action?" their energies will be directed into exciting channels. It makes no sense that they have no access to capital. If they are excluded from any such participation, their aggressive instincts may prompt them to look for trouble. A critical leverage factor in poor countries is ensuring that young people are trained to be entrepreneurs and that the bureaucratic roadblocks to entrepreneurship are removed.

Not all venture capitalists behave in the best way. Before the crash of 2000, American venture capitalists devised ingenious schemes for making themselves rich at the expense of everyone else. Often the founders of exciting new companies were "diluted" beyond reason. Many small companies failed when they could have been nurtured to success. Venture capital in poor countries must be of the nurturing kind. It may come from government agencies or charitable foundations as well as from profit-seeking investors. It makes sense for universities everywhere to encourage and teach entrepreneurship, have a venture capital arm to encourage student inventiveness and invest in student corporations that will eventually help build up the capital base of the university.

Over the last two centuries, countries that we now call First World went on a tortuous journey from quill-pen accounting to computers, from sailing ships to jumbo jets and from simplistic banks to the complex institutions of capitalism. It's difficult to invent such things, but once they exist, they can be put to use anywhere. Poor countries today don't have to repeat the journey of the last two centuries; they can leapfrog into the age of the Web, modern medicine, efficient corporations and opportunities for global trade. Today a country should, in principle, be able to make the journey from poverty to affluence much faster than in the past because the technology exists and efficient processes can be taught. Some countries have done so. Singapore and South Korea went from rags to riches in slightly more than two decades. Japan's "economic miracle" after it recovered from the Second World War was astonishing, and China's transformation will be even more so.

It's important to ask: Why is the takeoff of Singapore and South Korea so exceptional? Why are most undeveloped countries so dismally different? Why, from the same starting point, did South Korea boom while North Korea slipped into crippling poverty and starvation? Sometimes the poorest countries have had oppressive heads of state, prepared to murder, torture and use any other means to make themselves rich. They oppress the public and prevent genuine elections. This is a problem that hasn't been solved because the nations of the world have wanted to respect the sovereignty of other nations and not get involved in costly wars that do not directly benefit them.

Only if a corrupt regime is removed can measures like those of Jeffrey Sachs be taken to remove extreme poverty.

WORLD EDUCATION

There will soon be well over a billion teenagers on the planet. Informed teenagers everywhere seem highly concerned about the planet and its problems. They react strongly to topics in this book, and there should be much free education for this Transition Generation. As they mature, a billion teenagers using Internet chat rooms, worried about the planet, will become a great force for change.

Kids of all nations love the Internet if they have access to it. In one of the poorest shantytowns, we brought a simulated version of the Internet to a roomful of youngsters who had never seen anything like it. Their reaction was overwhelming. Children climbed on top of one another to touch the keyboard and wanted to play with it all day. In many poor countries, the

Internet isn't available in homes, but it can be available in schools, farms, offices and Internet cafés.

The UN Universal Declaration of Human Rights says that elementary education shall be *compulsory* and free. Technology gives new practicality to that profound goal. Much nonelementary educational material should also be available *free* to developing countries. A high-leverage use of the Internet could come from making a large body of education available free and making sure that its availability is well known to teachers.

In less-developed countries, there is an immense need for volunteers from elsewhere to teach the teachers. In 2004, I set out with Tom Benson to establish a nonprofit body called the World Education Corps (WEC) to do just that. Now based at Oxford University and linked to the 21st-Century School, it recruits volunteers from any country to go to other countries with digital-media products and use them to help teach teachers, showing them how to explore the huge body of educational material that is available. Today, the WEC operates in conjunction with iEARN (International Education and Resource Network), which operates in 110 countries. WEC volunteers anywhere can do a year of service. Those who prove to be exceptional may take an Oxford master's degree course after their experience in the field—the goal being to create future leaders. At a basic level, along with the teaching of reading, writing and arithmetic, is training in appropriate farming methods, sanitation, nourishment and health care. At a higher level, it can teach and promote professional skills. When we advertised for volunteers, we were flooded with so many applicants that we had to turn down many great ones.

Wonderful digital educational resources exist, but most people don't know about them. In order for this technology to be useful in the developing world, human guides are needed to show people what is available and how to use it. The volunteers help link the village schoolhouse to the World Wide Web. People don't necessarily need a computer to access Internet education; it can be made available everywhere on television sets and inexpensive handheld units. The Massachusetts Institute of Technology is developing a $100 personal computer. When capable children from any country become volunteers in another country and help use the leverage factors that can change the world, their lives become exciting and fulfilled. The best aid that rich nations can give poor nations is to help them put the most effective forms of education to work.

Sir Edmund Hillary, after he became the first person to climb Everest, returned year after year to the isolated valley in the high country of Nepal where the Sherpas live and helped them build schools. Of the primitive barefoot villagers, not previously taught to read, one became a commercial Boeing jet pilot, another became an executive of an Asian hotel group, and another directs the Asia conservation programmes for the World Wildlife Fund in Washington. iEARN is having similar experiences.

A goal of successful countries ought to be to help pull failed countries out of their nightmare. When affluent countries know how to grow food, cure diseases and create employment, it is tragic that this know-how is not transferred more readily to the billions of people who need it. In the long run, this transfer would benefit all countries—because the world's population would grow at a slower rate and potential future problems would be lessened.

It will not be a healthy planet if a large part of it is destitute, grossly overpopulated and destroying its environment, as is the case today. The rich world will not be insulated if, in two or three decades, there are billions of people living in appalling poverty and disease, not knowing how to feed their children.

TWO PICTURES OF THE WORLD

Two pictures of today's world should be put side by side—one of people living in rich countries with luxury houses, excessively fertilized lawns, stress, boredom, depression, drugs and three-car garages and one of people living in the unspeakable horror of Fourth World shantycities on less than one dollar per day, with unsanitary water, near-starvation, squalor, rape, rats, violence and uncontrollable diseases. If 1% of the income of the former were diverted to the latter, it would be too small to make a difference to the former, but it could eliminate extreme poverty in the latter.

Now fast-forward those images to 30 years into the future. There are more wealthy nations, and they are much wealthier than today, but the world is in the canyon. The difference between the highest earners and average people is much more extreme because of the technologies we have described. Uncontrollable hordes move to the poorest cities of the world, and the state of the Fourth World shantycities is worse. The police are shot if they try to enter them. The unsanitary conditions in the shantytowns are incubators of infectious diseases, and the world has quarantine plans, hoping to keep a pandemic from spreading globally. The combined population of China and India is around 2.8 billion, and their consumerist demand for better, perhaps more fashionable food has raised worldwide food prices to levels unaffordable in the poorest countries. Water is running out, and climate change has lowered the farm productivity of many poor countries. Muslim radicals advocate that jihad is the only answer and find endless converts. There are mass human migrations towards the wealthy countries and mass recruitment by terrorist organizations. Many Fourth World migrants are finding ways to get into First World countries.

Today's wealthy countries ought to see the wisdom of keeping such images from becoming reality. To do so, they need to act now, around the world. As with other major problems described in this book, we know what the great solutions are, but they are being applied on much too small a scale. At some time in the future, much extreme poverty will be curtailed, but the longer that time is delayed, the bigger the population in poverty will be and the more difficult the transformation.

REACTION CITIES

Certain places at certain times have developed an intensity of intelligent activity where brilliant people stimulate one another. Just as neutrons in a nuclear reactor cause the release of other neutrons, some special cities have become *reaction cities* where new ideas lead to other ideas. London in the days of Shakespeare was one such extraordinary place. Theatres were new; until then, there had just been roving bands of ragtag players. Around 1560, many plays were written and people joined in this activity as actors, costume designers and would-be playwrights. Even the plague didn't stop them. New ideas triggered other new ideas, as in a chain reaction. Four hundred years later, something similar happened in Silicon Valley with new technology. If an area starts to develop a reputation as a reaction area, bright people from around the world flock to it, as they did to Paris before the First World War or Prague in the 1990s. The 20th century saw the growth of new areas that encouraged creativity, such as filmmaking and advertising in Mumbai and elegant design around Milan. Reaction cities may concentrate on specific types of activity, as Vienna did with music, and Paris with the social interaction of the Belle Epoque.

There have been some spectacular examples of economic chain reactions in the Third World—for example, the phenomenal growth of Singapore, Hong Kong, Taipei and Mumbai, or Bangalore with its software industry. Shanghai has been the fastest-growing city of all time, stretching the imagination of architects. Some such cities have been governed so that they clean up poverty, as did Singapore; while others allow great wealth to coexist with devastating poverty, such as Mumbai.

Third World governments should deliberately try to create reaction cities, as Malaysia did with Kuala Lumpur. With emphasis on education and a high level of literacy, they would attract advanced corporations. The area may specialize in some type of new technology. Such a city may have a high-tech corridor between the city and its airport and an infrastructure for digital growth with high-bandwidth communications.

Many factors combine when reaction areas take off. An increasing innovation rate correlates positively with decreasing birthrate. Expenditure on education is essential. There are ways to attract leading-edge researchers and corporations from abroad. The pattern of intellect development, birthrate reduction and digital exports will be key to a better future for many poor parts of the world. A reaction city can grow faster with *knowledge* industries than with *physical* industries, as did India's software industry or the high-tech industries that took Singapore from rags to riches. If an area has certain knowledge industries as its target, it can offer high incentives to researchers and corporations in those fields. Biotechnology now offers numerous opportunities, especially in the new world of rejuvenative medicine.

Countries that are stagnant today should be encouraged to plot how they can create reaction cities. These cities may succeed if they are built around universities.

SINGULARITY CITY

While promising steps are being taken in some of the Third World to eliminate extreme poverty, the outlook for such programmes in much of the Fourth World is dismal. In most of the poorest countries, population is rising while earnings per person are falling, and the sorry state of the shantytowns is spreading.

Some countries in Asia jumped from agrarian societies to information societies, without going through the usual interim stages of industrialization. In every country, it should be asked how there can be a leap from today's no-hope communities to communities full of hope.

Consider the following project: In a Fourth World country, a site is selected for a city of the future. It is laid out with a 40-year plan, starting with a low budget but with plans for where funding will come from as it develops. Women or couples with children in destitute shantytowns are offered the chance to move to the new area and offered simple housing. The parents are taught to read and given jobs building the infrastructure of the city. It's to become a clean, sanitary, environmentally correct city without cars, but with excellent urban transportation, perhaps like Curitiba in Brazil. On the outskirts are car parks and, eventually, small forests of 5-megawatt wind generators. Within the city are large-field solar energy sites and the ability to generate fuel for fuel cells. High-bandwidth Internet capability goes into every building. Dramatic architecture is planned, but as is not the case in Brasilia, most of it will not be constructed until the city can generate enough money to finance it—perhaps by year 10 or 20.

Children go to Head Start–type schools from the youngest age. To help them learn English (as either their first or second language), they replay entertaining clips from films and are given iPod music in English. They grow up with BlackBerrys, Internet chat rooms, computer games and virtual reality. Good health care keeps pace with the population growth and makes sure that babies and children of all ages are well nourished. Initially, HIV-positive people are not admitted to the city. Appropriate behaviour for preventing AIDS is required, along with all possible anti-AIDS precautions. In the beginning, schools are concerned with the very young children. It is assumed that, by the time these children are 40, the Singularity will have occurred, and they will be educated in relevant material. After the Singularity, people will be employed very differently from today, and the education is aware of that. The city is called Singularity City.

Singularity City would necessarily have a social fabric utterly different from the failing fabric of the shantytowns. Children would grow up with family love and appropriate discipline but without their old world's tribal repressions and the widespread social mores that penalize a person for being an exception. As massive urban migrations grow in volume, selected people will be sent to Singularity City. The city will be designed to attract global attention and will have a top-quality board of advisers, including scientists from around the world.

Beyond their basic education, the young people will be trained in biotechnology, good Web design, hydroponics, the techniques of high resource productivity, regenerative medicine, nanotechnology, transhumanism and management methods for an age of ultracreativity. It will be an education of intense creativity, discovering applications of the new technologies. They will be taught about the problems described in this book and the solutions to those problems and will be made aware of how people may live to be 120. The youngsters will be taught that unethical behaviour is not an option. There will be a pervasive emphasis on creating lives that have value and meaning.

Eventually, Singularity City will have a virtual university, with a physical campus in the city and links to classrooms and research activities in some of the world's great universities. As the reputation of the city spreads, there will be numerous applicants wanting to move there, and the most promising of them can be selected.

As they grow up, the people of Singularity City will be taught how to be entrepreneurs and will interact with other entrepreneurs worldwide on websites. They will be taught about the endless new opportunities of the years ahead, opportunities that can come from anywhere because they relate to knowledge, ideas, computer intelligence and nanotechnology design. The city will have nanotechnology factories small enough to sit on a desk. Singularity City will be designed to become one of the centres of excellence of the "Flat World," astonishingly different from the terrible shantytowns where its founding children were born.

As happened in Ireland in the 1990s, the city will make itself as attractive as possible to foreign investors and to high-tech corporations that may set up laboratories, software factories and production facilities there. The city will try to meet the personnel needs of such corporations better than elsewhere by having lower taxes and a supply of young people trained in the new forms of creativity.

There can be many variations on the ideas of Singularity City. Many worldwide studies will be done of what a city should be like after the Singularity. Existing cities fall far short of that ideal. Singularity City will be focused on the future from the start, avoiding the vested interests of the past. It will host conferences on post-Singularity life. It will inevitably attract venture capitalists, leading-edge entrepreneurs and forward-thinking corporations.

The post-Singularity planet will have superstars everywhere, far more of them than there are today. Some of the best may be from an environment like Singularity City—children with no known father, lucky not to be born HIV-positive, plucked from destitution and boot-camped for a startlingly new human game. The goal is to demonstrate that we can take children of no hope and put them into a world of great hope. We can take kids whose future appears utterly tragic and give them a future of immense excitement.

23

RUSSIAN ROULETTE WITH
HOMO SAPIENS

THE MAGNIFICENCE of what human civilizations will achieve if they continue for endless centuries beyond this is past all imagining—so magnificent that it would be too tragic for words if humankind were terminated. To run the slightest risk of deleting *Homo sapiens* is the most unspeakable evil—the worst atrocity we are capable of. A view has grown among biocosmologists that it could be an extremely rare event for a planet to develop life with advanced intelligence like that on Earth. To exterminate humanity would, thus, be a spiritual crime of unparalleled magnitude. The public should understand this and have a feeling of utter outrage if a top scientist or politician accepts a course in which there's even a low probability that humanity's future would be erased.

Although a war in the late 20th century could have brought a nuclear winter, which would have caused starvation around the planet, an all-out world war in the 21st century could use biological weapons, both virus-based and bacteria-based, to exterminate the nuclear winter survivors. This is the first century that brings a variety of ways of terminating *Homo sapiens.*

This danger will be present for all future human centuries. We now have the power to wreck civilization or terminate humanity. That power will continue to increase; so, humanity needs to put safeguards in place.

THE INSTINCT OF THE SCIENTIST

Bill Joy, the former chief scientist of Sun Microsystems, has made a study of human activities that could threaten our existence. In my interview with him, he commented, "There's no proof—there's no a priori evidence—that what we can discover is conducive to our continued existence. We may discover things that are so powerful that we destroy ourselves."

When most people hear about atomic colliders that might be dangerous, runaway self-replicating nanotechnology, man-made pathogens for which nature has no protection or out-of-control gene modification, they react by saying, "This is insane!" They see it all as if it were lemmings charging towards a cliff. Whatever happens, stop it! While finishing this book, I participated in a seven-day conference of top-level physicists in Russia, all concerned with fusion and the "next" atom smasher. There was the most detailed discussion of alternate means of building that atom smasher but no discussion at all of what it might achieve. When we say "lemmings," we can also include brilliantly intelligent people, charging forward with an intense focus.

The instinct of the scientist is to deepen humankind's knowledge as far as it is possible to do so. The great astrophysicist Subrahmanyan Chandrasekhar paid tribute to the spirit of his mentor, Sir Arthur Eddington, by referring to Icarus, who flew too close to the sun on wings made from feathers and wax. Chandrasekhar said, "Let us see how high we can fly before the sun melts the wax in our wings."

Some experiments with atomic particle accelerators are very exciting to physicists. Atoms are composed of subatomic particles such as electrons, protons and neutrons. These are composed of smaller particles—and those of even smaller particles. Physicists proudly thought they knew all about matter when they just knew about electrons, protons and neutrons;

now, however, there are over 200 particles, some of which only last for a billionth of a trillionth of a second—a bizarre world. Particle physics seems like a tropical garden suddenly overrun with all manner of exotic entities.

A trillionth of a second after the Big Bang, the universe was a tiny, immensely hot and dense soup of such particles. Physicists hope to reproduce these conditions for the briefest moment. This would help them try to fathom more about the basic nature of matter, which seems ever more mysterious.

In order to do that, they would like to have an atom smasher immensely more powerful than the big ring-shaped one at CERN (the European Laboratory for Particle Physics), which circles across the border of France and Switzerland near Geneva. Instead, they would like to accelerate two beams of heavy atoms in opposite directions along an absolutely straight path—and when the atoms are travelling extremely fast, put the two beams into a head-on collision.

An atom of gold is very heavy. Each gold atom consists of 79 protons, 79 electrons and 118 neutrons. Physicists plan to accelerate gold atoms to speeds as close as possible to the speed of light and then smash them into one another. This spectacular smash-up would produce all manner of debris. The gold atoms would be shattered into protons and neutrons, and these would implode and break into tinier particles. The collision would produce a thousand quarks—subatomic particles that are the building blocks of protons and neutrons, which themselves would be shattered. Scientists hope to replicate, for an instant, a soup called a "quark-gluon plasma," which was what the universe consisted of shortly after it was created by the Big Bang.

A site in Germany was identified that is absolutely straight for 30 kilometres—a linear accelerator could be built there. The particle beams must not deviate by more than a fraction of the width of a hair from the straight line. Plans went ahead, but then detailed calculations showed that 30 kilometres might not be long enough to produce a collision of enough power. So the German government worked out how it could acquire land and demolish buildings to create an accelerator 40 kilometres long.

This atom smasher is referred to as "Next." It will create conditions that replicate what the universe was like a *billionth* of a second after the Big Bang. Next will be very expensive to build, but already scientists are sketching out how they could create an even bigger atom smasher, which they refer to as "Next Next." This one would replicate the conditions a *trillionth* of a second after the Big Bang. As the century progresses, physicists will want atom smashers that are more and more powerful, and are already jokingly using the terms "Next Next Next" and "Next Next Next Next." Nobody knows where they could find a dead-straight path hundreds of kilometres long—perhaps in some destitute country (and maybe they could build a Singularity City there).

Some physicists have expressed fear that Next might be dangerous. Whenever conditions are created that don't exist in nature, there is a slight risk that something could go wrong. It is possible that quarks might reassemble themselves into a very compressed object called a "strangelet," which has never been observed. If the strangelet had a negative charge, it would attract the atomic nuclei, which are positively charged. It could gobble up all the positively charged nuclei it encountered. A soup of strangelets could, in theory, attract atomic nuclei in a runaway process that consumes our entire planet.

A group of physicists reviewed the possible dangers of the collider at the Brookhaven National Laboratory and concluded that strangelets could be produced only under conditions of abnormally high pressure and low temperature. They said it's effectively impossible to produce them in the current machine where the atom smashing is being done.[1] A group of scientists at CERN, the biggest European atom smasher, concluded that if the experiments continued for 10 years, the probability of catastrophe was less than 1 in 50 million.[2] It sounds safe, but the odds are roughly the same as a person winning Britain's national lottery with a single ticket.

If you reproduce the conditions too close to the Big Bang, you might get another Big Bang. There are ways in which a tiny black hole might form and suck in everything around it. Whether such experiments could produce a reaction that gobbles up matter is something that only the rarest of mathematical physicists could assess. There might be only half a dozen people on the planet that could do the mathematics. If the probability is *almost* zero, the physicists might avoid mentioning it because they are so determined to build it.

The frontier of scientific understanding seems to recede further away the more we gain new understanding. Scientists will always want to pursue it. There are no plans for Next Next Next yet, but the mind of the scientist is such that it will want to go closer and closer to the Big Bang in the hope of learning everything it can about the fundamental nature of matter.

This is the first century in which humankind could terminate its own existence with technology that goes wrong—possibly with genetically modified pathogens, or with self-evolving nanotechnology, or with an atomic collider replicating the conditions close to those of the Big Bang—if not the "next" collider or the "next next," then a subsequent one sooner or later. Perhaps "next next next next" will trigger a subatomic event that consumes the atmosphere.

Lord Rees, the Astronomer Royal, describes scientific experiments in different fields that could lead to interesting results but have a very low probability of extreme danger—in plasma physics, nanotechnology, biological weapon research or genetic experimentation. He urgently believes that we must avoid any risk of such experiments going catastrophically wrong, banning them if the uncertainties are too great. We can't guarantee that other scientists would follow suit, however. I discussed this with Freeman Dyson, the legendary physicist and friend of the Astronomer Royal, and he quoted Shakespeare: "Dost thou think because thou art virtuous there shall be no more cakes and ale?"

ARTIFICIAL PATHOGENS

Nature has enormous numbers of pathogens. They get into our bodies all the time and generally don't do too much harm. Some cause serious illness, but they usually don't kill their victims. The reason for this is that nature, during millions of years of trial and error, has evolved protection from them. The immune systems of nature are of amazing complexity.

Unfortunately, if we modify the DNA of a pathogen, we can produce something that nature hasn't seen before; hence, there is no protection against it. Today we have the ability to modify DNA with a tool kit rather like a word processor. We can suddenly introduce a modified virus—a pathogen for which there is no protection. The tool kit for modifying DNA is becoming more widespread and easier to use. Some school science classes have students modifying DNA.

In just 24 weeks, the flu of 1918 killed far more people than AIDS has killed in the last 24 years. A total of 50 to 100 million people died from it—despite the fact that it killed only a small fraction of the people it infected. A man-made virus, for which nature has no defences, could kill 100% of the people it infects. Because Australia is overrun with mice, two researchers in 2000 tried to create a new mouse contraceptive. They found a mousepox virus and added one gene to it. To their surprise, it killed *every* mouse in the experiment. There were no survivors. Later, similar work was done in the United States that killed every mouse, including ones that had been vaccinated and ones that had been given antiviral drugs.

Mousepox can't be transferred to humans. Smallpox is, roughly, the human equivalent of it. Nature has given us some protection from smallpox so that not every human who has contracted the disease has died from it, and there has been a substantial survival rate. But smallpox can be modified, like Australian mousepox, so that it is a pathogen that nature hasn't seen before. Nature would then offer no protection. If a common highly infectious disease were artificially modified so that it contained a deadly toxin, we would have a dangerous killer on the loose.

Some types of virus, such as the common cold, are transmitted among people very easily. Don't go into the office if you have the flu, or you'll give it to lots of other people. If people are coughing and sneezing on an aeroplane, gargle with Listerine or another antiseptic.

Some viruses—like HIV—can hide for so long that people may be unaware that they have it until an opportunistic disease appears. Fortunately, HIV is not easily communicated, like the common cold or flu. Suppose that the 1918 flu had carried a pathogen with as long an invisible time as HIV. It would have spread almost everywhere on the planet before its presence was known. If, in addition, it was lethal, like HIV, most people around the planet would have been given a death sentence before it was known that anything was wrong.

Nature can protect itself from natural pathogens because it has had hundreds of years of trial-and-error mutations, but it can't protect itself from an artificial pathogen that it hasn't met before.

A truly deadly combination would be a virus or germ with the following characteristics:

- The virus is highly infectious, like the 1918 flu, so that it could spread worldwide.
- It carries a variant of a disease that was artificially created by gene engineering so that no protection against it exists in nature.
- It has a very long incubation period, like HIV, so that it could spread almost everywhere before anyone becomes sick.
- It is 100% lethal, like the Australian mousepox.

This combination might wipe out almost all humanity. It could not have happened until the present time because humanity has only recently acquired the capability to create artificial pathogens against which nature has no protection. Now it is becoming easy. Ron Jackson, one of the researchers who created the lethal Australian mousepox, said that it's not too difficult to create such a virus. A skilled person with the right equipment and appropriate training could do it in his basement.

In the world of today, the knowledge of how to make a weapon of mass destruction becomes almost as dangerous as the weapon itself. If people have the knowledge of how to make the weapon or acquire its components, the world is not safe. Because today's technology makes it possible for extreme individuals to trigger catastrophic events, dangerous information must be totally locked up and not be available on the Internet. *The knowledge must be locked up as thoroughly as the weapon.*

A detailed description of the creation of lethal Australian mousepox was submitted for publication in the *Journal of Virology*. Bill Joy and others protested that such publication was insane. Then the complete genome of the 1918 flu virus was posted on the Web so that it became immediately available worldwide.

As with weapons of mass destruction, access to such detailed information about how to make a lethal virus is almost as dangerous as the virus.

There must never again be instructions on how to make an atomic bomb on the Internet. Society should inflict the most serious penalties for any organization that allows that to happen.

SAFETY MEASURES

Risks that could terminate *Homo sapiens* are referred to as "existential risks". Genetically modified pathogens are one of these risks. Some decades in the future, we may have problems with self-replicating nano-technology devices too small to see. Various aspects of future technology could put humanity at risk. Marvin Minsky commented to me, "If we go on the way we are, we may not get through the next century at all. When there is a clear danger in the headlights, common sense says hit the brakes, but scientists often want to keep the foot hard down on the accelerator pedal."

When learning about the Holocaust, everybody recognizes the un-speakable evil that was perpetrated; we shudder even to think about it. Yet when discussing a futuristic scenario in which humankind happens to be "wiped out" by an engineered pathogen, the understanding of what this means does not sink in. This *must* change if we are to develop wisdom about the choices ahead.

The "Next Next" accelerator might give us a major step forward in our understanding of the universe, but the risk/reward ratio is absolutely unac-ceptable if the risk is a minute possibility of the termination of *Homo sapiens*.

There are defensive technologies to combat existential risks. For example, much better antiviral technologies will be developed. A rich set of antiterrorist measures is desirable. Solutions to existential dangers need to be thought about early enough so that they can be available in time. We need an appropriate level of research and investment on defensive measures. It's not appropriate today. A sudden fast-spreading pathogen would leave the world with a hopelessly inadequate quantity of vaccines, for instance.

You may live happily in your home never imagining that it might have a fire; nevertheless, you have fire insurance. It's vital to have insurance to ensure that *Homo sapiens* survives. What sort of insurance would work?

The best insurance is to study risks to our existence and to avoid them. With most risks, we can engineer the means for some humans to survive. By the end of the century, such risks and safeguards will probably be well understood—so, it is this century that presents the threat to *Homo sapiens*.

In the late 21st century, there will be independent human settlements on Mars or on large space stations far from Earth. If something happens that leaves no human survivors on Earth, there may be humans elsewhere to keep our species from permanent extinction. On Earth, we could have biological containment laboratories, like those in the Centers for Disease Control, large enough for long-term human survival. There are already habitats at the bottom of deep mineshafts, isolatable from whatever happens on the surface. If a global catastrophe happens, it's desirable that a few humans will crawl out of resilient hiding places. In that case, our culture may not be completely lost because the sobered survivors would have access to enormous digital libraries from before the event.

This is the first century in which we can genetically modify pathogens or create a subatomic reaction that might set fire to the atmosphere. A vital part of the meaning of the 21st century is understanding the possible risks to humanity's existence and establishing controls and defence technology that ensure that *Homo sapiens* survives. If we do survive the 21st century, we'll probably have the procedures to survive in the long-term.

BEYOND DR STRANGELOVE

Of all the possible threats to civilization, nuclear weaponry may still be the most dangerous. If all-out nuclear war happens, it will be combined with biological war. A vital part of the meaning of the 21st century should be that we make ourselves safe from wars that could end civilization.

The 21st century is thus confronted with one of the greatest paradigm shifts in history—*there will be either no all-out war between nuclear nations or no civilization.* Eventually biological weapons will become as lethal as nuclear weapons. We might ask, "Can today's politicians cope with a paradigm shift of such magnitude?" If yes, then today's young people will live their lives without experiencing anything like the 20th century's world wars.

Henry Kissinger observed that the greatest danger of nuclear war lies not in the deliberate actions of wicked men but in the inability of harassed men to manage events that have run away from them. This surely describes the future. There are many ways to stumble into a disaster that neither side intends. The potential paths to disaster need to be meticulously researched so that we can think ahead and eliminate, as far as we can, the possibility of harassed men blundering into unspeakable catastrophe.

In the days when there were movies like *Dr Strangelove* and *Fail-Safe,* nuclear bombers took 12 hours to reach the USSR from the United States and vice versa. The United States built the SAGE (Semi-Automatic Ground Environment) defence system to warn President Kennedy if Soviet bombers were on their way.

Ten years later, ICBMs (InterContinental Ballistic Missiles) could make the trip in 25 minutes; so, NATO built the BMEWS (Ballistic Missile Emergency Warning System), which had radar on many mountaintops. It was designed to detect Soviet ICBMs and funnel this information to a computer centre carved deep in the granite of Cheyenne Mountain in Colorado. It would take half of the 25 minutes to confirm that the Soviets had really launched an attack (and that it was not a computer error) and then a frenzied few minutes in which nuclear bombers take off, ICBMs are activated, and President Nixon could decide whether to launch a counter-attack.

Then we got submarines like *Red October.* That class of Soviet "boomer" carried 20 missiles each with 10 independently targeted warheads; so, it could vaporize 200 cities with a much shorter flight time than ICBMs making the long journey around the planet. There would barely be time to wake up President Reagan.

In the 1980s, the game speeded up even more. Cruise missiles were built, like those that slid under the Baghdad radar defences in the Gulf War of 1991. Such missiles were designed to carry a nuclear warhead immensely more powerful than the Hiroshima bomb. They could be hidden undetected in an innocuous-looking ship approaching the coast near

Washington, DC. Although there was massive publicity for SDI, the so-called Star Wars, there was no publicity at all for nuclear command-and-control systems designed so that, at the very highest level of alert, nuclear retaliation would happen *automatically* with preprogrammed missiles. Both the United States and the USSR implemented automatic "Launch On Warning" capability because, if they were attacked, there would be no time for the national leader to make a decision to press "the button." So, the computers had to be set to launch *automatically* if they detected that an attack was under way.

The situation becomes like that of the gunfighters in the classic westerns. Each wants to be "the fastest gun in the West," instantly ready to fire. The leisurely pace of *Dr Strangelove* changes to computerized hair-trigger Launch On Warning.

The two sides built an immensely complex computerized nuclear system. It was chained up with superb security. It had many safety catches that could be successively released as a crisis grew (DEFence CONdition), e.g., DEFCON 5, DEFCON 4, DEFCON 3.

The world came shockingly close to nuclear war in 1962 with the Cuban Missile Crisis. The events at the brink were described with some accuracy in the film *Thirteen Days,* but reality was even more alarming because, unbeknownst to the filmmakers or to the United States "Excom" (executive committee) who called the shots, four Soviet submarines had nuclear weapons. In the defining moment of the crisis, the US blockaded the Soviet fleet to prevent it from going to Cuba. At about 5 p.m. on 27 October 1962, an American ship depth-charged a Soviet submarine, unaware that the Soviet sub had a nuclear weapon on board. The depth charge exploded next to the hull but didn't penetrate the hull. The Russian captain felt honour-bound to retaliate and ordered that a nuclear weapon be launched at the Americans. To do so, two other officers had to agree to the firing and turn their keys simultaneously. At the last moment, the second captain, Vasili Alexandrovich Arkhipov, refused. If he had not done so, there would have been a devastating nuclear war.

Fortunately, there was no Cuban Missile Crisis in the 1980s. No crisis caused either side to put its forces on an alert as high as that of 1962.

The nuclear confrontation of the Cold War illustrates the inexorable logic of the computer age. Computers become steadily more powerful and become linked to a nervous system. The system has senses that are constantly alert. Computers acquire intelligence of deepening capability. It is inevitable in war or business that complexity increases and reaction times decrease until we have systems with electronic intelligence confronting each other in real time.

The control system for nuclear war might be likened to a giant creature. Radars and sensors are its nerve endings. Telecommunications are its nerve impulses. The creature has two types of brains: an automatic brain and a thinking brain. The automatic brain uses computers to tell the creature if it is being attacked and ensure that it is ready to respond. The thinking brain assesses the situation like a leopard at dawn in Africa surveying the threats and opportunities. The creature's nuclear claws are drawn when it is not on alert. If it is attacked, it must respond with extreme speed.

A major problem with the control of nuclear forces, however, is that the brains will be the first thing to be attacked. What happens if Washington and the US president are destroyed? What happens if the command-and-control computers are destroyed? When the top levels are destroyed, lower-level commanders have to take over, but then how are unauthorized firings prevented? The US and Russian missiles were designed with no self-destruct mechanism; they couldn't be destroyed after they were launched. If a mistake occurred, the missiles couldn't be stopped.

In the public folklore, only the US president can "press the button" to initiate the use of American nuclear weapons. The reality has to be very different because the US president could be target number one.

The NATO and USSR nuclear creatures were built to monitor each other with sophisticated electronic eavesdropping. In one sense, this can actually facilitate escalation. For example, if Side A goes to a higher state of alert, Side B knows and takes precautions. Then Side A reacts to these precautions, and Side B observes the reaction. These actions and reactions can act like a ratchet, reinforcing one another. In a high level of alert, there is a mutual ratcheting of actions and reactions that becomes difficult to control.

MAD

From about 1965 to 1985, there was a balance between the nuclear creatures of the United States and the USSR represented in a philosophy called
MAD, a memorable acronym for "mutually assured destruction." Each
side was deterred from attacking because it could expect to be totally
destroyed in return. It is sometimes argued that mutually assured destruction has helped to avoid a Third World War, but it resulted in the building
of nightmarish computerized systems.

The nuclear creatures watched each other every second of every day,
always suspicious and always ready to lash out with unthinkable destruction. Each side wanted to be sure it could destroy the other side even
though the other side's defences were improving. Each side built more and
more nuclear weapons, always trying to make its own arsenal more powerful and the other side's more vulnerable. This led to the unbelievable number of some 75,000 nuclear weapons by the mid-1980s. Every time I go to
Russia, I reflect that Russians and Americans are people who tend to like
each other. They get along. They are natural partners. There is an insanity
in the affairs of man. In the 21st century, the insanity has to be tamed.

In the early years of MAD, the nuclear creatures were relatively stable,
but technology changed, as it always does. The time between the detection
of an attack and the latest moment for retaliation steadily shrank. Nuclear
submarines, trying to be silent and undetected, moved closer to the coasts
of their opponent, eventually carrying as many as 200 independently targeted warheads.

A dangerous logic applies to nuclear systems. Neither side wants
nuclear war, but neither side trusts the other. Therefore, each side makes it
disadvantageous for the other side to strike by building the capability to
retaliate, but this capability to retaliate is seen by the other side as being a
capability for attack.

The nuclear systems progressed from being relatively simple and stable
in the 1960s to being outrageously complex and difficult to control in the
1980s. They were chained up with the tightest security and had layers of

elaborate safety catches, but if a crisis happened the safety catches would be taken off layer by layer. The innermost safety catch was a hair trigger with which a simple order could launch many thousands of nuclear missiles towards preprogrammed targets.

Robert McNamara instituted the doctrine of MAD as the declared policy of the United States. Two decades later, he advocated "the return, by all five nuclear powers, insofar as practicable, to a non-nuclear world."[3] He comments, "People don't understand the risk that, quite unintentionally, we maneuver ourselves into a position where these things will be used. They don't understand the fog of war. You make mistakes in war. That's the lesson of the Cuban Missile Crisis. Unless you've been in it and have been responsible at the upper level, you don't recognize how often you make mistakes and how serious they can be. The indefinite combination of human fallibility (which we can never get rid of) and nuclear weapons carries the very high probability of the destruction of nations."[4]

COLLAPSED SOCIETY

A megaton warhead exploding on a city would cause a firestorm greater than any so far experienced on Earth. If it were exploded 8,000 feet above Manhattan, the searing heat of the fireball would last about 30 seconds. The initial blast would destroy, flatten or gut every building within a diameter of about 10 miles. The blast wave would surround and crush whole buildings. The ravines of New York would fill with debris from buildings collapsing. At a distance of 2 miles from ground zero, the winds would reach 400 miles per hour. At 4 miles, they would reach 180 miles per hour. The intense heat would melt the asphalt of streets and the steel skeletons of buildings. Nine miles away in Brooklyn and New Jersey, mass fires would break out. At that distance many people would be in appalling agony—battered, crushed, burned and irradiated, with no hope of finding nurses or hospital beds. The raging fire would suck in the surrounding air, causing the wind to blow towards the city. A vast roaring column of fire would rise above Manhattan. The fires over an area of 100 square miles would converge into a single fire. Immense amounts of radioactive debris would be swept up in the column of heat. Lethal radioactive material would rain down on an area 15 miles wide and some 150 miles long, depending on the wind speed. Much of the population exposed to this would die.

The Soviet Union's SS-18 missiles each had *10* independently targeted warheads. The Soviets planned that after an attack with thousands of nuclear warheads, they would use biological warheads carried in similar missiles. The Soviets tested a variant of the SS-18 that could carry 10 refrigerated warheads, protected from the heat of reentry, each of which released large numbers of small bomblets spraying smallpox, anthrax and deadly plagues on a massive scale to wipe out any shattered survivors of the nuclear devastation.[5]

Most food in the United States comes from hundreds of miles away, but the distribution systems would be gone. The 1990s' just-in-time techniques designed to minimize inventory holding costs would ensure crippling shortages of food, gasoline and other essentials. People desperate for food and medical supplies might not have money that could buy such things.

Twenty years from now, the infrastructure of an advanced society will be far more complex than today. Supercomputer intelligence, quite unlike human intelligence, will have applications in all aspects of running society, but the programs will be far too complex to be written by hand; they will be generated by computers—highly interdependent systems of NHL (non-human-like) artificial intelligence. In 2003, a small incident caused much of the power grid in the eastern half of the United States to fail for more than a day. Excessive complexity of interdependent systems in society creates a metastable situation. Such an interdependent society works well if not perturbed too much, but if its computer centres, networks, telecom switching centres, airports, railway yards, ports, distribution systems and banks were destroyed, the infrastructure would collapse like a house of cards.

When thinking men write about the most wicked things of the 20th century, they describe the purposeless First World War, the Holocaust,

Stalin's purges and so on, but the United States and the USSR built a high-tech system for killing (with a single command) far more people than the worst despots killed. If an international confrontation had spun out of control, nuclear war could have been started by a US president or by Boris Yeltsin stressed to the edge of sanity.

A small number of men had built something that was immoral beyond belief. Yet the great forces of morality had nothing to say about it. The Catholic Church had much to say about not using condoms. The Dalai Lama wrote books about how to be happy. The Anglican Church argued about gay marriages. One might have expected total outrage from the great religions about making possible an atrocity unsurpassed in human history, but they were silent.

As we think about the 21st century, we must ask how humankind reached a situation where there were 75,000 nuclear warheads and command-and-control systems that were playing Russian roulette with civilization. How could this ridiculous situation possibly have come to pass? Perhaps we have an excuse in that nothing like it had ever happened before. It was beyond our wildest imagination. It was the result of thousands of decisions, most of them shrouded in secrecy, and the public in general couldn't see where it was leading and couldn't confront the issue.

After Hiroshima and Bikini, we knew what nuclear warheads could do, but we allowed 75,000 of them to come into service—enough to destroy every city on the planet a hundred times over. If there was an argument for building 75 nukes, there was no conceivable sense in building 75,000—it was runaway, circular "logic" at best.

Historians looking at the nuclear systems of the 1970s and 1980s will say, "Mankind was lucky." The nuclear command-and-control mechanisms of that period were an accident waiting to happen. Fortunately nothing triggered the accident. Then the USSR caved in, and humankind was given time in which it could rethink its ways.

Now that we see the whole story in retrospect, we must address the vital question: "How can we stop something like that from happening again?"

2040 CONFRONTATION

Imagine a scenario: It is 2040, and the United States and China have approximately the same number of nuclear weapons. China is now affluent but bulging at the seams. Its population density is 20 times that of Russia and 50 times that of Canada. In a world stressed for resources, China is short of oil, uranium, copper and water and can't come close to growing enough food for its people. Suppose that it peacefully annexes outer Mongolia, and then its military invades Myanmar, as it did Tibet. There are many harsh words, but nobody takes military action because the downside of a nuclear war is far larger than any upside. And what else could save Myanmar? China sets out to build a smart coastal city near Yangon with the architects who re-created Shanghai. Emboldened, China now appears ready to invade Kazakhstan and the predominantly Muslim states of Tajikistan, Uzbekistan and the astonishingly beautiful Kyrgyzstan, all of which could be defeated in a short, violent blitzkrieg. The United States issues a warning—if China invades these countries, it will take military action. China tells the United States to mind its own business.

By 2040, the reaction times of nuclear systems have become much shorter and the means of delivering weapons more diverse. The weapons have become small and easy to hide. There are numerous Hiroshima-plus "stealth" weapons systems that can slip through radar undetected. Small cargo boats can have small pilotless aircraft hidden in them, for a two-minute flight to coastal cities like Shanghai or New York, and the nuclear weapons have pinpoint accuracy. Their first targets are the national leadership and the command-and-control systems. Major powers have designed their nuclear creature with the assumption that war begins with "decapitation" (destroying the head of state, the Pentagon, the nuclear command-and-control centre, etc.). The computerized creature must react fast; it can't wait for the US president to press "the button." A future nuclear nation has to design its defence to survive decapitation. The only attack strategy that is safe in future nuclear war may be an attack that is a complete surprise, which destroys the leadership of the enemy.

There is something even more destabilizing. When an attack is about to occur, the activity can be detected; so, the defence logic says "launch before attack" instead of "launch under attack." In other words, you launch a preemptive strike that prevents you from being attacked—one designed so that, once it is launched, the enemy's response capability is pinned down in any way possible. If a Muslim state is about to nuke Israel, Israel would probably attack first and vice versa. Future nuclear command-and-control systems will be designed for preemptive strikes. Each side will want to counteract that by having a large number of hidden, highly distributed weapons so that some will survive the preemptive strike.

Because the other side could fire at any instant, a country designs its nuclear weapons so that it can fire almost all of them while under attack—in other words, while the other side's weapons are in flight towards it. "Use them or lose them." Both sides can launch while under attack. If one weapon were fired because of a misunderstanding, there would be a sudden spasm of mutual destruction. Once they start to go, they all go.

Computers with NHL intelligence will explore all military options. Their very short reaction times reduce the likelihood of human wisdom prevailing. At the highest level of alert, the computers might warn the head of state that a preemptive nuclear strike is essential. Imagine a command committee, like the Excom of the Cuban Missile Crisis, so tense that they can barely think straight, faced with an ultraintelligent computer that says, "Attack with nuclear weapons immediately, or you will be annihilated."

Weapons and computerized war fighting will become increasingly complex throughout the 21st century. The reaction time before attack will become drastically short. Nuclear confrontations won't have the clear-cut nature that they had in the 20th century. There will be fanatical politics in some parts of the world. The Cold War command-and-control systems were designed with knowledge of who the enemy was. In the future, a major power may not have that luxury. It may need to protect itself from large terrorist organizations not associated with one country.

HOW DO WE STOP IT?

A most urgent 21st-century question is, "How do we stop this madness?"

This is a complex question, but there are some obvious components to the answer. First, if we have any future technology that could wipe out civilization, it must absolutely be eliminated. Second, open, intelligent discussion is necessary on any subject that could terminate *Homo sapiens*. All thinking people of the human race must know what's going on. Third, simulation is needed of all situations that could lead to disaster, long before anything bad happens, exploring alternatives in relentless detail. Fourth, comprehensive study of the techniques that can diffuse high-tech war situations before they become too dangerous must become a well-researched scholarly discipline.

The elimination of nuclear weapons is not an impractical dream, but it cannot happen quickly. There needs to be a carefully designed journey towards elimination, with safeguards at every step. The journey would take time and would encounter many difficulties, but any other course is insanity.

The arguments for and against elimination were studied in great detail by Jonathan Schell for his 1998 book *The Gift of Time: The Case for Abolishing Nuclear Weapons Now*.[6] He interviewed the people most responsible for the nuclear systems of the Cold War era and found an intense belief that such weapons could and must be eliminated. One might have expected this belief to be that of civilians rather than warriors, but the most intense arguments for elimination came from top generals and military nuclear strategists. Robert McNamara sums it up: "Put very simply, the risk in *not* eliminating nuclear weapons is totally unacceptable."[7]

An all-out drive is needed to make ourselves safe from end-of-civilization weapons. That isn't happening yet; so the disturbing thought is, "Do we have to have a nuclear war before we take the subject seriously?" Is this a catastrophe-first situation?

24

REVOLUTION

HOW MANY PEOPLE?

Some authorities have attempted to calculate how many people the Earth can support in the second half of this century. They have models of future water shortages, farm capability and increase in food prices, and they have also determined what the needs of the Earth's control mechanisms are if our pleasant climate is to be maintained. Their conclusion is that the population figures now forecast would be too high. Lester Brown thinks 7 billion is the limit. Some other calculations show a lower figure. Even that may be optimistic because the calculations underestimate the quantity of water that will be diverted to the new cities.

James Lovelock states that a population of 6 billion people is wholly unsustainable in the present state of Gaia. Lovelock states that it would be wise to aim for a *stabilized population of about half to one billion.*[1] They could have very diverse living styles.

Lovelock says that Gaia will do the culling and eliminate those that break her rules—as she always has done. We have a choice, he says, between this grim fate and somehow gaining control of our own destiny.

The fertility rate in Hong Kong is 0.91. If the world had a fertility rate like that, the Earth's population would drop. In a century, it would become less than a billion. Pandemics might accelerate the decline. If somehow our population does become lower, or if its activities become less harmful, then our task would be to ensure that in the future we live within what the Earth's control mechanisms can handle.

However, we don't have much time. At the current rate of increase of carbon in the atmosphere, the Earth's control mechanisms will reach runaway positive feedback long before our population reaches an acceptably low figure. It seems likely that humankind will reach figures beyond those that Gaia can handle. There will be too much carbon and methane in the atmosphere, too much farmland, too much desertification, Arctic ice melting too fast and not enough forests to absorb the carbon dioxide. The models of earth system science show that we are in trouble. Perhaps we should have realized that the Earth has complex control mechanisms— which would go wrong if we used its facilities as a source of products to be plundered endlessly.

When zoologists study an endangered species, they talk about the number of breeding pairs that are left. James Lovelock warns that by the end of the century there may be a relatively small number of human breeding pairs, most of them in the Arctic regions where the climate remains tolerable.[2]

In my view, this is much too grim a conclusion. If we are heading towards this state, we will get plenty of warning, and we'll do something about it. Humanity, once it wakes up to such a demise, will take extreme action to rescue the situation.

Catastrophes can wake up governments snarled in bureaucracy and the falsification of science. A month after the Pearl Harbor attack, President Roosevelt gave the most extraordinary address to the nation. He announced huge targets for armaments production. "Our task is hard. Our task is unprecedented. And the time is short. We must strain every existing armament-producing facility to the utmost. We must convert every available plant and tool to war production. That goes all the way from the greatest plants to the smallest—from the huge automobile industry to the village machine shop." He stopped car production almost immediately. For nearly three years there were essentially no cars produced in the USA.

Imagine a similar campaign to stop manufacturing petroleum cars and replace them with advanced hybrids and, eventually, fuel-cell cars. At the same time, coal power stations are told that either they sequester their carbon or they will be closed down. There is a massive campaign to make the public cut down on its energy wastage. Airfares are drastically increased, and as much physical travel as possible is replaced with spectacular video conferencing.

THE TAPESTRY OF BIG ISSUES

The story of what is happening to humanity and its home planet can largely be put in place now. It is not sharply in focus; our knowledge in many areas is sketchy. Yet the tapestry of big issues is visible. We are damaging our future in diverse ways, but there are solutions to these problems—numerous solutions in different disciplines. A massive transition is needed, and the agenda can be created for the generation that is going to bring about that transition. Broadly speaking, we know what needs to be done. It involves all nations. The issues are global. There is no place to hide.

There is a problem, however. Much of what needs to be done is not happening. The grand-scale transition of the 21st century could occur gently. There could be step-by-step replacement of carbon-based fuels, steady improvement in food-growing capability, measures to conserve water, a lockdown of sources of fissile uranium, growth of antiterrorist measures, a drive for eco-affluent lifestyles and so on—it's a long list, but today's computer models make it clear that we are not changing our ways fast enough. We are drifting towards irreversible climate change faster than we are taking any actions to keep that from happening. Water is essential for food production, but we are taking water from aquifers at a rate that will cause many of them to run dry. In addition to water depletion and soil degradation, food production will also be seriously lowered in some countries by drought and heat waves caused by the rising quantity of greenhouse gases. As if that weren't bad enough, we are diverting huge amounts of water from farms to cities, and that trend will grow because there are massive migrations of people from the countryside to cities.

There are many solutions to our problems, but it is common to find corporations resisting the solutions and governments that seem incapable of taking the actions needed. If we continue to allow roadblocks to impede correct behaviour, negative events will overtake us. The planetary damage caused by the successful countries will breed anger among the unsuccessful. It will enhance the recruiting capability of al Qaeda and terrorist protest squads. This will occur at a time when the technology of nuclear and biological weapons is becoming more easily available, getting out of control and costing less.

There is ultimately no way of avoiding the 21C Transition. Humanity can't go on for ever using more water than is replenished, or increasing its population while damaging its food-growing capability. If steady, gentle change does not happen, the world will end up in situations where only revolutionary change will work. If governments continue to take almost no action, the transition, when it happens, will be traumatic, expensive and often violent. Large-scale catastrophes will trigger change, but the problem with the catastrophe-first syndrome is that the potential catastrophes are getting bigger. The worst famines and the worst pandemics in human history are yet to come. A war with the anger of the Second World War and modern technology could set civilization back centuries.

To make the 21C Transition as painless as possible, we need to make positive changes as early as possible, before the problems become too bad. In almost all areas, however, this is not happening. Where steady transitions are possible—for example, with the change to noncarbon fuels—almost nothing is being done. The US Environmental Protection Agency tries to repair environmental damage after it has occurred, but it seems to have almost no ability to change the economic practices that cause the damage in the first place. World summits on sustainable development have had agendas of immense importance but have taken almost no action on anything. Their follow-up can be described as studied avoidance of any changes that are controversial. The acronym UNCTAD (United Nations Conference on Trade And Development) is often said to stand for Under No Circumstances Take Any Decision.

In many areas, the situation is even worse than if nothing were being done. There are enormous subsidies being paid for the oceans to be over-fished. The subsidies for fuels that damage the environment are massive, but subsidies for fuels that would help the environment are small. As I commented earlier, Norman Myers catalogued $2 trillion per year of "per-verse" subsidies—subsidies that do more harm than good. They leave the environment or the economy worse off than if the subsidy had never been granted. If voters and taxpayers were given a listing of subsidies they pay, along with the net harm from those subsidies, they would be in revolt. Not surprisingly, governments tend to hide this information.

The longer the transition is delayed, the more difficult it will be. The Transition Generation will not sit idle and see their world go down the drain. The 21C Transition could become revolution, not evolution. As often before in history, revolution will be the consequence of compla-cency. Craig Venter, the legendary genome mapper, commented in my interview with him that the danger to our society is not science—it's apathy.

DIVERSION OF WATER

Demographers have detailed computer models of the Earth's population. Assumptions fed into the models lead them to predict a peak population in midcentury of about 8.9 billion. But the furious growth of the massive new consumer class in China (and somewhat later in India) is redrawing the world economic map. The First World has a population of 1 billion. China and India will have a joint population rising to 3 billion. There is great excitement among young people in the new consumer classes and almost no thought about their impact on the rest of the world.

Cities need large amounts of water, both for hygienic living and for their new industry. A thousand tons of water can produce one ton of wheat, which is worth $200 or less. Lester Brown quotes situations in China in which one ton of water can increase industrial output by $14,000—70 times as much. Because of this, water is increasingly being diverted to where such industry is, and this leaves already water-short farmers with even less water. Cities are prepared to pay much more for water than farmers can afford to pay.

The magnitude of the difference between what farmers could pay and what cities will pay for the same water ensures that much water will be diverted from farms to cities. The largest human migration in history is China's current migration from country areas to new cities, and the diversion of water to cities in China will become massive. A similar pattern is developing in India and in other countries. The economic forces behind this are immense. Many city governments have signed contracts to buy water for the next 50 years or more. San Diego, for example, has bought the rights to 247 million tons of water per year for the next 75 years.

As China's aquifers decline and it increasingly diverts its river water to cities, it will be able to grow less grain. It needs more grain, however, because the hordes of young people in its new consumer societies want to eat meat instead of rice—and they want high-quality meat, often pork, from well-fed animals. Meat production is very expensive in terms of grain and, hence, in terms of water. The amount of water needed increases as the quality of the meat goes up.

When China and India run so short of water that they can't grow enough grain, they will buy grain on the world markets. Buying a ton of grain is equivalent to buying 1,000 tons of water, and buying 18 pounds of meat is equivalent to buying roughly a ton of grain.

China buying huge amounts of grain will cause the price of grain to go up. India is likely to follow a similar pattern—its population will become larger than China's in two or three decades. The price of grain will almost certainly exceed what poor countries can afford, and global warming will make this situation worse by lowering farm productivity in many water-stressed countries. The poor will not be able to grow the grain they need, or purchase it.

Unless very strong action is taken by wealthy countries to move destitute countries to the first rungs of the ladder of economic development, there will be massive death tolls in such countries, afflicted as they are by AIDS, malaria and diseases caused by unsanitary water. The newly rich classes who buy grain will probably have no concern for the 3 billion or so people who can't pay the market price for grain.

In Egypt, it hardly ever rains, and for 5,000 years, the people have been entirely dependent on the Nile. Today's 73 million Egyptians use almost all the water of the Nile. So for part of the year, the great river is only a trickle by the time it reaches the sea. Now, however, Ethiopia and Sudan, countries upstream of Egypt on the Nile, are rapidly growing in population and are dependent on the river for growing food. These countries want to divert the water to their cities, and Ethiopia is planning to build a dam on the Blue Nile, which will lessen the flow of water to Egypt. The population of these two upstream countries is projected to grow from 106 million today to 231 million by 2050. Two decades from now, the Nile will be dried up before it reaches Egypt. Then Egypt will have no water. This is an impossible situation, but there seem to be no plans in Egypt to deal with it. If Egypt's population grows at its present rate, it will be massive by the time the water runs out. Will the country that was, for 3,000 years, the grandest in the world become like Somalia?

If the Earth's maximum sustainable population is 2 billion or so less than projected, it is desirable that population reduction be achieved by human choice rather than by the nightmare of starvation or the most lethal human violence—2 billion is, indeed, a very large number.

CATASTROPHE-FIRST DEVELOPMENT

Catastrophes can wake up governments snarled in bureaucracy.

When China's pollution finally becomes too much, its all-powerful government may put China on a warlike footing, switch all vehicle production from petroleum to fuel cells, terminate further building of coal power stations and announce Henry Ford–like mass production of pebble-bed reactors, ban energy-intensive forms of air-conditioning and so on. There will be a nonpolluting nuclear industry using fuel that cannot be used for atomic bombs. New codes will be enforced so that buildings use less energy and have good insulation, benign forms of cooling, low-energy light bulbs and large-scale solar generation. There will be plans for "green cities," with bicycles and scooters, large car-free areas and public transportation going into malls and campuses. The warlike urgency may ration the sale of meat and set quotas for fish production from freshwater ponds.

There will be a drive for mass exports of fuel-cell cars, radically redesigned so they are pleasant to drive. The fuel cells will power the home as well as the car. Japan may sell millions of hydrogen-storage devices for cars and homes. Wind generators and gas-cooled nuclear reactors will generate the hydrogen needed. Farmers worldwide may sell hydrogen as well as food, and use large wind generators to act as windbreaks in soil-conservation schemes. There will be a world market for green home technology. Mass production of large-field solar panels will make them cost-competitive.

One subject will become of great importance: Gaia-friendly farming. Cattle and sheep take up far more land than pigs and chickens. It's much better, ecologically, to eat like the Chinese—pigs, chickens, ducks and pond fish, all of which consume mainly vegetable waste. It will be essential in the future to have forested land for the needs of Gaia.

Feeding 9 billion people from agriculture would use too much of the planetary surface; not enough would be left for Gaia's needs. Hydroponic farms could grow food in high-rise glass city blocks, close to the customers. An important subject will be the production of synthetic food. This would require carbon, nitrogen and sulphur, perhaps from sequestered power station effluents, water and trace elements. Humanity would eat farmed foods as well as synthetic foods, but synthetic foods would release enough land to Gaia for the regulation of the climate and chemistry of the Earth.

LARGE-SCALE SOLUTIONS

A vitally important aspect of the Gaia-friendly future is to stop the carbon input to the atmosphere in time to prevent a runaway positive feedback of processes that heat up the planet. When a pond is becoming stagnant, we have to catch it in time. Otherwise, we can't bring it back to good health. Gaia today has a fever, and we need to act quickly to bring it back to health, hoping that we are not already too late.

There are various ways to lessen the sunlight reaching the Earth. Cooling could be achieved deliberately. One way is to generate artificial clouds. Our industry has released a massive quantity of aerosol particles into the atmosphere, and these particles can reflect sunlight back into space—causing global cooling of 2 to 3 degrees Celsius. John Latham, from the National Center for Atmospheric Research in Colorado, proposed large numbers of aerosol devices on the oceans that use seawater to generate mist just above the ocean surface.

Sunshades in space have been designed by two engineers in the US Lawrence Livermore Laboratory, Lowell Wood and Ken Caldiera. Between the sun and the Earth is a location called the Lagrange point where the gravitational pull of the sun and Earth are equal and opposite. At this location, little effort is needed to keep a space object in position. A disk 7 miles in diameter—if placed there—would block a few per cent of the sunlight reaching the Earth and would be barely noticeable from most parts of the Earth. Such a disk would be deployed by spinning it in space. It would weigh about 100 tons and would not be prohibitively expensive to build. If it works well, there may be many disks.

The same engineers proposed using many small balloons in the stratosphere of the Earth. They would block a similar amount of sunlight.

Ken Caldiera also proposed that carbon dioxide from power stations be pumped into a suspension of chalk in water. This would produce a solution of calcium bicarbonate, which can be disposed of relatively easily.

Klaus Lackner, an American scientist, proposed making a powder from the alkaline rock, serpentine, and making it react with the carbon dioxide from the air. This creates a product called magnesium carbonate, which can be used as a building material.

In 1991, the volcano Pinatubo, near Manila in the Philippines, erupted and injected a massive amount of sulphur dioxide into the stratosphere. It oxidized to form an aerosol of sulphuric acid droplets, which lessened global warming for several years. Large numbers of jet aircraft fly at that altitude. It has been proposed that they could burn fuel containing a small amount of sulphur. Fuel suppliers today *remove* sulphur-containing compounds from aviation fuel. They could do the opposite.

Given the grand-scale problems that are looming, engineers will dream up grand-scale ways of solving them. There are diverse ways of bringing about global cooling—as opposed to global warming. These may be used to prevent runaway heating of the Earth for a few decades, while we put into place Gaia-correct long-term solutions—fusion, population reduction, non-carbon cars, benign energy sources, carbon sequestration, pebble-bed reactors and a civilization based on eco-affluence.

THE 21ST CENTURY REVOLUTION

As we commented, the Industrial Revolution started the avalanche that led to our current situation. The avalanche will go thundering on into the future, probably for centuries. To make human civilization work well with such technologies and exist at peace with Gaia, we need another revolution, putting into place the desirable management, laws, controls, protocols, methodologies and means of governance. This is a complex and absolutely necessary transition—the 21st Century Revolution. The Industrial Revolution was not violent. The Luddites who smashed mill machinery didn't amount to much. Similarly, the 21st Century Revolution need not be violent. But it might be. It might be very violent. It might be utter mayhem if we cause Gaia to go spinning out of control.

Whether the revolution happens smoothly depends on the education that is put into place and how widely it is acted upon. Much of the Transition Generation will, presumably, become educated about the subject matter of this book. They will be taught the results of computer models that relate to their future, showing predictions of climate change, decline of water sources, food availability and population growth in different countries. Just as there are well-known measures of GDP per capita for different countries, there may be such well-known measures of good citizenship as controls on population, carbon emissions, aquifer depletion, proliferation of fissile uranium, drug traffic, war-inducing behaviour and tariffs that block trade.

A freight-train trend of the 21st century will be the growth of wealth. In the First World, real wealth is likely to go up by at least an order of magnitude (10 times) in the course of the century because of improving productivity and better management—and because wealth will relate increasingly to intellect and digitized knowledge rather than to physical goods. I commented earlier that the world's *increase in wealth will be very much greater than its increase in population.* This offers hope that the world will be made a more decent place for most of humanity.

The new riches raise questions about how society will spend its money. The big unknown is this: Will members of the Strong Nations Club spend part of their money pulling the poorest nations out of their trap? Or will the poor nations be left to fester? If the latter, it seems likely that many will not survive as their farm capability declines, food prices rise and diseases spread.

The 21st century will witness a gigantic dichotomy. Part of it will be inventing civilizations with the potential to become magnificent, while part of it will be hell on Earth. How many people in the First World will live to be 120? When they happily read *The No. 1 Ladies' Detective Agency,* do they know that life expectancy in Botswana will soon be only 27 years? The water needed for the poor to survive will be made scarcer by the rest of the world's having enormous quantities of pointless consumer goods sold in air-conditioned glass shopping malls. The Strong Nations Club may grow to almost 4 billion people, but it will have a fertility rate that can halve its number in 50 years. As that happens, it may allow a substantial fraction of the rest of humanity to join the Club.

The main factor in achieving a Gaia-friendly world will be the many diverse forms of eco-affluence. Machines will run the factories, and there will be much less need for people to work, but avocations will be complex and extremely challenging. Mobile phones have evolved into cameras and will evolve into the most eclectic assortment of virtual-reality environments. Technology is evolving so that most of us spend much of our time with low-energy devices. Games will embody diabolical complexity, some capable of being played for weeks by people around the world. Sailing, gliding, exploring the Earth's lush forests, love of theatre, music, computer analysis of golf swings and endless new forms of creativity—endless eco-affluence will take no real toll on the Earth's control systems.

THE TRUE PURPOSE OF WHAT WE ARE

A great building has to have a sound foundation, and the foundation of our world should be a planet with a stable climate, environmental correctness, food security and the Compassionate World scenario. Creating that foundation is a tough task, but not beyond our capability. But what is built on top of the foundation is just as important—it invokes profound questions.

Surely our destiny is to build something better than a society with endless, mostly trivial consumer goods. We have created consumer societies on a grand scale. In such societies, a growing number of people are unhappy and seeking therapy, and most of them complain of feelings of emptiness and pointlessness. Extreme consumer societies become devoid of deep values.

Václav Havel, the Czech Republic's deeply philosophical ex-president, refers to this dominance of efficient persuasive consumerism as "a new totalitarianism." He wrote, "It is necessary to change our understanding of the true purpose of what we are and what we do in the world. Only such a new understanding will allow us to develop new models of behaviour, new scales of values and goals, thereby investing the global regulations, treaties and institutions with a new spirit and meaning."[3]

How many people will want to be part of a world of high civilization? Can we bring back a sense of great beauty into our lives and cities? Is the building of a civilization with great music, literature, gardens and cities what it's all about? Is a high-culture civilization a top-level goal, or is there something much more important and fundamental?

Anything we can describe today is only the surface appearance, like cosmetics on the face of a woman. There is something much deeper. We cannot know the depths when we look at the face. We will learn about them only as we get to know her. Having reached this point, the journey of *Homo sapiens* will be extraordinary, though with major setbacks, like the journeys of ancient epics.

We are sowing seeds of enormous change that will take centuries to mature. When Johann Gutenberg invented his printing press, he had no idea about what it would make possible. He couldn't have imagined Rupert Murdoch or modern academia. Similarly, we can have no idea what culture will emerge from artificial intellect, artificial genetics and the extreme-bandwidth wiring of the planet. Future artilect culture is as alien to us as Shakespeare's world would have been to Gutenberg. The time elapsed between Gutenberg's invention and Shakespeare's first play was 139 years. A similar journey in the artilect world might take 20 years. The consequences of such activity will be incomparably deeper than anything we can describe today.

What should we become as technology enables humanity to reinvent itself? It appears now, for the first time, that we can change human nature, but what should we change it to? The authorities on transhumanism talk about its methodologies but not its goals. Again, we are debating how to make the train run efficiently but not about what its destination should be.

How will we use near-infinite computing? Can we make the Singularity an asset to humanity? The increasingly strange science of this century will lead to extraordinary creativity. Evolution will change from being glacially slow to being ultra-parallel applications of millions of supercomputers wired together across the planet. Automated evolution will become "infinite in all directions." Hugo de Garis says we'll produce Godlike machines. If so, what will they do? When we have mapped the genomes of every living thing, where will that understanding lead us? Craig Venter believes we'll invent biological things. Will we eventually regard life as a clay for human creativity?

Is extreme evolution the purpose of our existence? When Rome fell, barbarians swept across its empire, burning libraries and purging monasteries. The Dark Ages lasted for six centuries, and books were banned. If the vigilance of the Dark Ages had continued, we would have had no cathedrals like Chartres, no Michelangelo, no science and what it makes possible. From today's perspective, this would have been a tragedy of immense proportions. If we fast-forward through the 21st century, we might accuse intellectual Luddites of wanting a postmodern Dark Age— of wanting to stop what humankind could become capable of with the evolution of machine intelligence linked to transhuman enhancements. Humankind would then never develop the culture that could be made possible by such changes, and this culture might be as far from today's culture as Beethoven's Ninth is from jungle drums.

There are overwhelming reasons to stop evil technology—to banish the fissile material with which atomic bombs can be made, for example. There are areas where technology must be held back until the toughest safety engineering and associated laws are in place. Until we understand it thoroughly, human gene enhancement should be restricted to changes that are not automatically inherited. But that leaves us lots of leeway.

University courses have become increasingly complex. A student who briefly falls in love with Tolstoy or Bach is torn away from them and has to come to grips with calculus or quantum chemistry. Relaxed education to become civilized has been replaced with intense education for the professions. We need both. There needs to be emphasis on different types of education—curricula developed for high-culture civilization and curricula with the synthesis of topics necessary to understand what is happening to humankind and what its options are. We must foster in young people a passion for multidisciplinary excellence.

The Transition Generation everywhere needs to be taught about the meaning of the 21st century, as described in Chapter 13, "The Awesome Meaning of This Century," and the challenges that are part of it. This subject needs to be taught to today's schoolchildren. When I have tried to teach it to young people, they have reacted with immense enthusiasm. In a sense, education for survivability is the most important subject we can teach. The subject matter of this book can be both a stand-alone course and can pervade many other courses. It can be built into courses about engineering, manufacturing, accounting, environmental science, geopolitics, architecture and city planning.

An alien professor from a planet far away might observe us and say, "Now, here's a situation worth watching. The people of Earth have reached a point where they could destroy themselves. Amazingly, they have no end of business schools but no school for survival."

A business school studies factors that affect the running of business. A *civilization school* should study factors that affect civilization. The business school concentrates mainly on short-term horizons—issues that concern today's investors. The civilization school would focus on a longer-term horizon. The new 21st Century School at Oxford University is concerned with the problems described in this book. It researches diverse subjects that are important to the future and, where useful, synthesizes the results.

WHEN WOULD YOU WANT TO BE ALIVE?

Young people have sometimes asked me, "If you were to pick any time in history to be alive, which time would you pick?" There have been moments in history when there was excitement in the air and people vying with one another for creating better ideas—London in the Shakespeare years, or Paris of the Belle Epoque. Would I want to live in the Athens of Pericles or the Florence of Michelangelo?

When I reflect on such times, it seems to me that none can compare with the lifetime of today's young people. We don't have the high civilization of Athens or Florence, but this is a far more extraordinary time to be alive. Debates about the meaning of the 21st century and the global evolution of human affairs will be immensely rich in content. Technology will reach a self-evolving chain reaction, with ubiquitous computer intelligence racing ahead of people.

I commented that South Africa's magnificent "floral kingdom," the fynbos, took 5 million years to evolve. When automated evolution is in top gear, with massively parallel supercomputers, evolution will happen a billion times faster. This implies that fynbos-like complexity could evolve in two days.

Our future wealth will increasingly relate to knowledge in the broadest sense of the term. We might use the term *knowledge capability* to refer to the quantity of available knowledge multiplied by the power of technology to process that knowledge. The quantity of *usable* knowledge is rising fast (for example, precisely mapping the genome of everything biological), and the power of technology to process that knowledge capability is increasing at roughly a Moore's Law rate. Combining these, knowledge capability is approximately doubling every year. It seems likely that this doubling will go on throughout the century (if there is no catastrophic disruption). That means that, during the 21st century, knowledge capability will increase by two to the power of 100—an unimaginably large number—a thousand billion billion billion. The individual is immersed in such an expanding ocean of capability to process knowledge. That makes the 21st century radically different from any other century so far.

Reflecting on all this, I reply to their question, "If I could choose any time to live, I would want to be a teenager now (in a country where great education is available)." There is excitement in the air—perhaps more excitement than at any other time. Many people react to this by saying, "You've got to be crazy. We're heading into the canyon. We're destroying the planet, and there'll be extreme tensions caused by overpopulation, overconsumption and water and other resources running out. There'll be corporations larger than nations, runaway nanotechnology and devastating famines. We have terrorism, weapons of mass destruction and the possibility of a biotechnology war. Why not pick a benign, safe time like the 1950s?" But we were then terrified by the hydrogen bomb. The Russians exploded a 50-megaton bomb and talked of a "doomsday machine." Why not pick 100 years ago? But then, First World teenagers were drifting toward the trenches of the First World War.

The most important reason I would choose today is that, more than at any other time, young people will make a spectacular difference. The 21st Century Revolution is absolutely essential, and today's young people will make it happen. There needs to be an absolute crusading determination to bring about the changes we describe. I am old enough to remember the Battle of Britain. Today's young people will collectively determine whether civilization survives or not. It will be a time of revolution establishing the processes by which humankind can achieve levels of greatness never dreamed of before.

With technologies that are infinite in all directions, what can humanity become?

The Chessboard of Growing Computer Power

Moore's Law was concerned with the number of transistors on a chip. The speed of computers has also doubled approximately every year and a half. The first massive vacuum-tube computers at the end of the Second World War (one in Britain and one in the US) could manage about a hundred floating-point operations per second (referred to as FLOPS).

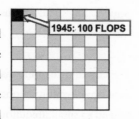

Computers were so expensive and unreliable that they were not of much practical use until about 10 years later. By 1960, IBM had a high-selling transistorized computer for commercial operations, with a power a thousand times greater than the first monstrous computers.

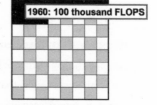

With the Moore's Law rate of growth—doubling in power every year and a half (one chessboard square)—the power of computers increases by a factor of a thousand every 15 years. By 1975 it was 100 million FLOPS.

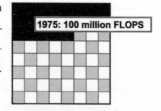

As sales increased, computers dropped in cost. The cost became low enough to make personal computers. The massive market for PCs and the intense competition in that market brought major drops in the cost of the technology. At the opposite end of the scale there was a drive for ever more powerful supercomputers, reaching 100 billion FLOPS by 1990.

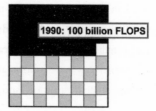

Speeds of parallel computers are quoted as the number of processors multiplied by the power of each processor. A parallel computer with 10,000 processors, each of 10 billion FLOPS in power, would be rated as 100 thousand billion FLOPS (although all 10,000 processors would almost never be operating simultaneously).

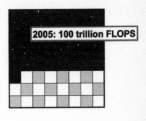

The curves of exponentially advancing technology will take us in two directions at once, to ultra-powerful supercomputers and to small ubiquitous microsystems. As microsystems spread in vast numbers they will have very low unit cost. Powerful supercomputers can be massively parallel machines built from these low-cost chips.

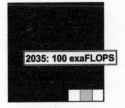

The rate of improvement of silicon chip technology may slow down well before 2020, and give way to nanotechnology with techniques that include carbon nanotubes and quantum dots. Massively parallel supercomputing will often be achieved by linking petaFLOPS computers over optical-speed networks.

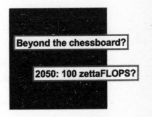

Three-dimensional nanotechnology is possible because nanotechnology will have low heat dissipation. After 2035, improvement will continue at a roaring pace as three-dimensional nanotechnology gains in power and drops in cost. What will the world be like when advanced nations have machines with a million times human brain power? When that happens, the

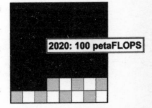

world may have a population of almost 9 billion, and 2 billion might be living in appalling destitution, and terrorism might be reaching epidemic proportions.

NHL (Non-Human-Like)
Intelligence Techniques

The following are common techniques of NHL intelligence:

- *Software that tries vast numbers of combinations* in an attempt to find an optimum solution.
- *Evolutionary* techniques, in which software is designed to evolve on the basis of results. Evolution may continue for a long time and take place at high speed.
- *Breeding* techniques, in which genetic-like breeding in software has a measurable target in mind, and software progresses towards that target, often going through many thousands of "generations" before it reaches the target. We can "breed" software that we could not design.
- *Learning* techniques—a variety of techniques with which software can learn, often learning things that humans would not learn, and steadily improving its ability.
- *Neurocomputing,* in which software "neural networks" emulate in principle, but not in detail, the neural networks of our brain. They can be set up so that we can "train" them, for example, to recognize patterns in data, or to constantly adapt their behaviour on the basis of feedback. Neural networks can be set up so that they learn and improve their own capability.
- *Pattern recognition* techniques other than neurocomputing, where software can be trained to recognize patterns.
- Techniques in which large numbers of small software objects execute rules of behaviour independently. This process is used to study complex systems that constantly adapt their behaviour and, often, unpredictable results are found.

- *Inference* techniques, in which software chains together many independent rules to derive inferences. Software can produce complex logic from an established collection of rules.
- *Data mining,* referring to a variety of techniques, such as those above, for deriving useful insight from data in a data warehouse.

NHL thought processes have various characteristics that cannot be emulated in human thought. First, they may employ *vast quantities of data,* sometimes trillions of bytes, which would be overwhelming to humans. Second, they may execute *highly parallel* procedures. Many procedures, each by itself simple, may be linked in ways that make the overall result complex. The results of the parallel processing may be surprising and completely unpredictable. Third, the process may have *feedback* that modifies the process itself. It may examine its own results and adjust its own behavior so that it does a progressively better job. It may search systematically for alternate patterns of behaviour. Fourth, the behaviour of different components may be *competitive,* with each component competing to have more influence than the others. Fifth, the formulations may be complex and *sprawling.* They may incorporate many thousands of rules rather than the small number that are in diagrams or equations for human use. Sixth, the behaviour is steadily adjusted on the basis of results so that a process may, in effect, *learn* at electronic speed.

NOTES

1. THE TRANSITION GENERATION

1. Michael Porter interviewed by James Martin, Harvard Business School, Cambridge, Mass., May 2004.

2. Rees, Martin, *Our Final Century: Will the Human Race Survive the 21st Century?* Heineman, London, 2003.

2. WHAT GOT US INTO THIS MESS?

1. At www.worldwatch.org.

3. RICH KIDS AND THEIR TRUST FUNDS

1. Costanza, R., R. d'Arge, R. de Groot, S. Farber, M. Grasso, B. Hannon, K. Limburg, S. Naeem, R. V. O'Neill, and J. Paruelo, "The Value of the World's Ecosystem Services and Natural Capital." *Nature* 387 (15 May 1997): 253–60.

2. Daly, Herman, *Beyond Growth: Avoiding Uneconomic Growth.* International Society of Ecology Economics, 5th Bien Conference, Santiago, Chile, 1998.

3. Worldwide Fund for Nature, *A Third of the World's Natural Resources Consumed Since 1970.* Agence France-Presse, Paris, October 1998.

4. Daily, G. C., ed., *Nature's Services: Social Dependence on Natural Ecosystems.* Island Press, Washington, DC, 1997.

5. Roodman, D. M., *Paying the Piper.* Worldwatch Institute, Washington, DC, May 1996.

6. Myers, Norman, and Jennifer Kent, *Perverse Subsidies: How Tax Dollars Undercut the Environment and the Economy.* Island Press, Washington, DC, 2001.

7. Roodman, D. M., *Getting the Signals Right: Tax Reform to Protect the Environment and the Economy.* Worldwatch Institute, Washington, DC, May 1998.

8. Myers, *Perverse Subsidies.*

9. Hillel, D., *Out of the Earth: Civilization and the Life of the Soil.* Free Press, New York, 1991.

10. Wackernagel, M., and W. Rees, *Our Ecological Footprint: Reducing Human Impact on the Earth.* New Society Publishers, Gabriola Island, British Columbia, Canada, 1995.

11. At www.earthday.net/goals/footprintnations.stm.

12. Ibid.

4. Too Many People

1. "World in Figures: Men and Women, Most Male Populations." *The Economist,* London, 2005.

2. Franke, Richard W., and Barbara H. Chasin, *Kerala: Radical Reform as Development in an Indian State.* Food First, Oakland, Calif. 1994.

3. Freire, Paulo, *Pedagogy of the Oppressed.* Translated by Myra Bergman Ramos. Seabury Press, New York, 1973.

4. Franke, Richard W., and Barbara H. Chasin, "Is the Kerala Model Sustainable? Lessons from the Past—Prospects for the Future," 1999. In Parayil, Govindan, ed., *The Kerala Model of Development: Perspectives on Development and Sustainability.* Zed Books, London, 2001.

5. For extensive Kerala bibliographies and links to Kerala sites: http://chss.montclair.edu/anthro/kerala.html.

6. "World in Figures." *The Economist,* London, 2000.

7. At www.populationmedia.org/popnews/popnews.html. The story of Mexican television soap operas is in Brown, Lester R., *Eco-Economy: Building an Economy for the Earth.* Norton, New York, 2001.

8. Henderson, Kathy, "Telling Stories, Saving Lives: Hope from Soaps." *Ford Foundation Report,* Fall 2000.

5. THE GIANT IN THE KITCHEN

1. Sen, Amartya, *Poverty and Famines: An Essay on Entitlement and Deprivation*. Oxford University Press, Oxford, 1982.

2. Gleick, Peter, *The World's Water 2002–2003*. Island Press, Washington, DC, 1998. Gleick produces a report on freshwater resources biennially.

3. Eisenberg, E., *The Ecology of Eden*. Knopf, New York, 1998.

4. Stuart, K., and H. Jenny, "My Friend the Soil." *Whole Earth* 96 (Spring 1999): 6–9.

5. E. Eisenberg quoted in Hawken, Paul, Amory Lovins, and L. Hunter Lovins, *Natural Capitalism: Creating the Next Industrial Revolution*. Little, Brown, New York, 1999.

6. Ibid.

7. Lal, R., J. L. Kimble, R. F. Follet, and V. C. Cole, *The Potential of U.S. Cropland to Sequester Carbon and Mitigate the Greenhouse Effect*. Sleeping Bear Press, Chelsea, Mich., 1998.

8. Eisenberg, in *Natural Capitalism*.

9. Hawken et al., *Natural Capitalism*, quoting David Pimentel of Cornell University.

10. FAO, *Yearbook of Fishery Statistics: Aquaculture Production 1998*. Vol. 86, no. 2, Rome, 2000.

11. Goldberg, Rebecca, and Tracy Triplett, eds., *Murky Waters: Environmental Effects of Aquaculture in the US* Environmental Defense Fund, New York, 1997.

12. Benyus, J. M., *Biomimicry: Innovations Inspired by Nature*. Morrow, New York, 1997.

6. DESTITUTE NATIONS

1. Homer-Dixon, Thomas Fraser, "On the Threshold: Environmental Changes as Causes of Acute Conflict." *International Security* 16, no. 2 (Fall 1991): 76–116.

2. "Angola: Measuring Corruption." *The Economist,* 26 October 2001.

3. De Soto, Hernando, *The Mystery of Capital: Why Capitalism Triumphs in the West and Fails Everywhere Else*. Bantam, London, 2000.

4. Ibid., figure 6.3, p. 203.

5. Ibid., figure 2.1, p. 19.

6. Reuters, reprinted in *Financial Review,* 11 May 1992, p. 45.

7. CLIMATE CATASTROPHE

1. Michael B. McElroy quoted in Shaw, Jonathan, "The Great Global Experiment." *Harvard Magazine,* November–December 2002.

2. Byrnes, Michael, "Scientists See Antarctic Vortex as Drought Maker." Reuters News Service, 23 September 2003.

3. At www.cnn.com, December 2003.

4. Gelbspan, Ross, *Boiling Point: How Politicians, Big Oil and Coal, Journalists and Activists Have Fueled the Climate Crisis—and What We Can Do to Avert Disaster.* Basic Books, New York, 2004, p. 112.

5. Stevens, William K., "Experts Doubt Greenhouse Gas Can Be Curbed," *New York Times,* November 3, 1997, pp. A1 and A12.

6. Blair, Tony, and Göran Persson, letter to the European Council. London and Stockholm, 25 February 2003. At http://www.defra.gov.uk/environment/business/envtech/pdf/blair-persson.pdf.

7. Millais, Corin, "European Wind Energy Achieves 40% Growth Rate." Press release, European Wind Energy Association (EWEA), Brussels, 13 November 2002.

8. For example, one produced at the Hadley Centre in England.

9. Shaw, "The Great Global Experiment."

10. McElroy, Michael B., *The Atmospheric Environment: The Effects of Human Activity.* Princeton University Press, Princeton, N.J., 2002.

11. At www.whoi.edu.

12. Gagosian, Robert G., "Abrupt Climate Change: Should We Be Worried?" Prepared for a panel on climate change at the World Economic Forum, Davos, Switzerland, 2003. At www.whoi.edu/institutes/occi/hottopics_climatechange.html.

13. Gelbspan, Ross, *The Heat Is On.* Basic Books, New York, 1997.

14. Ibid., "Boiling Point," p. 61.

15. A phrase used by von Weizsäcker, Amory B. Lovins, and L. Hunter Lovins, *Factor Four: Doubling Wealth, Halving Resource Use.* Earthscan Publications, London, 1999.

16. *Solar Living Sourcebook,* 11th edition. Chelsea Green Publishing, White River Junction, Vt., 2001.

17. Lester Brown video interview with James Martin, May 2004.

18. Figures from *Solar Living Sourcebook,* which sells these products.

19. Statement by Lewis Strauss, chairman of the US Atomic Energy Commission, to the National Association of Science Writers. New York, 16 September 1954.

20. "Cost of Closing Reactors Crucial to Privatization." *Independent,* London, 5 July 1988.

21. Hansard, HMSO, London, 24 July 1989.

22. "Nuclear Site Clean-up Costs More Than Double to £8.2 billion." *Financial Times,* London, 18 June 1994.

23. "Russia: Forgotten Victims of Chernobyl Taking Their Own Lives." IPS/*Moscow Times,* 12 January 1993.

24. "Chernobyl Cost $55 Billion in Medical Aid." *East European Energy Report,* February 1993.

25. A 10-megawatt pebble-bed reactor, HTR-10, which became operational in 2004, built by Tsinghua University's INET (Institute for Nuclear and New Energy Technology) in Beijing.

26. Clean Air Task Force, *Power Plant Emissions: Particulate Matter-Related Health Damages and the Benefits of Alternative Emission Reduction Scenarios.* June 2004. At www.cleartheair.org/dirtypower.

27. At www.crispintickell.org.

28. Rajendra Pachauri, chief of the U.N. 191-nation Intergovernmental Panel on Climate (IPCC), interviewed by Worldwatch, March–April 2003.

29. Lovelock, James, *The Revenge of Gaia: Why the Earth is Fighting Back—and How We Can Still Save Humanity.* Penguin, London, 2006.

8. INVISIBLE MAYHEM

1. Sharpe, Richard M., and Niels E. Skakkebaek, "Are Oestrogen Involved in Falling Sperm Counts and Disorders of the Male Reproductive Tract?" *The Lancet* 341 (29 May 1993): 1392–95.

2. Auger, Jacques, et al. "Decline in Semen Quality Among Fertile Men in Paris During the Past 20 Years." *New England Journal of Medicine* 332, no. 5 (2 February 1995): 281–85.

3. Carlsen, Elisabeth, et al. "Evidence for Decreasing Quality of Semen During Past 50 Years." *British Medical Journal* 305 (1992): 609–13.

4. Sharpe, "Are Oestrogen Involved."

5. Kaiser, Jocelyn, "Scientists Angle for Answers." *Science* 274 (13 December 1996): 1837–38.

6. COMPREHEND—COMmunity Programme of Research on Environmental Hormones and ENdocrine Disrupters. At http://www.ife.ac.uk/comprehend.

7. Sumpter, John P., "Feminized Responses in Fish to Environmental Estrogens." *Toxicology Letters* 82–83 (December 1995): 737–42. See also Purdom, C. et al.,

"Estrogenic Effects of Effluents from Sewage Treatment Works." *Chemistry and Ecology* 8 (1994): 275–85. Also Jobling, S., and J. Sumpter, "Detergent Components in Sewage Effluent Are Weakly Oestrogenic to Fish: An In-Vitro Study Using Rainbow Trout (*Oncorhynchus Mykiss*) Hepatocytes." *Aquatic Toxicology* 27 (1993): 361–72.

8. Hines, M., "Surrounded by Estrogens? Considerations for Neurobehavioral Development in Human Beings." In Colborn, T., and C. Clement, eds., *Chemically Induced Alterations in Sexual and Functional Development: The Wildlife-Human Connection.* Princeton Scientific Publishing, Princeton, N.J., 1992.

9. Muir, D., R. Nordstrom, and M. Simon, "Organochlorine Contaminants in Arctic Marine Food Chains: Accumulation of Specific Polychlorinated Biphenyls and Chlordane-Related Compounds." *Environmental Science and Technology* 22, no. 9, 1998.

10. Colborn, Theo, Dianne Dumanoski, and John Peterson Myers, *Our Stolen Future.* Plume, New York, 1997. Some victims of industrial accidents may have denser concentrations.

11. Eighteen scientists, "Statement from the Work Session on Environmental Endocrine-Disrupting Chemicals: Neural, Endocrine, and Behavioural Effects." International School of Ethology at the Ettore Majorana Centre for Scientific Culture in Erice, Sicily, November 5–10, 1995.

9. GENETICALLY MODIFIED HUMANS?

1. Ridley, Matt, *Genome: The Autobiography of a Species in 23 Chapters.* HarperCollins, New York, 1999.

2. Campbell, John, and Gregory Stock, *Engineering the Human Germline,* Part I: "A Vision for Practical Human Germline Engineering." Oxford University Press, New York, 1999.

10. NANODELUGE

1. IBM has a $100 million research effort to build a supercomputer of 1 petaFLOPS, called "Blue Gene," by 2005. Blue Gene's massive computing power initially will be used to model the folding of human proteins. At www.rs6000.ibm.com.

2. Estimates by the late Richard Smalley, who won the Nobel Prize in 1996 for his work in nanoscience.

11. AUTOMATED EVOLUTION

1. Cowling, Richard, and Dave Richardson, *Fynbos.* Fernwood Press, Vlaeberg, South Africa, 1995. (A spectacular book.)

2. Dawkins, Richard, *The Blind Watchmaker*. Norton, New York, 1987. In this wonderfully written book showing that Darwin's views were largely right and anti-Darwin views are largely wrong, Dawkins describes his software universe, which he calls "Biomorph Land."

3. Genobyte home page: http://www.genobyte.com.

4. Dennett, Daniel C., "How Has Darwin's Theory of Natural Selection Transformed Our View of Humanity's Place in the Universe?" In Purves, W. K., et al., eds., *Life, the Science of Biology,* 7th edition. Sinauer Associates, 2003.

5. Term used by de Garis.

6. Kurzweil, Raymond, *The Singularity Is Near*. Penguin, New York, 2005.

7. Ibid.

12. THE TRANSHUMAN CONDITION

1. At http://transhumanism.org/index.php/WTA/faq.

2. Raymond Kurzweil interviewed by James Martin, Boston, May 2004.

3. McKibben, Bill, *Enough: Staying Human in an Engineered Age*. Holt, New York, 2003.

4. Ridley, Matt, *Genome: The Autobiography of a Species in 23 Chapters*. HarperCollins, New York, 1999.

5. West, Michael D., *The Immortal Cell*. Doubleday, New York, 2003.

6. "Telomerase: Role in Cellular Aging and Cancer." At www.geron.com.

7. Kramer, Peter D., *Listening to Prozac*. Penguin, New York, 1993.

8. Reus, Victor, Owen Wolkowitz, and Brian Knutson, "Antidepressant Can Change Personality Traits in Healthy People." *American Journal of Psychiatry,* March 1998.

9. "Transcendental Medicine." *The Economist,* 21 September 2002.

10. Raymond Kurzweil interviewed by James Martin, June 2004.

11. Dennett, Daniel C., *Consciousness Explained*. Little, Brown, New York, 1992.

13. THE AWESOME MEANING OF THIS CENTURY

1. James Lovelock. *Independent,* London, May 2004.

2. Julian Filochowski, director of the Catholic agency for Overseas Development, in *Time,* 7 October 2002.

3. Sachs, Jeffrey, *The End of Poverty: Economic Possibilities for Our Time*. Penguin, New York, 2005.

14. A PERFECT STORM

1. Friedman, Thomas L., *The World is Flat: A Brief History of the 21st Century.* Farrar, Straus & Giroux, New York, 2005.

2. Hayek, Friedrich, "The Use of Knowledge in Society," *The American Economic Review,* September 1945.

3. von Weizsäcker, Amory B. Lovins, and L. Hunter Lovins, *Factor Four: Doubling Wealth, Halving Resource Use.* Earthscan Publications, London, 1999.

15. THE VITAL ROLE OF CORPORATIONS

1. Rugman, Alan, *The End of Globalism: A New and Radical Analysis of Globalism and What It Means for Business.* Random House, London, 2000.

2. Stiglitz, Joseph, *Globalism and Its Discontents.* W. W. Norton, New York, 2002.

3. At http://www.ifsia.com.

16. CULTURE'S CRUCIBLE

1. McLuhan, Marshal, *Understanding Media: The Extensions of Man.* Harper & Row, New York, 1964.

2. Huntington, Samuel P., *The Clash of Civilizations and the Remaking of World Order.* Touchstone, New York, 1997.

3. Huntington, Samuel P., "Conflicts and Convergences," Keynote Address. 125th Anniversary Symposium: Cultures in the 21st Century, Colorado College, Colorado Springs, Colo., 4 February 1999.

17. A COUNTER-TERRORIST WORLD

1. Abadie, Alberto, "Freedom Squelches Terrorist Violence." *Harvard Gazette,* Cambridge, Mass., August 2004.

2. Michael Scheuer, the ex–senior intelligence analyst who created and advised a secret CIA unit for tracking and eliminating Osama bin Laden, interviewed on *60 Minutes,* 14 November 2004.

3. Allison, Graham, *Nuclear Terrorism: The Ultimate Preventable Catastrophe.* Times Books, New York, 2004.

4. Bunn, Matthew, Anthony Wier, and John P. Holdren, "Securing Nuclear Weapons and Materials: Seven Steps for Immediate Action." Managing the Atom Project and Nuclear Threat Initiative, Harvard University, Cambridge, Mass., May 2002.

5. Bukharin, Oleg, and William Potter, "Potatoes Were Guarded Better." *Bulletin of Atomic Scientists,* May–June 1995.

6. Crenshaw, Martha, *Terrorism in Context.* Pennsylvania State University Press, University Park, Penn., 1995.

18. WORLD SCENARIOS

1. *The National Security Strategy of the United States of America.* The White House, Washington, DC, September 2002.

2. Sir John Vereker, private correspondence, 2005.

19. A GREAT CIVILIZATION?

1. von Clausewitz, Karl, *On War.* Viking, New York, 1983.

2. U.N. International Commission on Intervention and State Sovereignty, *The Responsibility to Protect.* International Development Research Center, Ottawa, Ontario, Canada, 2001.

3. The Earth Charter, "Values and Principles for a Sustainable Future." Brochure c/o University for Peace, P.O. Box 319-6100, San José, Costa Rica, 2000. At www.earthcharter.org.

4. Ibid.

20. VALUES OF THE FUTURE

1. *National Vital Statistics Report.* Washington, D.C., 2001.

2. Sapolsky, Robert M., "Will We Still Be Sad 50 Years from Now?" In Brockman, John, ed., *The Next Fifty Years: Science in the First Half of the Twenty-first Century.* Vintage, New York, 2002.

3. Martin, James, and Adrian Norman, *The Computerized Society.* Prentice-Hall, Englewood Cliffs, N.J., 1970.

4. Suggested by Leon R. Kass, M.D. In *Life, Liberty and the Defense of Dignity: The Challenge for Bioethics.* Encounter Books, San Francisco, Calif., 2002.

21. CATHEDRALS OF CYBERSPACE

1. Zubrin, R., S. Price, L. Mason, and L. Clark, "An End to End Demonstration of Mars In-Situ Propellant Production." 31st AIAA/ASME Joint Propulsion Conference, San Diego, Calif., July 10–12, 1995.

2. Ibid.

3. Stoker, C., et al., "The Physical and Chemical Properties and Resource Potential of Martian Surface Soils." In Lewis, J., M. Mathews, and M. Guerreri, eds., *Resources of Near-Earth Space*. University of Arizona Press, Tucson, Ariz., 1993. Includes table comparing plant nutrients in soil on Mars and on Earth.

22. RICH AND POOR

1. "Please Adjust Your Set." *The Economist,* 20 November 2004.

2. Bell, Clive, *Civilisation*. Rupa, New Delhi, India, 2002. First published by Random House, London, 1928.

23. RUSSIAN ROULETTE WITH *HOMO SAPIENS*

1. Alpert, Mark, "Apocalypse Deferred: A New Accelerator at Brookhaven Won't Destroy the World After All." *Scientific American,* December 1999.

2. Dar, A., A. de Rujula, and U. Heinz, "Will Relativistic Heavy Ion Colliders Destroy Our Planet?" *Applied Physics Letters,* 1999, pp. 142–48.

3. McNamara, Robert, with Brian VanDeMark, *In Retrospect: The Tragedy and Lessons of Vietnam*. Times Books, New York, 1995.

4. Schell, Jonathan, *The Gift of Time: The Case for Abolishing Nuclear Weapons Now,* chapter 2. Owl Books, London, 1998.

5. Alibek, Ken, with Stephen Handelman, *Biohazard: A Description of the Massive Soviet Program to Create Secret Biological Weapons, by the Man Who Ran It*. Hutchinson, London, 1999.

6. Schell, *The Gift of Time*.

7. Ibid.

24. REVOLUTION

1. Lovelock, James, *The Revenge of Gaia: Why the Earth Is Fighting Back—and How We Can Still Save Humanity*. Penguin, London, 2006.

2. Ibid.

3. Havel, Václav, *The Art of the Impossible: Politics as Morality in Action*. Fromm International, New York, 1998.

INDEX